CONVERGED MULTIMEDIA NETWORKS

CONVERGED MULTIMEDIA NETWORKS

Juliet Bates, Alcatel, UK
Chris Gallon, Fujitsu Telecommunications, UK
Matthew Bocci, Alcatel, UK
Stuart Walker, Leapstone Systems Inc, UK
Tom Taylor, Nortel, Canada

John Wiley & Sons, Ltd

Other Wiley Editorial Offices

John Wiley & Sons Inc., 111 River Street, Hoboken, NJ 07030, USA

Jossey-Bass, 989 Market Street, San Francisco, CA 94103-1741, USA

Wiley-VCH Verlag GmbH, Boschstr. 12, D-69469 Weinheim, Germany

John Wiley & Sons Australia Ltd, 42 McDougall Street, Milton, Queensland 4064, Australia

John Wiley & Sons (Asia) Pte Ltd, 2 Clementi Loop #02-01, Jin Xing Distripark, Singapore 129809

John Wiley & Sons Canada Ltd, 6045 Freemont Blvd, Mississauga, ONT, L5R 4J3, Canada

Wiley also publishes its books in a variety of electronic formats. Some content that appears
in print may not be available in electronic books.

Library of Congress Cataloging-in-Publication Data:

Converged multimedia networks / Juliet Bates . . . [*et al.*].
    p. cm.
  Includes bibliographical references and index.
  ISBN-13: 978-0-470-02553-6 (cloth : alk. paper)
  ISBN-10: 0-470-02553-0 (cloth : alk. paper)
1. Multimedia communications. 2. Convergence (Telecommunication) I.
Bates, Juliet.
  TK5105.15.C66 2006
  621.382′1 – dc22

                                                            2006020556

British Library Cataloguing in Publication Data

A catalogue record for this book is available from the British Library

ISBN-13: 978-0-470-02553-6
ISBN-10: 0-470-02553-0

Typeset in 10/12pt Times by Laserwords Private Limited, Chennai, India.
Printed and bound in Great Britain by Antony Rowe Ltd, Chippenham, Wiltshire.
This book is printed on acid-free paper responsibly manufactured from sustainable forestry
in which at least two trees are planted for each one used for paper production.

Contents

Foreword   xi

Preface   xiii

Acknowledgments   xv

1 Introduction   1
  1.1 Motivation for Network Convergence   1
  1.2 The Core Network   2
  1.3 Legacy Service Requirements   4
  1.4 New Service Requirements   5
  1.5 Architectures   6
  1.6 Moving to SIP   7
  1.7 Growing Revenue   7
  1.8 Network Operators – Dealing with Convergence   9
      1.8.1 Scenario 1 – A Cable Operator   9
      1.8.2 Scenario 2 – A Video-on-demand Service Provider   10
      1.8.3 Scenario 3 – A High-speed Internet Service Provider   11
      1.8.4 Scenario 4 – A Mobile Operator   11
      1.8.5 Scenario 5 – A Fixed Network Operator   13
      1.8.6 Scenario 6 – The PSTN Operator   13
  1.9 Enabling Technologies for Converged Networks   14

2 Call Control in the NGN   15
  2.1 NGN Network Architectures   16
  2.2 The Operation of Call Control   24
  2.3 Call Processing in the Legacy PSTN   25
  2.4 Call Processing in an NGN Call Agent   28
  2.5 The Basic Call State Model   30
      2.5.1 The IN CS-2 Originating BCSM   31
      2.5.2 The IN CS-2 Terminating BCSM   34
  2.6 Call Signalling in the NGN and the Role of SIP   35
      2.6.1 A Brief Discussion of the SIP Architecture and Network Elements 36
      2.6.2 A Simple Call Set-up Using SIP Signalling   38
      2.6.3 Simple Call Clearing Using SIP Signalling   52
      2.6.4 SIP Redirection Servers and SIP Forking   54
      2.6.5 Privacy CLI and the SIP P-Asserted-Identity Header   56
      2.6.6 SIP Registration Procedures   58
      2.6.7 Routing SIP Messages, Record-route, Route and Via Headers   60

    2.6.8 SIP Routing in Real Networks   64
    2.6.9 The P-Charging-Vector Header   64
  2.7 The SDP Protocol   65
    2.7.1 An Example Session Description   66
    2.7.2 The v =, o =, s = and t = Lines   66
    2.7.3 The m = Line (Media Announcement)   67
    2.7.4 Static and Dynamic RTP/AVP Payload Types   68
    2.7.5 SDP Attribute Lines   69
    2.7.6 Building an SDP Answer to and SDP Signalling Conventions in SIP   69
  2.8 Media Transport Using RTP and RTCP   71
    2.8.1 The RTP Header   71
    2.8.2 The RTCP Protocol   72
    2.8.3 RTCP Reports   73
    2.8.4 RTCP Extended Reports   74
    2.8.5 RTP Port Numbers and Symmetric RTP   75
  2.9 Addressing Issues   75
    2.9.1 The SIP "tel-URI"   76
    2.9.2 Locating Telephone Numbers, ENUM   76
  2.10 Summary   77
    References   78

3 Securing the Network and the Role of Session Border Gateways   **81**
  3.1 General Principles of Security and the NGN   81
    3.1.1 Security Assets   82
    3.1.2 Risk Analysis   82
    3.1.3 Common Pitfalls   83
  3.2 The Problem of Secrets   83
    3.2.1 Passwords   83
    3.2.2 Shared Secrets   84
    3.2.3 Public Key Infrastructure (PKI)   84
  3.3 IPsec   85
    3.3.1 Key Management   85
    3.3.2 Key Distribution   85
  3.4 Session Border Controllers and Session Border Gateways   86
    3.4.1 Functions of a Session Border Controller   86
    3.4.2 Session Border Gateways   89
    3.4.3 Gates and Pinholes   90
    3.4.4 Preventing Denial of Service Attacks with Session Border Gateways   90
    3.4.5 Additional Functions of Session Border Gateways and Session Border Controllers   92
  3.5 Protecting the PSTN Call Control Platforms in the NGN   92
    3.5.1 The Importance of Customer Access Type on Security   93
    3.5.2 SIP Security Mechanisms   94
    3.5.3 The Impact of the Threat Model on Control Plane Security   96
  3.6 Summary   97
    References   98

4 The NGN and the PSTN   **99**
  4.1 Circuits and What they Carry   99
  4.2 Signalling and Supervision   101
    4.2.1 Signalling and Supervision on the Access Link   102
    4.2.2 Inter-exchange Signalling and Supervision   106
  4.3 The Birth of the Call Agent   108
    4.3.1 History of an Idea   108
    4.3.2 Applying the Architecture   110
  4.4 Media Gateways   116
  4.5 A Look at Media Gateway Control Protocols   121
    4.5.1 SGCP and MGCP   121
    4.5.2 The Megaco/H.248 Protocol   126
  4.6 The Sigtran Protocols   142
    4.6.1 The Stream Control Transmission Protocol (SCTP)   142
    4.6.2 User Protocol Adaptive Layers   144
  4.7 Summary   146
    References   146

5 Evolution of Mobile Networks and Wireless LANs   **149**
  5.1 Introduction   149
  5.2 1G and 2G Mobile Networks   151
  5.3 Development of 3G   151
  5.4 Release 99 UMTS Architecture   152
  5.5 General Packet Radio Service (GPRS)   153
  5.6 Enhanced Data Rates for GSM Evolution (EDGE)   154
  5.7 Release 4 UMTS Architecture   154
    5.7.1 Circuit Switched Domain   156
    5.7.2 Packet Switched Domain   157
  5.8 Wideband Code Division Multiple Access (W-CDMA)   158
  5.9 Introduction to IMS   159
    5.9.1 The Proxy Call Session Control Function (P-CSCF)   160
    5.9.2 The Interrogating Call Session Control Function (I-CSCF)   162
    5.9.3 The Serving Call Session Control Function (S-CSCF)   163
    5.9.4 IMS Subscriber Identities   164
    5.9.5 The Breakout Gateway Control Function (BGCF)   165
    5.9.6 The Media Resource Function (MRF)   166
  5.10 GPRS Access to IMS   168
    5.10.1 Creating a Session   171
    5.10.2 Authorisation and Reservation of an IP bearer   173
    5.10.3 Storage of Session Paths   173
  5.11 Broadband Data Wireless Access   174
  5.12 Wireless LAN Interworking   175
    5.12.1 3GPP Release 6 Integration of Wireless LANs   176
  5.13 Mobile TV and Video   177
  5.14 Related Work in other Standards Bodies   179
    5.14.1 ETSI   179

    5.14.2 ITU-T   181
    5.14.3 ATIS   182
    5.14.4 IETF   182
    Summary   184
    Appendix   185
        3GPP Specifications   185
        3GPP Technical Specifications for MBMS   186
        MBMS Bearer Service (Distribution Layer)   186
        MBMS User Service (Service Layer):   187
        IETF Specifications   187
    References   188

6 Value-added Services   **191**
  6.1 Introduction   191
  6.2 Service Creation and Delivery Technologies   191
      6.2.1 Service Delivery in the PSTN   191
      6.2.2 SIP Application Servers   194
      6.2.3 Parlay   199
      6.2.4 Parlay X   201
  6.3 Service Orchestration   205
      6.3.1 IMS Model   206
      6.3.2 MSF Model   210
  6.4 Service Orchestration Examples   217
      6.4.1 Service Combination Example – IMS   217
      6.4.2 Conflict Resolution Example – MSF   221
  6.5 Service Delivery Platforms   226
      Summary   228
      References   228

7 Core Network Architecture   **231**
  7.1 The Convergence Layer: Multiprotocol Label Switching   231
      7.1.1 Quality of Service in IP Networks   233
      7.1.2 MPLS Traffic Engineering and Traffic Management   234
      7.1.3 Signalling and Routing in MPLS Networks   234
      7.1.4 Protection, Restoration and Service Assurance in MPLS   236
  7.2 Virtual Private Networks   243
      7.2.1 Layer 3 Virtual Private Networks   244
      7.2.2 Layer 2 Virtual Private Networks   245
  7.3 Summary   263
      References   264

8 Guaranteeing Quality of Service in the NGN   **267**
  8.1 Introduction   267
  8.2 Defining QoS   268
  8.3 QoS in IP Networks   269
  8.4 Traffic Engineering in the MPLS Core   273

8.5 Video Services   276
8.6 Business VPN Services   278
8.7 Extending QoS for VPN Services across Multiple Providers   279
8.8 QoS and the PSTN   281
8.9 QoS Architectures for PSTN Services   283
     8.9.1 A Simple DiffServ-based QoS Solution   283
     8.9.2 A Session Border Controller–based Solution with Explicit
            Reservations   287
     8.9.3 Bandwidth Manager–based Architectures   288
8.10 The MSF Architecture for Bandwidth Management   291
     8.10.1 The Bandwidth Management Layer and Scaling the Network   294
     8.10.2 Interactions with the Underlying Network   295
     8.10.3 Handling Network Interconnect   299
     8.10.4 An Alternative Approach to Network Interconnect   300
     8.10.5 Signalling QoS Requirements   300
     8.10.6 Signalling QoS with SIP Preconditions   303
     8.10.7 Bandwidth Reservation Using the Diameter Protocol   306
     8.10.8 Challenging Cases and Responding to Network Failures   312
     8.10.9 A Call Set-up with Guaranteed QoS Using Bandwidth Managers   314
     8.10.10 End-to-end QoS, Spanning Multiple Networks   316
     8.10.11 Supporting IMS-based Networks   318
8.11 Protecting the Network from Application Layer Overload   320
     8.11.1 Principles of Control Plane Overload   321
     8.11.2 Control Plane Overload Control in the PSTN   322
     8.11.3 An Overview of Control Plane Overload Control in the NGN   324
     8.11.4 Congestion Control Mechanisms Required for Black Phones and
             Access Gateways   325
     8.11.5 Trunking Gateway Overload Protection Mechanisms   327
     8.11.6 A Framework for SIP Overload Control   327
     8.11.7 A Protocol-independent Approach (GOCAP)   328
8.12 Summary   328
     References   330

Index   333

Foreword

I am delighted to have the opportunity to introduce this book on Converged Multimedia Networks.

Around the world, both network operators and service providers are experiencing massive changes brought about by the growth of Broadband, the Internet and the rapid adoption of mobile communications.

Unprecedented cooperation across the industry has enabled the establishment of 21st Century Networks, ensuring an inherent capability for the rapid introduction of new services, while empowering customers with greater control, as well as driving down costs.

Years of discussion with technology experts in the MultiService Forum, and other standards bodies, are finally coming to fruition. BT's 21st century network, the largest cross BT transformation process ever, provides a single multi-service, multi-entity network where both fixed and mobile services are supported through an application-layer approach to service creation and delivery.

The authors of this book have provided a readable, holistic account of the definition of the open architecture of the Next Generation Network, carrying a world-class communication service, minimising execution risks, maximising the opportunity for innovation and reducing complexity.

Mick Reeve
BT Group Technology Officer

Preface

In 1998, the MultiService Forum (MSF) started on a unique journey. Supported by BT and other major carriers, the MSF set out to create a collaborative working environment where engineers from all backgrounds could explore and discuss their ideas. The forum's aim was to take the specifications of a twenty-first-century network to the implementation stage. Building on agreed architectural concepts, the work would take the design right through to practical implementation and testing. Cost saving was a factor, carriers were beginning to see the need to make economic savings and multi-service platform was an attractive concept. But the technical challenge, as always, was the main driving force.

Today, almost eight years later, the success of MSF is evident. Numerous detailed implementation agreements have been published and interoperability tests have taken place across the globe, to prove end-to-end solutions. The MSF membership will testify to the very many hours spent in problem solving but also that the journey has been memorable and worthwhile.

And now at last a book! This has come about by the determination of some of the MSF contributors to provide a readable and well-referenced account of the essential elements involved in the future Next Generation Network. All the major topics are covered from Call Control, the implementation of Value-added Services, the relation to Mobile/IMS/Wireless technologies and the importance of the MPLS Core. In conclusion, there is a short section on the achievement of Quality of Service and the role of Bandwidth Manager, which, of course, reminds us that the pace of development and innovation in our industry remains undiminished.

Roger Ward
MSF President

Acknowledgments

Our ideas on Converged Multimedia Networks have been helped by many stimulating conversations with colleagues and friends within the MultiService Forum (MSF), and we would particularly like to thank Roger Ward, President of the MSF, for his support and for allowing us to publish sections from some of the MSF's specifications.

We would also like to thank the following colleagues who very kindly provided valuable inputs to the book: Dal Chohan for technical assistance with Chapter 2, Ian Batten for contributing the section on NGN security in Chapter 3, Mustapha Aissaoui for his contribution to Chapter 7, Jim Guillet for contributions to Chapter 1 and Chapter 7, and Hans de Neve, Olov Schelén and Erik Lundgren for their help with Chapter 8.

Finally, we would like to thank those who have kindly reviewed various sections of the book, suggesting many useful improvements, particularly Peter Chahal, Chris Liljenstolpe, Albrecht Schwarz and Alistair Urie.

The publishing team led by Richard Davies at John Wiley & Sons have helped us keep to the demanding schedule involved in the publication process and we are grateful to them for all their professional guidance.

We would also like to thank our families for all the support that they have given us during the project.

The authors welcome any comments and suggestions for improvements or changes that could be implemented in possible new editions of this book.

1

Introduction

The telecommunications industry is currently changing at a rapid pace. This is driven to a very large degree by the emergence of new technologies that are rewriting the business cases and cost models upon which telephony has been based for years. Faced with such an environment, carriers are changing their strategies, and convergence has become a keyword in the industry. These new technologies mean that it is now possible for a carrier to move away from running separate networks for services such as mobile telephony, fixed telephony, data and broadband and, instead, to deploy a converged network which can support all of these services. Furthermore, the deployment of a converged network lends itself to converged services with all the opportunities for revenue that they bring. This book looks at the issues and provides detailed coverage of the key protocols and architectures that will support Converged Multimedia Networks.

1.1 Motivation for Network Convergence

Convergence aims to enable both new and legacy services to be delivered by one core network, minimising the number of network layer protocols and combining the transport of all types of network traffic across a single multi-service common core network. Already a subscriber can roam transparently between wireless, mobile and fixed networks. Soon, the subscriber will no longer be able to clearly identify the services being carried over a Public Switched Telephone Network (PSTN) from those which are being carried over a mobile cellular network or a Broadband Internet connection. New services will assume seamless access to the Internet and the PSTN, and a multimedia session will be able to transfer across different types of access network, without any obvious change, or interruption, in the services being offered.

The fast, "always on" Broadband Internet is a key driver for these changes, pushing forward and enabling the delivery of multimedia applications in all types of networks. The Session Initiation Protocol (SIP) is another key driver, because SIP provides the ability to combine different multimedia flows in a single session, and simplifies the management of parallel user services. A third key driver is the industry's agreement on a standard architecture, the Internet Multimedia core network Subsystem (IMS) which enables many different types of access network to interwork across a converged core.

Converged Multimedia Networks   Juliet Bates, Chris Gallon, Matthew Bocci, Stuart Walker and Tom Taylor
© 2006 John Wiley & Sons, Ltd

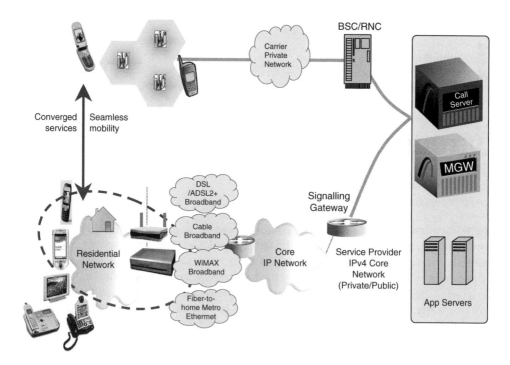

Figure 1.1   Towards Triple and Quad Play

Figure 1.1 illustrates how new Triple Play Services will provide Broadband Internet, combined with Voice and Video/Television in an "all inclusive" package. The so-called "Quad Play" will add mobile services. Imagine starting a voice call on your home phone line and transferring it seamlessly to your mobile as you drive to work. Or watching a movie on a TV, pausing it in mid-show and then watching it on a wireless Personal Digital Assistant (PDA) as you relax in the garden. Imagine having a cell phone conversation with two or three friends and simultaneously sharing a video of the football match you are attending. Then imagine that all of these things can be done with a single account, on a single login, with multiple devices over different types of access networks. These are only a few examples of the seamless multimedia services that can be accessed by users "anywhere" at "anytime".

1.2 The Core Network

For the core network, convergence means using a common network layer protocol between the edges of the core to carry all services. This protocol must be flexible enough to meet all the requirements of the current and foreseeable future services. The use of a common protocol is a major element in developing a fixed cost network infrastructure, and generating revenues from the broadest range of traditional and emerging services over this is the underlying argument for convergence. This vision is fundamentally business driven, because service providers must achieve profitability objectives while exposing themselves to the least risk. For example, the deployment of any new service has associated risk, which is increased if the operator introduces new communications protocols or deploys new network elements in the core network in order to deliver the service.

In addition to reducing risk, a converged network may reduce the number of deployed network elements that need to be managed and maintained as well as the number and complexity of network management and operations support systems. This is because network, nodes and management systems have historically specialised in supporting one or only a few protocols. For example, Time Division Multiplex (TDM)-based private line and voice services required TDM switches, while Ethernet services required the deployment of Ethernet switches, and IP services required IP routers. Such nodes would typically be interconnected by separate transport network connections and managed by different specialised divisions within a network operator. An objective of convergence is, therefore, to support all these protocols and their associated interfaces on a single platform, managed by one network management system.

Reducing the number of physical overlay networks, the associated network elements and management systems will naturally reduce the number of skilled personnel required to operate and maintain the network. It can also improve the utilisation of equipment since the consolidation of all services on a single network reduces the inefficiencies due to asset fragmentation; that is, inefficiencies associated with having similar equipment deployed in multiple, service-specific networks. Cost is therefore removed from the network, both in terms of Operating Expenditure (OPEX) and Capital Expenditure (CAPEX).

Network convergence not only simplifies operations but also improves an operator's multi-service capability. A single, converged network provides continuity of service for the operator's complete service portfolio. It also introduces the concept of service convergence into the operator's business; the ability to offer new, innovative services is facilitated by having different service traffic physically integrated over the same network infrastructure. This can also improve the time-to-market deployment of new service offerings and increase the responsiveness to customer service requests.

These are compelling arguments that help explain why operators and equipment vendors are driving towards this vision. However, it should be noted that convergence also introduces uncertainties and compromises in its own right. While it may be obvious, a point that cannot be over-emphasised is that for network convergence to be a viable solution today, and in the future, it must support the breadth of services currently offered in addition to those planned.

As introduced above, the need to eliminate risks associated with forecasting service mix and growth rates is a key driver for a converged network. Operators need to be able to develop an investment evolution strategy that allows them to adapt their infrastructure to support service changes. However, it is challenging to anticipate and forecast future service requirements at the best of times. A single, converged network has the potential to satisfy this requirement only if the current and forecast service mix is readily supported by current technology. It is important to assess this assumption on a case-by-case basis because mitigating forecasting risks with a single, converged network, by definition, implies that there is only one network upon which to support all services.

Service breadth is perhaps the main issue in determining whether network convergence using the available technology in today's solutions is an appropriate deployment option for service providers. Each service provider must therefore evaluate the business requirements and individually assess the applicability of a converged network.

1.3 Legacy Service Requirements

Legacy Services include the following:

- Traditional Public Switched Telephone Network (PSTN) voice services Plain Old Telephone System (POTS)
- Mobile voice services, for example, GSM and UMTS
- Leased line services such as TDM, Frame Relay and ATM
- Next Generation Network (NGN) services including Internet (both access via xDSL, WiFi, WiMAX, dial-up and peering)
- IP Virtual Private Networks (IP VPNs)
- Ethernet Private Lines and Ethernet Virtual Private LAN Services (VPLS).

Traditionally, each legacy service has been supported by a separate network, which was specifically engineered to meet the requirements of its service. For example, the PSTN was originally only designed to support voice services and, thus, required a network that could support a single quality of service (QoS) with strict delay and loss guarantees, but the same set of guarantees for all users and the same bandwidth for all calls.[1]

Private line services, such as Frame Relay, require more flexibility in bandwidth and delay, and are thus often supported using ATM networks that use flexible traffic management schemes to support this service breadth.

These networks have also begun to support more and more voice traffic as legacy PSTN networks are being replaced. The current TDM narrowband voice switching system has proved very effective and reliable but is coming towards the end of its economic life. The future solution will be packet voice, which, for broadband customers, will mean voice in the broadband service payload but there will be a large community of users who will wish to retain their PSTN service using their existing handset and without having to purchase additional equipment. Unless an operator wishes to run the rump of the PSTN, for many unprofitable years, a way has to be found to bring these legacy users seamlessly onto the new network. While this is a major concern for a fixed network operator, a still bigger worry is the increasing mobility of workers in the business environment. This means that mobile handsets are now often preferred as the most convenient communication device, and this presents a challenge for fixed-line operators seeking to retain customers. Market trends bear this out, as revenues for fixed voice services run relatively flat, while mobile voice revenues continue to show significant growth.

However, while mobile subscribers are using more services from their mobile phones and mobile network traffic has increased significantly, sharp competition on pricing has caused both fixed and mobile service providers to experience a cut in the Average Revenue Per User (ARPU) and both fixed and mobile Service Providers are being forced to differentiate themselves by offering new Multimedia Services to increase profits and, at the same time, also find ways to reduce the cost of provisioning all services on their networks. Convergence of fixed and mobile traffic can provide a way to make much better use of a service provider's available capacity; for example, the busy hour for voice traffic is often not the same as the busy hour for data traffic. Furthermore, the greater bandwidth

---

[1] Differences may exist at the call level, for example, emergency service calls may be prioritised over non-emergency calls.

available for fixed-line access (even if the last hop is delivered via a wireless router) means that fixed network operators who have successfully integrated support for mobile traffic into their architecture can differentiate themselves from pure mobile operators by offering high bandwidth multimedia services to mobile subscribers when they are in the range of a fixed network access point.

Services must also meet strict, and in many cases regulated, levels of availability. For example, the PSTN voice platform provides a very important lifeline service and provides a very high degree of service protection, which a replacement PSTN must at least equal. Typical Service Level Agreements (SLAs) specify that a user must not experience an unplanned outage of the service more than 0.00001% of the time, equating to an availability of 99.999%, or "five nines". NGN services compound the requirement for service flexibility and availability; Triple and Quad Play services require that voice, video and data and television be delivered reliably and with appropriate levels of QoS to potentially the same user and eventually using the same converged network.

1.4 New Service Requirements

The new converged network must not only meet the service capabilities of existing networks but also enable new services to be deployed more economically than today. This must be achieved without forcing operators to build a separate network for each service, thus requiring that the network is more flexible, scalable and cost-effective than today.

Transparency to both existing and new services is required if operators are to maintain and grow revenues from the existing services while reaping the CAPEX and OPEX benefits of convergence.

There are two elements in the evolution of a converged core network. The first is where existing services are consolidated onto the same infrastructure as newer IP services. The second is where existing services are extended alongside the new services, utilising the same infrastructure. From an operator's perspective, the future vision is a network that allows operators to reduce CAPEX though shared functionality and reuse of infrastructure for multiple services and, at the same time, provides for the reduction of OPEX through simplified architecture and reuse of the same infrastructure for multiple services:

- A network that allows operators to mix and match services to address specific market segments and enables the rapid deployment of new products.
- A network that allows operators to open up their networks to third parties in order to enhance tailored services to their customers and limit loss of customers to competitors.

A business case showing the long-term savings in OPEX must be developed that shows the savings both in cost reduction for existing services and for cost-reduced deployment of new services. Indeed, it is arguable that the greatest justification for, and requirement on, a new converged core network is the ability to support new services with lower operating costs than today.

The first major implication of these general requirements is that the network must be able to cost-effectively cope with traffic growth. This means that it must be possible to allocate network resources to rapidly growing new services, without impacting the QoS of existing services, while remaining sensitive to the incremental costs of increasing the available bandwidth in the network. The underlying network architecture must enable the operator to adapt the utilisation of the infrastructure to both the current and future changes in service diversity and demand. This task is made easier if a certain amount of intelligence is built into the edge of the network, allowing service-aware policies for routing customer traffic at the edge. The network edge must have the visibility of, and the capability to, select specific core network resources. Simplifying the core of the network reduces the cost and increases the scalability of the core. These capabilities assist operators by enabling them to offer different services with different performance objectives (e.g. Virtual Leased Lines, Internet access) or multiple grades of the same service (e.g. Gold, Silver) and thus generate new revenue streams from the converged network. A performance objective might specify an allocation of bandwidth for a downlink (network to subscriber) and a different allocation for the uplink (subscriber to network). Part of that allocation in each direction can be reserved for high-priority traffic and the rest for non-priority traffic. Network operators can also charge for policy upgrades.

In order to satisfy the requirements of the specified availability and performance objectives which have been specified in SLAs, flexibility in the way that resources are allocated needs to be complemented with flexible levels of protection and restoration. Existing "legacy" services have well-defined availability objectives. New services must be differentiated from these to generate new revenue opportunities, requiring enabling new services with different availability commitments (e.g. Best Effort residential Internet versus Premium guaranteed services for peering Internet Service Providers). There must also be flexibility in the way in which an operator can deliver that commitment. For example, the operator should be able to provide local protection (link by link, or node by node) or end-to-end protection, as required by their network design and service delivery model.

Operations and Maintenance (OAM) procedures are key contributors to enabling flexible levels of service assurance. Proactive OAM procedures alert operators to, for example, network faults, thus allowing remedial action to be taken, minimising or eliminating any impact on the SLA. Reactive OAM procedures allow fault localisation and diagnosis to take place.

Seamless interworking with the new core network is the key to the migration of existing services. This interworking is needed at the user, control and management planes. Bocci *et al.* (Ref. 6 of Chapter 7) provide an in-depth review of the development of these interworking techniques in international standards bodies, and Chapter 7, of this book, explains the principles underpinning the design of the converged core network architecture.

1.5 Architectures

The IMS developed by the Third Generation Partnership Project (3GPP) provides a network infrastructure to support fixed and mobile convergence. The IMS is an open-systems architecture designed to support a range of IP-based services employing both wireless and fixed access technologies. The IMS model adds call session control to a network to enable peer-to-peer real-time voice, video and data services over a packet-switched domain. Chapter 5 explains the evolution of the IMS and shows how the IMS architecture will be integrated with different types of access networks. Progress towards a full IMS core solution is likely to take another two to three years, but recent significant changes in access networks already provide much greater interoperability and by 2008/2009, we may have the same IMS implementation for fixed, mobile and wireless networks.

1.6 Moving to SIP

While IMS is a significant architecture for the implementation of multimedia services, undoubtedly the most influential enabler is SIP. SIP (RFC 3261) is a client–server protocol used for the initiation and management of communications sessions between users and can run over a variety of transport protocols, including Transmission Control Protocol (TCP), User Datagram Protocol (UDP), and the Stream Control Transmission Protocol (SCTP). In SIP, Internet endpoints, called *User Agents* (UAs), discover one another and agree on a characterisation of a session they would like to share. SIP UAs register with Registrars and use these in addition to Proxy Servers to help them locate and send invitations to other prospective session participants. Typically, multimedia architectures will include other protocols as well as SIP, such as the following:

- The Session Description Protocol (SDP) (RFC 4566) for describing multimedia sessions
- The Real-time Transport Protocol (RTP) (RFC 3550) for transporting real-time data and providing QoS feedback
- The Real-time Streaming Protocol (RTSP) (RFC 2326) for controlling delivery of streaming media
- The Media Gateway Control (MEGACO) Protocol/H.248 protocol for controlling gateways to the PSTN.

SIP is a principal component of multimedia architectures and the viability of converged networks will depend upon the economical implementation of SIP-signalling servers that are capable of meeting real-time performance constraints. Chapter 2 expands further on the role of SIP in call control in the PSTN and in converged networks. The nature of any packet network, with its widespread use of Network Address Translation (NAT), and related security risks, provide additional challenges to a voice service provider. Chapter 3 covers some of the measures that are required to secure the network from denial of service and theft of service attacks. Designers of the NGN have accepted the necessity for the NGN and the PSTN to coexist at least for the next decade. The description of how this is achieved is covered in Chapter 4 which divides naturally into three topics: PSTN interworking, PSTN emulation and PSTN simulation.

1.7 Growing Revenue

Adoption of a converged network infrastructure can provide cost savings for service providers, thereby increasing their competitiveness and profitability. However, increased competition for communication services continues to drive down service prices, resulting in lower and lower margins for service providers. In the not-so-distant future, service providers in competitive markets will be faced with the choice of either becoming the low-cost "communications" utility (the so-called Cheap Fat Pipe) or attempting to grow revenue by providing additional value to their customers.

The initial defence of most service providers against revenue erosion is the bundling of services. The Triple and Quad Play business models are attractive because delivery through a converged network is more efficient and the customer benefits from the convenience of a single bill. However, this approach alone does not allow service providers to differentiate their service offering from that of competitors; in this scenario the only differentiation becomes price resulting in another downward price and margin spiral, as competitors strive to take market share from each other.

Service differentiation and revenue growth are ultimately to be found in the form of Value-added Services delivered over and above the basic communication services. Much attention has been given to the service layers of converged network architectures and, in particular, how to enable the delivery of the best-in-class applications in a timely and cost-effective manner. Since there is little indication that the application vendor community is gravitating towards a single technology for the creation of applications and services, most service providers will require the ability to deliver services from application platforms based on different technologies, in order to provide themselves with the widest set of application vendors, from which to select the best-in-class applications to deliver to their customers.

Next generation service architectures must also support two other key enablers for service provision, these being *Service Velocity* and *Service Agility*. Service Velocity is quite simply the ability to get new service offerings into the market in much shorter timescales than was possible with traditional service creation and delivery technologies and techniques. This provides two benefits to service providers. Firstly, and most obviously, the time to revenue is reduced; service providers receive revenue from services much earlier. Secondly, the service provider's competitor may launch a successful service before the service provider does; service velocity enables service providers to react rapidly and release their version of that service quickly, curtailing the period of market exclusivity enjoyed by their competition and thereby reducing any impact in terms of subscriber and revenue churn.

Service Agility is the ability to maximise the utilisation of any application a service provider deploys. Typically this means enabling the inclusion of any deployed application into different service bundles aimed at different market demographics. Service Agility distributes the cost of an application across multiple service propositions. Together, Service Velocity and Service Agility can reduce the cost, time and effort needed by service providers to deliver new services. This in turn can extend the "economic reachability" of the value-added services market, and open up new market opportunities for service providers.

Traditional service deployment methods, being relatively both costly and time consuming, have restricted service deployment to those services that could be sold to a large proportion of the subscriber base and would be active in the network for an extended period of time. Reduction in cost and increase in the time-to-market speed allow service providers to profitably launch services with limited penetration (targeted at a particular subscriber niche) or a limited lifespan (e.g. a service associated with a reality TV program). Each individual service may yield less revenue than a traditional service; however, the service can still be profitable in its own right, and there are likely to be many more of these. It is also important to consider that services with the traditional penetration-duration demographics can still be deployed, but at a lower cost, and the new niche market and ephemeral services represent additional revenue over and above that provided by these traditional services. The reduction in cost to deploy new services also permits service providers to take a much more entrepreneurial role with new services. Service providers can put new services into the marketplace with considerably less market planning and testing since the business consequences of unsuccessful services are considerably reduced.

Chapter 6 provides a detailed insight into the new architecture that service providers are turning to, in order to economically deliver new and innovative services to their sub-scribers. Chapter 8 reminds the reader that the user's perspective of satisfactory QoS is an essential consideration, but the complexity required to manage and enforce differentiated levels of QoS, with sufficient granularity, must be weighed against a previous remedy, which is over-provisioning the network. Chapter 8 sets out different strategies for the implementation of QoS.

1.8 Network Operators – Dealing with Convergence

This last section of this chapter deals with several real-life scenarios, which are intended to illustrate why, and how, Converged Multimedia Networks will be implemented, and should also help you to find your way around this book.

1.8.1 Scenario 1 – A Cable Operator

A cable operator currently provides "Triple Play" bundled services, including a range of broadband and dial-up Internet services, local, long-distance and international telephone services, and digital and analogue cable television, to residential customers, and provides a range of retail and wholesale voice, data and Internet products and services to the business market.

The Cable Operator is considering merging with a Mobile Network Operator (MNO) or a Mobile Virtual Network Operator[2] (MVNO) to provide Quad Play Services. The mobile operator should be able offer a reduced tariff for mobile calls, which originate in the "home" cell, to compete with those charges applied by a fixed PSTN operator (approximately 30% of calls from home are made on a mobile). A further aim of the Cable Operator is to provide one phone, one phone number, one bill and one mailbox, and additionally they also would like advice on the best way to protect the QoS of each of the services.

The Cable Operator should deploy a single service layer for both the fixed-line access and mobile access, using the emerging IMS architecture as described in Chapter 5, which explains how the IMS model facilitates the one phone concept, and this chapter also looks at how broadband data wireless technologies can provide a "bridge" between fixed and mobile networks.

The introduction onto the network, of mobile traffic with its characteristically tight delay requirements, means that the Cable Operator will need to ensure strict separation between delay-sensitive real-time services, and congestion-aware non-real-time services. In cases of failure, careful network design can contain less critical traffic, in favour of higher revenue premium services. Chapter 7 deals with Core Network Architectures and shows how a common network protocol can be used to support a broad range of services, while achieving the required performance objectives.

Chapter 8 on QoS and Bandwidth Management describes how Connection Admission Control (CAC) can guarantee QoS for mission-critical services, both within a single service provider's domain as well as across the multiple domains of different service operators.

---

[2] A mobile virtual network operator is a company that does not own a licensed frequency spectrum, but resells wireless services under their own brand name, using the network of another mobile phone operator.

1.8.2 Scenario 2 – A Video-on-demand Service Provider

A Video-on-Demand (VoD) network connects regional server centres to the national PSTN operator's local exchanges. Local Loop Unbundling (LLU) and Asymmetric Digital Subscriber Line (ADSL) technology is used to deliver digital television and video services to customers, over a normal telephone line, without affecting the customers' existing telephone service.

VoD The Service Provider would like to extend their service offering to include telephone services. The VoD company wants a single authentication process for all services as well as the ability to provide telephone services and to include personation of all services. They also want communication regarding incoming calls to appear on the subscriber's TV and the ability for the subscriber to divert the calls to another terminal/telephone or to voice mail (Figure 1.2). The VoD company should look to provide a voice solution based on the SIP protocol, and Chapter 2 describes how SIP user registration and authentication is implemented and how peer-to-peer voice services can be supported using SIP. Chapter 6 has examples of how applications can be grouped together into Service Capability Features (SCFs) which can export Call Control and Charging and provide for centralisation of authentication and billing services.

If the VoD company is looking to provide a complete PSTN replacement service (rather than a second-line service), they should deploy an access platform that is capable of handling any xDSL data in addition to terminating the analogue phone line and converting it to voice over IP. The access platform must include an access gateway, which will be controlled by a call control application, using a protocol such as H.248. Chapter 4 looks at access gateways and the H.248 protocol and also describes how an operator can interconnect back into the TDM PSTN in the core of the network, to allow a cost reduction in calls, originating or terminating in the legacy PSTN.

Figure 1.2   Combining Video Conferencing with Telephony

Providing a full PSTN replacement service implies having a very robust solution for QoS. Chapter 8 describes how to protect the network against the kind of overload caused by emergencies or television phone-ins. The network must be able to connect calls to both analogue phones as well as SIP terminals, and a Session Border Gateway should be deployed to secure the network against potentially misbehaving PC based SIP clients. Chapter 3 discusses the role of the Session Border Gateway in securing a network as well as describing some other measures that the company could take to protect themselves against attack.

A key service, which the VoD company should consider offering, is the ability to present a call to any number of possible terminals and to provide appropriate notification of an arriving call. This can be achieved using the capabilities of the SIP protocol. Chapter 2 explains how SIP Forking and its redirection services can be used to send incoming calls to more than one device and how to provide an efficient location service. Chapter 6 includes an example of the Whisper connect service which allows the multimedia subscriber to determine who is calling him before deciding whether to interrupt his current call in order to take the new call.

1.8.3 Scenario 3 – A High-speed Internet Service Provider

A new high-speed Broadband Internet Service Provider uses Local Loop Unbundling (LLU) to provide residential customers and businesses with a high-speed Internet service. The Digital Subscriber Loop (DSL) access infrastructure is in place but the company wants to understand the process of interconnecting to the PSTN to find out which other components must be added to their network. Multimedia conferencing is a service they would eventually like to offer (Figure 1.3).

Gateways are needed to interconnect to the PSTN, and SS7 signalling, used in the PSTN, has to be converted to SIP. Chapter 4 explains the principles behind gateways, describing the functions of a Media Gateway (MGW), a Media Gateway Controller (MGC) and a Signalling Gateway (SGW), which are needed to interconnect the Internet to the PSTN.

In the NGN, multimedia sessions are supported by multiple RTP streams, for example, one for video and one for the soundtrack. Chapter 4 describes how Multimedia Conferencing uses RTP to take the contributing sources that make up a media stream and mixes them together while ensuring that each stream maintains its own synchronisation source. Voice services require low delay and delay variation because they are interactive, but voice can usually withstand a limited number of voice samples being lost by the network, whereas Video services require a low packet loss, but they are typically not interactive and so packet delay and delay variation can be higher. Chapter 8 is dedicated to the provision of QoS for multimedia services.

1.8.4 Scenario 4 – A Mobile Operator

A Mobile Operator is currently upgrading the network from 3GPP Release 99 to 3GPP Release 4. Access to the Circuit Domain for standard GSM calls can continue, but eventually more and more circuit-switched services will be offered from the packet domain. The mobile operator is interested in the potential for expanding its service offerings, through a partnership with a fixed operator. The company is also evaluating new wireless technologies such as Unlicensed Mobile Access (UMA) and would like to understand how Broadband wireless might benefit its business (Figure 1.4).

Figure 1.3   Multimedia Conferencing

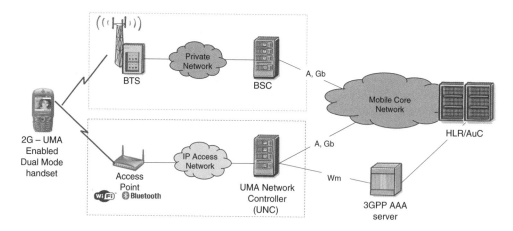

Figure 1.4   Mobile upgrade from Release 99 to Release 4

The Mobile Network Operator (MNO) is planning a package of Rich Services:

- Availability – "on line" (objective), "don't disturb" (personalised), "only urgent calls" (indicative)
- Reachability – "In a call" (network state), "on a broadband connection" (network capabilities), "In the airport" (network location)
- Location – "abroad" (roaming network), "in the station" (place), "00°12'23"N– 21°40'11"W" (coordinates)

A variety of rich services can be offered by Application and Media Servers. Chapter 6 shows that by insulating the applications from one another, it is possible for a Service Provider to mix and match application offerings from different vendors, without requiring the vendors to make custom changes, in order to get a particular application to interwork with other applications.

1.8.5 Scenario 5 – A Fixed Network Operator

A Fixed Network Operator hopes to integrate Enterprise Private Automatic Branch Exchanges (PABXs) connected to wireless access points, with mobile access, through setting up a partnership with a MNO or by becoming a Mobile Virtual Network Operator (MVNO) (buying airtime from a MNO) (Figure 1.5).

Wireless "Hotspots" can be used to transfer a call from a mobile to a fixed network and recent advances in the development of wireless are a hugely significant player in the advance towards Converged Multimedia Networks. The fixed operator would like to understand the benefits of wireless access. Chapter 5 includes information on the shorter-range wireless protocols such as WiFi and Bluetooth, as well as information on Wireless Local Area Networks (WLANs) and Broadband wireless.

1.8.6 Scenario 6 – The PSTN Operator

A PSTN operator is facing the need to upgrade their existing PSTN switches; in recent years, they have rolled out broadband services successfully, and rather than invest capital

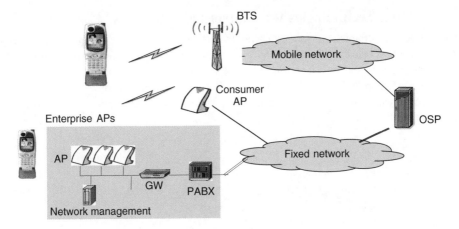

Figure 1.5   Combining fixed wireless access and mobile

in TDM technology, they have decided to migrate to a single converged network based on their broadband infrastructure, while still supporting the PSTN (which is a critical national infrastructure and is regulated).

To achieve this, the network operator must upgrade their xDSL infrastructure to deploy multi-service access nodes that support access gateways. Customer lines must be gradually migrated onto the access gateways from their TDM local exchanges. The carrier needs to upgrade their call control to a SIP-based infrastructure, possibly using SIP with ISUP (SIP-I) (see Chapter 4) for PSTN transparency, and provide a control interface to the access gateways, using the H.248 protocol with the Stimulus Analogue Line Package which guarantees support for the full range of European Telecommunications Standards Institute (ETSI) PSTN services. Because this company currently offers Integrated Services Digital Network (ISDN), they must cater for ISDN user terminals by making use of the Internet Engineering Task Force SIGTRAN architecture, using the ISDN User Adaptation Layer. Chapter 2 looks at some of the strategies that may be used by an operator migrating from broadband to a full PSTN replacement service and describes how the SIP protocol can be used to support voice services over a packet core. Chapter 4 describes the role of access gateways, H.248 and SIGTRAN, and shows how the SIP-I variant of SIP assists in migrating services from the PSTN.

To transport PSTN voice over the same network, as broadband and business services, the operator may look to deploy a technology such as Multiprotocol Label Switching (MPLS), with its support for facilities such as fast re-route, its ability to support Layer 2 private wire services as well as Layer 3 VPN services and its relatively sophisticated QoS technologies such as Traffic Engineered tunnels. Chapter 7 describes the resilient multi-service capabilities of MPLS and Chapter 8 describes how the MPLS network can provide service separation and QoS through MPLS traffic engineering. PSTN replacement, while complex, is probably just the start and the carrier will eventually want to develop new services to take advantage of SIP-based call control, using many of the techniques described in Chapter 6. The carrier could bring mobile communications into their portfolio, and may look at moving their call control to an IMS architecture eventually, as described in Chapter 5.

1.9 Enabling Technologies for Converged Networks

This introductory chapter has looked at the drivers for network convergence and described some example scenarios to show what this really means to network operators and to provide pointers to the key technologies that the network operators will need to use in these cases. The remainder of this book will look at these enabling technologies for converged networks in detail, describing the problems that they solve and characterising the solutions they provide.

2

Call Control in the NGN

While much has been made of Voice over Internet Protocol (VoIP) and the advantages
of migrating to an IP-based infrastructure, it remains the case that voice support in the
converged multimedia network relies on a call control function that remains fundamentally
unchanged from the Public Switched Telephone Network (PSTN) based Call Agents
deployed in networks today. In part, this is due to the fact that the complexity of call
control and the vast array of service intelligence and features that has built up in the
PSTN makes it impractical to start from scratch in the new networks and also due to the
fact that changing the network infrastructure from the Time Division Multiplexed (TDM)
circuit-switched PSTN to a packet-based converged IP network does not change the fact
that the voice service many people wish to deploy is largely unchanged at the application
layer.

This chapter examines the evolution of networks towards the converged multimedia
Next-generation Network (NGN) and considers the properties of call control in both the
traditional PSTN and in the new converged networks. It also examines how the approach
to network design differs between those operators looking to provide a low-cost voice
over Internet service and those operators looking to deploy a complete PSTN replacement
service.

Having outlined the basic architecture and network elements that make up many of
today's VoIP solutions, the chapter looks in detail at a number of areas.

Section 2.2 provides an overview of call control, Section 2.3 describes how it was
implemented in the PSTN and Section 2.4 considers how call control implementations
must change to run over a converged multimedia network. The Basic Call State Model
(BCSM) that is used in some form by all call control platforms is described in Section 2.5.

In order to allow users to communicate with each other, call control requires a peer-to-
peer signalling protocol and in the core network, this is primarily the Internet Engineering
Task Force (IETF) Session Initiation Protocol (SIP) [1]. Section 2.6 looks at the SIP
signalling required to set up a peer-to-peer voice call in detail. It considers the basic
messages involved, the routing used and some of the advance features that the protocol
supports including how the SIP protocol allows the NGN to support a degree of nomadicity
by allowing individual users to register their location with their chosen voice provider.
In addition to controlling the call state, NGN signalling must also be able to describe the

Converged Multimedia Networks   Juliet Bates, Chris Gallon, Matthew Bocci, Stuart Walker and Tom Taylor
© 2006 John Wiley & Sons, Ltd

type of bearer that two users wish to use to communicate, as there are many types of voice and video codecs that users may wish to select. In the NGN, the protocol chosen to do this is the IETF Session Description Protocol (SDP) [2] which is described in detail in Section 2.7. In the NGN, voice and video is transported within IP packets, which, given the connectionless nature of IP networking, requires additional information to be passed end to end to assist in clock recovery and depacketisation. This is achieved with the IETF Real-time Transport Protocol (RTP) [3] which is described in Section 2.8.

In order to support this infrastructure, it is necessary to allow individual users to be addressed using either fully qualified domain names or other mechanisms such as traditional telephone numbers. A major issue that providers of voice services face is how to move between the world of telephone numbers and the world of IP addresses and domain names. Section 2.9 discusses the mechanisms chosen to resolve these issues and looks at the role of telephone number mapping services (ENUM services) [4] in supporting this.

In this book, it is not possible to describe in detail all the solutions and all the protocols that are used to support voice and peer-to-peer video services over the NGN. However, the most important protocols SIP, SDP and RTP are looked at in detail because they are so important. This includes examples that describe the messages used in typical exchanges down to the actual parameter values. It is not necessary for the casual reader to dwell on this information if they are just seeking an understanding of the framework; however, it is hoped that providing such detail will enable any user wishing to debug or analyse such networks with some useful examples that will make the process of understanding the messaging considerably easier.

Throughout this chapter, reference is made to standards, but particular attention is paid to the work of the MultiService Forum (MSF) and its Implementation Agreements (IAs), which have been used for the real-world interoperability testing of NGN solutions at Global MSF Interoperability events. These IAs bridge the gap between the traditional industry standards bodies such as the IETF and the International Telecommunication Union Telecommunication Standardisation (ITU-T), and network deployments. This is achieved by identifying issues and ambiguities in the standards and by offering more tightly defined standards profiles to allow network operators to get a head start in implementing NGN solutions. In addition, many IETF and ITU-T standards are referenced and while this chapter hopes to assist the user in accessing these standards and understanding them through many detailed examples, the standards must always remain the primary point of reference for any engineering project.

2.1 NGN Network Architectures

Call control exists to enable peer-to-peer voice and multimedia communication between users over a network. Traditionally, this has been achieved using a TDM circuit-switched network consisting of local exchanges (known as *Class Five exchanges* in North America), which provided service intelligence and predominantly analogue signalling to the customer's terminal equipment. These local exchanges were interconnected by a network of transit switches which had the function of concentrating inter-exchange traffic from a large number of local exchanges onto a mesh of large TDM-based links known as *trunks*. These transit exchanges typically support fewer services but were much larger than the local exchanges (in the latter days of the TDM switching, call rates of up to 6.8 million busy-hour call attempts could be supported in a single exchange), in North America they were known as *Class Four switches*.

Individual network operators built large networks of local exchanges, connected by trunk exchanges and then entered into interconnect agreements with other PSTN operators to link these networks together over a network of international tandem exchanges. When mobile communications evolved, these mobile networks were linked into the same architecture, allowing universal voice and circuit-switched data communication between network subscribers. A typical example of this architecture is shown in Figure 2.1.

The diagram shows how customer phones are connected into the local Class Five exchange (Digital Local Exchange i.e. DLE) by Remote Concentrator Units (RCU). In this network, each DLE is connected to two transit switches (XIT) for resilience and these transit switches are meshed using a core network of high-capacity TDM trunks. In addition, the transit network is connected to two International GateWay exchanges (IGW), which provide a connection into the international network, allowing worldwide communication.

The emergence of the Internet in the 1990s led to a new type of network where customers subscribed to an Internet Service Provider (ISP) who gave them access to the loose confederation of networks and content that is the Internet. Typically, users dialled

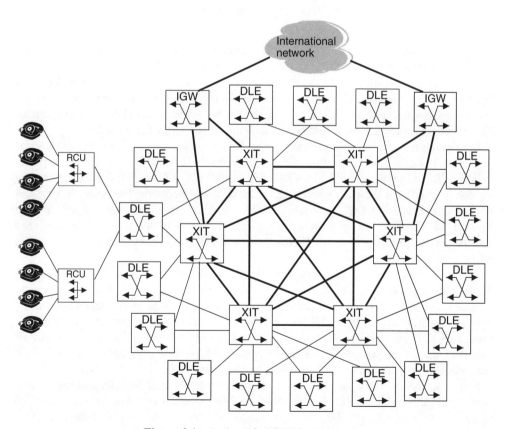

Figure 2.1 A simplified PSTN architecture

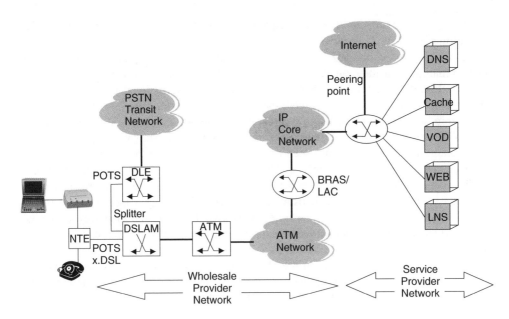

Figure 2.2   An early broadband DSL architecture

into a Radius Access Server using their PSTN landline and established a connection using the Point-to-Point Protocol (PPP) and were then authenticated by their service provider. As the Internet became content rich and the take-up increased, there was a drive for higher-speed access and subscribers moved away from the dial-up access to the PSTN and instead subscribed to cable modem or Digital Subscriber Line (DSL) based services. This architecture for a DSL-based access network is shown in Figure 2.2.

In this architecture, the customer gets the DSL and phone services from the same copper pair but the Digital Subscriber Line Access Multiplexer (DSLAM) uses a passive splitter unit to split out the baseband POTS signal and hand it over to the PSTN while forwarding the data carried over the DSL interface into the backhaul network. The network is typical of an early-generation DSL deployment with an ATM backhaul into a small IP core network to which service providers are connected. The network between the customer and the service provider typically belongs to a wholesale operator, usually a large public telecommunications operator.

In this example, the wholesale operator provides a combined Broadband Remote Access Server (BRAS) and an L2TP Access Concentrator (LAC), which is shared amongst many service providers. There are two modes of operation that can be used, depending on the network operator's preference. The first is to just provide a LAC; in this case, each service provider's PPP traffic is concentrated and passed through the core IP network in an L2TP tunnel to the service provider's interconnect point. The service provider is responsible for authenticating the customers and terminating the PPP using their L2TP Network Server (LNS). The second mode of operation is to provide a full BRAS facility to the service provider. In this case, the BRAS terminates the PPP on behalf of the service provider and routes the IP packets over the core network to the service provider. A radius interface is provided to the service provider to allow the authentication of the customer and the remote configuration of network information such as the customer's IP address and the service provider's Domain Name Services (DNS) addresses.

The service provider offers caching, web servers for locally hosted content, possibly services like Video On Demand and critically a high bandwidth IP interconnect or peering point out to the wider Internet. Although the customer receives the POTS and the DSL service over the same copper pair, the POTS and data networks are completely separate, with the end user's phone service being entirely supported from their local exchange.

The PPP-based architecture was limited in terms of Quality of Service (QoS) compared with the PSTN network but it allowed rapid data transfer, and most of the services used by customers such as web surfing, email and file sharing did not require bounded delay and jitter but instead required significant bandwidth. As technology improved, access and core network speeds increased such that current xDSL services of 50 Mbps access speeds are not unknown in Asian markets and today these markets are starting to move away from high-speed copper access towards fibre to the home which will enable speeds of 100 Mbps and beyond to be achieved [5].

As bandwidth became cheaper and packet technology improved, it became clear that peer-to-peer voice, previously the preserve of the PSTN, could be seen as just another service on the packet network. Hardware and software-based codecs can packetise voice into relatively high bandwidth streams equivalent to the quality seen in the PSTN (the G.711 codec) or into lower bandwidth streams that offered equivalent quality to early mobile networks (the G.729 codec) and a number of points in between these two extremes. Although packetised voice carries a premium on bandwidth compared to the TDM (a 64 kbps voice stream typically requires bandwidth of the order of 120 kbps on a packet network, based on a G.711 codec with Ethernet encapsulation), these bit rates are acceptable when compared to the access speeds available on DSL and fibre access networks. Furthermore, because the cost per bit carried is very much less in a packet network than in the TDM network, it rapidly became clear that a business case could be made for low-cost voice services carried over "the Internet".

In order to provide voice services over a traditional packet network, a service provider typically needs to invest in a number of specialised voice platforms as follows:

- A Call Agent that provides the call control application for establishing peer-to-peer voice communication as well as billing and other capabilities essential for operating a voice service (e.g. in many territories it must be possible for lawful intercept facilities to be available on any public voice network).
- A Session Border Controller (SBC) which handles issues arising from the widespread use of Network Address Translation (NAT) in packet networks and also provides a gate functionality to ensure that only media streams associated with authorised voice calls are permitted into the network.
- A media gateway that is able to provide a mechanism for interconnect with the TDM-based PSTN to allow customers of the packet voice service to communicate with the vast majority of people who remained on the traditional PSTN. This type of media gateway is referred to as a Trunking Gateway (TGW) because it typically connects to class 4 transit switches in the PSTN and these are often referred to as *trunk switches*.
- Residential voice gateway equipment that allows a customer to plug a telephone into their broadband connection. The residential voice gateway is responsible for ensuring

that the phone rings, provides dial tone and supports features such as call waiting and caller display while using only the broadband data path to connect to the network. In addition, the voice gateway is responsible for packetising the voice for transmission over the data network. Alternative solutions using PC-based clients rather than black phones and gateways can also be used but typically appeal to a more tech savvy and, hence, less mass market audience (they also typically cannot provide an exact replication of the PSTN service set).

- A key requirement for network operators providing a voice service is getting the Operational Support Systems (OSS) and billing systems in place to ensure that the business can charge for calls and thus operate profitably. These aspects of a voice service become particularly important when interconnecting with the PSTN because network operators charge for terminating or transiting other networks calls.

A typical network architecture employed by an operator or ISP looking to provide a low-cost voice over the Internet service is shown in Figure 2.3. This type of Internet-based VoIP service is sometimes referred to as a *voice on net* service, as it is VoIP running over a customer's basic Internet service.

The network architecture shows two customers connected to the ISP's voice service. They are each supplied with (or purchase) a Residential GateWay (RGW) device which, for example, has ports for two normal phones on one side and an Ethernet port for connecting to a home network on the other. Customers are connected to the ISP's voice service through their SBC, which handles any NAT that has taken place. The Call Agent

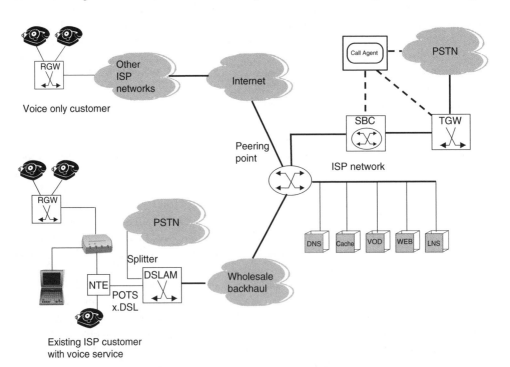

Figure 2.3   Adding a VoIP service to an ISP's portfolio

controls the RGWs, responding to local events, such as a subscriber going off hook and dialling a number, and also alerting the subscriber to incoming calls. It provides the essential call-routing function to identify the called subscriber and route the media path accordingly. All media is passed into the ISP network and reaches the SBC where (depending on the Call Agent's instructions) it is either passed back out into the packet network to terminate on another subscriber's RGW or sent to the PSTN using a TGW to convert the packetised voice back into TDM. In this case, the Call Agent will send call set-up signalling to the PSTN (usually using Signalling System No 7).

The key point of this infrastructure is that the use of the Session Border Controller (SBC) enables the solution to be independent of any NAT that has taken place between the point at which the customer has connected to the network and the edge of the voice service provider's network. This ensures that subscribers can take their RGW anywhere in the world, plug it into a broadband connection and receive exactly the same telephone service that they would get at home; furthermore, they will not suffer any additional billing, penalty despite the fact that they are calling from another country because the access portion is effectively free on a per-call basis (the customer still has to pay for the basic broadband access). For example, in the diagram, both customers connect to the PSTN at the same TGW (and incur the same PSTN call charges, regardless of where in the world they are). Similarly, regardless of where in the world the customer is located, somebody dialling his or her phone number will always be connected to him or her and without the customer having to pay the incoming call costs associated with mobile roaming scenarios.

However, this type of solution is not a replacement for the PSTN in reality, although in many countries, it has had a sharp impact on PSTN revenues, because it suffers from a number of limitations. The key limitation is that there is a lack of QoS. In most circumstances, the customer will receive low packet loss and latency and, hence, acceptable voice quality; however, if, for whatever reason, the packet network becomes congested (either because of unexpected demand or because of a network failure), all the calls on the network, without discrimination, are likely to suffer significant degradation and there will come a point where users will experience too much packet loss or delay to carry on an effective telephone conversation. While this type of user experience is not by any means common, the nature of networks and people is that the load on networks tends to increase significantly when events on the ground dictate that they are most needed; furthermore, once a network becomes congested, customers retry calls, thus leading to an ever-increasing load. While the PSTN employs sophisticated overload control and call gapping technologies to ensure that the most vital calls get through, this type of voice on net service lacks the required controls and so may collapse completely in extreme circumstances. Therefore, it is not a suitable architecture on which to build critical national infrastructure but is adequate for normal usage patterns.

Another limitation of this type of service is that the same lack of location awareness that allows the customer to receive calls anywhere in the world without additional billing, also makes identifying the physical location of the customer difficult. This has implications for the emergency services, which typically like to be able to receive verifiable information about the origin of a call. Some network offerings therefore specifically exclude access to emergency services, while others permit it, preferring to hope that well-trained emergency centre operators can cope with a lack of location information.

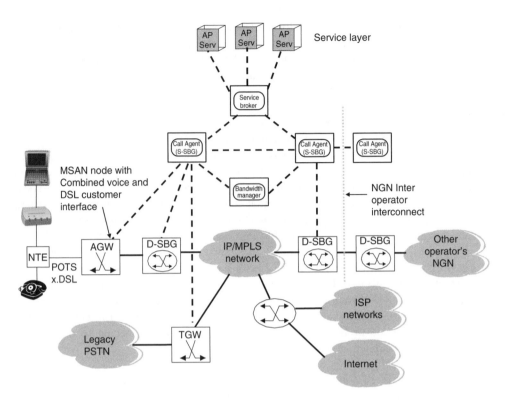

Figure 2.4 A converged NGN based on the MSF R2 architecture

As PSTN operators struggle to come to terms with the impact of low-cost voice on net VoIP services eroding the margins of the TDM network and the need to support new services to compete with the multimedia capabilities of packet-based voice services, a new type of NGN is starting to emerge, one that attempts to wholly replace the PSTN. This type of architecture has to be able to offer all the QoS and resilience to overload that the existing PSTN can provide while allowing voice to be carried over a lower cost and multimedia-capable packet network infrastructure. The issues of QoS and reliability are difficult problems to solve because of the limitations of the existing IP router-based infras-tructure, and significant work has been undertaken within the many telecoms standards organisations to try and address these issues, in particular, by the ITU-T and by ETSI TISPAN (which is the ETSI competence centre looking at converged networks) in Europe and ATIS (the Alliance for Telecommunications Industry Solutions) in North America. A good example of what such a network might look like is the MSFs Release 2 architec-ture, as shown in Figure 2.4, which concentrated on how the physical boxes required to provide a PSTN capable NGN would be deployed and connected [6, 7].

This network architecture is an evolution from the voice on net solution in that it contains additional elements to tightly integrate the voice and multimedia services with the underlying network, thus allowing QoS, security and overload control to be implemented. The key elements of this architecture are as follows:

- The Call Agent performs the same functions as the VoIP voice on net architecture; however, it has additional capabilities to transport legacy PSTN call signalling within the NGN. In addition, it is capable of supporting a full PSTN service set to customers who are connected to the network via Access Gateways. The Call Agent may peer with neighbouring networks' Call Agents to enable true IP interconnect of voice services. It should be noted that the terms Call Agent and Call Agent are often used interchangeably in the industry to refer to this network entity.
- The access gateway allows customers with traditional POTS phones to connect to the network with their existing equipment and without requiring additional equipment. This type of access gateway is in reality a part of a Multi-service Access Node (MSAN) that is able to take POTS, ISDN and broadband services and connect them to the core network while providing the QoS and overload control functions required. An NGN Call Agent using H.248.1 [14] controls the access gateway; in Europe this may follow the V5 [8] model as supported in H.248 using the ETSI Stimulus Analogue Line package (Stimal) [9]. See Chapter 4 for a fuller discussion on Media Gateways and Access Gateways and the use of H.248 packages.
- The SBG effectively performs the same role as the SBC did in the voice on net VoIP architecture. However, in this type of NGN, it may be decomposed into a gateway portion (the D-SBG) and a controller portion (the S-SBG). It is used to secure the boundaries of the network for packet-based terminal access and also the boundaries of the network for inter-carrier interconnects. In fact, the gateway portion of the SBG, that is responsible for carrying the data packets, is best thought of as an enhanced IP edge router, which may be required to scale upwards of 160 Gbps of throughput.
- The bandwidth managers provide a mechanism for guaranteeing QoS over the packet network by allowing the Call Agent to request bandwidth to be reserved for individual calls or for aggregate groups of calls by the Call Agent (and other application servers that wish to use the network to deliver services to end users). In the event of network congestion, the bandwidth managers provide a Connection Admission Control (CAC) function that protects calls in progress from being impacted by new call attempts. In addition, bandwidth managers also provide mechanisms to allow individual calls to receive preferential treatment in the event of an unforeseen network outage, thus ensuring that not all calls are degraded and that key government and emergency service communications can be maintained. A detailed discussion on QoS and bandwidth managers is provided in Chapter 8.
- The TGW provides the same functions as the voice on net VoIP service, allowing the NGN to interconnect with any PSTN networks that remain on the old TDM infrastructure; however, over time the number of these gateways are likely to decline as more networks move to an IP infrastructure.
- At the centre of the IP network are a number of very large core routers providing the high-speed backbone of this multi-service network. These are able to support at a minimum class-based queuing. They ensure that traffic that has been marked by access gateways and SBGs as high priority (e.g. voice and real-time peer-to-peer traffic) receives preferential treatment compared to that marked as lower priority (e.g. web surfing or file sharing) traffic. In the case of an MPLS network, traffic-engineered tunnels may be provisioned in the network to provide hard bandwidth guarantees; see Chapter 7 for further information about the IP/MPLS core network.

- Above the Call Agents, the Service Broker provides a mechanism to arbitrate services between the Call Agents and the application servers that support legacy Intelligent Network (IN) services or new multimedia services, possibly built using Parlay and Parlay-X APIs. The service broker provides a powerful tool for handling the problem of feature interaction that has long plagued telephony service designers.

Although these converged voice and multimedia networks are very much in their infancy, they are now starting to be deployed in the real world, in the UK BT's visionary, 21st Century Network is one such example. However, the world of telecoms standards does not sit still and a further wave of convergence is being proposed on the basis of the emerging mobile IP Multimedia Subsystem (IMS) architecture. As a result of this, the MSF have published their Release 3 Architecture [10] which evolves the architecture shown in Figure 2.4 to integrate it into the IMS core. The 3GPP-defined IMS is described in detail in Chapter 5.

2.2 The Operation of Call Control

The function of call control is to enable peer-to-peer voice communication, typically between two users, and to provide a framework for the addition of value-added services which allows features such as multi-party conference calling, call forwarding services and call completion services such as ring back when free (otherwise known as Call Completion to Busy Subscriber (CCBS) [11]). At the heart of a call control system is the Basic Call State Model (BCSM), which is typically split into two halves, the originating BCSM (O-BCSM), which maintains the call state from the calling party's point of view and the terminating BCSM (T-BCSM), which maintains the call state from the called party's point of view.

Call control maintains these state models for each individual call and provides the facilities needed to set up, maintain and tear down the call. These facilities are typically as follows:

- Digit and number analysis, which is used to determine the number dialled by the user against a national dialling plan and then analyse the number to provide an entry into the second key component of call control.
- Call Routing, which selects an outgoing route for the call, effectively the next network hop to where the call is to be sent. In the PSTN, this might resolve to a SS7 trunk group (if the next hop is a class 4 or transit exchange) or a port on a line card in a class 5 local exchange (if the next hop is an end user).
- Call Signalling, which is used to communicate with the identified peer call control entity. This might be another exchange using a signalling system such as ISDN User Part (ISUP) [12] or an end user terminal using a signalling system such as Q.931 [13] for ISDN terminals or V5 [8] if the call is being sent to an analogue phone terminated on a traditional access concentrator. In the NGN, signalling protocols such as SIP [1], MGCP [14] and H.248 [15] are utilised but their function remains fundamentally the same as the PSTN protocols they replace.
- Bearer Control, which is used to allocate and maintain an appropriate bearer to carry the voice communication to the next hop. In the PSTN, this was a TDM timeslot on a

trunk or port, while in the NGN, this is typically an IP-based Real-time Protocol (RTP) session [3].

- Billing, which is used to record the information required to bill for a call, such as origin, destination, tariff and duration and which is typically passed to an offline platform such as a call-rating engine for processing.

At certain points in the call, it is possible for the user to invoke additional services, for example, by keying a digit on getting subscriber engaged to invoke the CCBS service. These points in the call are represented as trigger points in the BCSM and the trigger point at which a feature is invoked forms a key entry into the service processing. Initially, such services were embedded in the call-processing logic of the telecoms switch; however, in the late 1980s and early 1990s, the Intelligent Network (IN) standards were developed which allowed this feature processing to be formalised and permitted off-board feature servers called *Service Control Points*, thus allowing network operators to buy services from third parties and deploy them on stand-alone computer platforms. With the introduction of the NGN, additional flexibility has been added to service processing, allowing greater openness and faster time to market for new services. The handling of services in the NGN is described in detail in Chapter 6.

2.3 Call Processing in the Legacy PSTN

In order to understand the mechanisms behind call control in both the legacy PSTN and the emerging NGN, it is useful to examine how a Call Agents might be implemented. Figure 2.5 shows an abstract model of a PSTN switch identifying key components and their interactions.

Central to the operation of the switch are the two Basic Call State Models, one for the originating call half and one for the terminating call half. These state machines drive the call set-up, supervision and teardown and are described in detail later. Above the two BCSMs, the service feature–processing function ties the two halves of the call together and provides the processing required to insert additional service logic into the call (the service logic blocks are shown above it). The invocation of service logic is triggered when a particular point in the call state model is reached and therefore an active service is invoked or at a point in a call set-up where a user enters a particular keystroke. The point in the basic call state model that has been reached is known as a trigger point, because it may trigger additional service invocation. For example, in the case of the CCBS service, invoked from a trigger point in the O-BCSM, the service feature–processing function would identify which service has been invoked and select the appropriate service logic for CCBS to insert into the call chain. The problem of which service should be selected and how it interacts with other services that may already have been invoked for the call is known as *feature interaction* and has long been a difficult problem for PSTN networks. This issue remains critical in the emerging NGNs and is discussed in detail in Chapter 6. An example of the type of problem that feature interaction processing must resolve is what should happen if the called party is engaged in another call and has set up call forwarding on busy to a personal assistant, who is also engaged in another call. Should the calling party be able to invoke CCBS when they hear subscriber engaged and if so, on which line should it set?

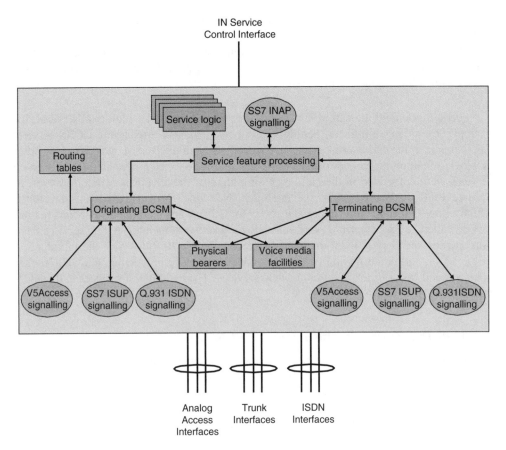

Figure 2.5   An abstract model of a legacy PSTN switch

During call establishment, the O-BCSM uses the call-routing tables of the switch to determine where the call should be directed. These routing tables consist of a large amount of call-routing data, statically defined, and are often complex and feature rich. For example, the entry point into the call-routing tables is determined by a number of factors of which called *party number* is just one and it is not unusual for the called party number to be modified as part of the analysis. To illustrate this, consider the case of a user dialling the emergency services, for the United Kingdom the dialled number is 999 or 112. This will result in a routing number, which corresponds to an emergency centre for handling the call; this routing number will replace the originally dialled 999 and a route will be selected to progress the call. If it is not possible to find a free route to the chosen emergency centre, it is possible that re-translation will take place, resulting in a new routing number (corresponding to an alternative emergency centre) being selected. The use of number analysis is, by its nature, proprietary to the switch vendor and the network operator handling the call. However, in North America, regulation has dictated that certain routing lookups are performed by off-board call databases. This results in a trigger being hit at the number analysis stage of the basic call state model and a query being sent to the correct routing database. By separating the call-routing information from the call control, this regulation has allowed greater competition by permitting smaller operators to process calls without the overhead of maintaining (for example) emergency services call routing because they are able to query a known database maintained by another organisation. In NGNs, this separation has been continued with the introduction of ENUM, which provides mechanisms for resolving number analysis via external servers. ENUM servers and their uses are discussed later in this chapter.

In order to perform call set-up and processing functions, legacy switches have a number of facilities for communicating with other network entities. These include signalling protocol stacks and of course the physical bearers required to actually carry the call. A number of signalling protocols are typically supported by switches depending on whether the call is originating or terminating on a Network-to-network Interface (NNI) or a User-to-network Interface (UNI). For a typical local exchange (a Class Five exchange in North America), the following signalling protocols would be used.

- ISUP for interfaces to other Call Agents, either trunk exchanges (known as *class 4 exchanges* in North America) or other local exchanges. ISUP is a peer-to-peer protocol and has many country-specific variants, for example, ANSI ISUP in North America, UK ISUP in the United Kingdom and J-ISUP in Japan.
- Q.931 for interfaces to ISDN user terminals. Q.931 is another peer-to-peer protocol and provides call set-up and bearer control functions over the access interfaces to ISDN terminals.
- V5 for interfaces to access concentrators. V5 is a master slave protocol whereby the Call Agent acts as the master to slave access concentrators. These access concentrators have limited intelligence and are responsible for translating Call Agent instructions into electrical conditions on lines to analogue telephony equipment. In North America, the GR.303 [16] protocol performs a similar function. A detailed discussion of V5 and access networks is provided in Chapter 4.
- Intelligent Network Application Part (INAP) [17] for communication between the Call Agents and IN platforms supporting value-added services.

Although these are the major protocols, the nature of the evolution of the PSTN and the different rates at which services and solutions were developed in multiple countries by large, often state owned, network operators mean that there is literally a plethora of signalling protocols that have actually been deployed. Within the United Kingdom, for example, most services today are supported using the BTNUP (British Telecom National User Part) signalling rather than the ISUP because ISUP was not available at the time the then British Telecommunications company was deploying its digital exchanges. Because BTNUP was able to support all the services required in today's PSTN, there was never a business justification to migrate to ISUP.

In North America, solutions such as Robbed Bit Signalling were developed which carried signalling information in-band with the bearer traffic by setting bits within the T1 bearer frame that would normally be used to carry the encoded voice (hence the name robbed bit). This approach is known as *Channel Associated Signalling*, and there are many subtle variations of it depending on the type of CPE that is using it.

This book concentrates on covering those protocols that form the basis for delivering call control over an NGN and, therefore, it is beyond the scope of the book to look in detail at the many signalling protocols that have been used in voice networks. At the fundamental level, all these protocols perform the same function, to enable call control to manipulate the state of a call at either the customer interface or the network interface to peer exchanges. It is important to note, however, that these legacy protocols have not truly gone away and, in many cases, support must be provided to tunnel them through the NGN or to interwork them at the boundary of the NGN (and in some cases, both). This problem of legacy solutions and protocols poses one of biggest challenges for network operators wishing to migrate their PSTN to a packet-based network and is often the least well documented.

The final components of a legacy switch are the media facilities. These are the trunks that carry the voice traffic and also the announcement machines, for playing recorded announcements and conference bridges to allow multiple parties to have their media merged so they may take part in audio conferences or three-way calls. The decision as to when to play an announcement, and when to connect a call to a conference bridge, is driven by the call processing, the events that take place during a call and the features invoked by the calling or called party.

2.4 Call Processing in an NGN Call Agent

Having examined how a legacy PSTN server performs call processing, it is interesting to look at how a next-generation Call Agent operates. The key functionality is clearly the same; however, the separation of call control from the bearer path in the NGN changes some of the components required in the Call Agent from those seen in a legacy switch.

As can be seen from Figure 2.6, the core call-processing capability remains fundamentally unchanged. However, what has changed from a legacy switch is that there are no longer any media resources resident on the Call Agent and that all of the interfaces that now connect to the Call Agent are IP based.

The originating and terminating BCSMs operate similar to a legacy PSTN switch; however, some of the states in the call state model exhibit different behaviour because the physical bearers are now separated and pass through other network equipment instead, such as media and access gateways. These differences are highlighted in Section 2.5, which discusses the originating and terminating BCSMs in detail.

The service feature-processing function is also fundamentally unchanged; Call Agent–based features can be inserted into the call logic as before. However, the INAP interface has been replaced by a service-level interface based on SIP. This service interface may be to a feature server or a service broker, which is able to arbitrate between different feature servers and potentially remove some of the feature interaction burden from the Call Agent. The MSF has defined a SIP interface between the Call Agent and the service broker [18], and the role of the service broker is discussed in detail in Chapter 6.

The routing function in the NGN Call Agent is enhanced to take advantages of the capabilities offered by off-board ENUM servers [4]. The ENUM framework provides a mechanism to allow a Call Agent to query a database with a telephone number and receive one or more DNS Naming Authority Pointer (NAPTR) resource records in return. These can be used to resolve the telephone number to an IP address, which would allow the call to be forwarded to, for example, the customers' serving Call Agent. ENUM and its implications are discussed later in this chapter.

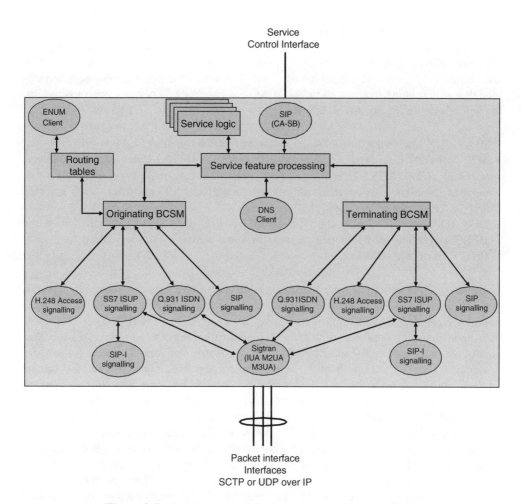

Figure 2.6   An abstract model of a next-generation Call Agent

The major difference between an NGN Call Agent and a legacy PSTN switch is at the signalling layer. Legacy access protocols such as V5 are replaced with the new ITU-T H.248.1 protocol, and the IETF SIP is widely used for signalling to end terminals and also as a transport mechanism to enable ISUP signalling to be carried over the network; in this role, it is known as SIP-I [19]. Because the NGN Call Agent is connected to a packet network and not a traditional TDM network, it follows that a mechanism must be found to transport traditional telephony signalling, such as Q.931 for ISDN customers and ISUP for interconnect to legacy class trunk and local telephone exchanges over an IP network. In order to solve this problem, the IETF has developed the Sigtran (Signalling Transport) suite of protocols [20].

Sigtran provides a number of tools to support legacy-signalling protocols over an IP network. For transport, it uses Stream Control Transmission Protocol (SCTP) [21] which offers resilience and recovery mechanisms above and beyond those provided by TCP so as to support the greater reliability and real-time resilience requirements of protocols originally developed to run on TDM infrastructure. Running above SCTP, a number of user part adaptation layers have been defined; these adaptation layers provide the protocol information required to allow the application (e.g. ISUP or ISDN signalling) to be encapsulated within an SCTP payload in such a way that the application can behave as though it was running over a traditional TDM network. For PSTN replacement networks, the key adaptation layers are as follows:

- IUA (ISDN User Adaptation Layer) to support ISDN signalling [22].
- M2UA (MTP2 User Adaptation Layer) [23], M3UA (MTP3 User Adaptation Layer) [24] and M2PA (MTP2 User Peer-to-peer Adaptation Layer) [25] which are all variations of solutions to support the transport of traditional SS7 signalling such as ISUP and TCAP over a packet network.
- DUA (DPNSS User Application Layer) [26] is used to support the transport of the DPNSS-signalling protocol that is used in some networks for signalling to Private Branch Exchanges (PBXs).

The roles of the various Sigtran adaptation layers in the NGN are discussed in Chapter 4.

Because an NGN Call Agent is connected directly to a packet network, often using multiple gigabit speed links and because the packet network runs IP (and is both more open and therefore more insecure than the TDN network it replaces), security and overload control are major issues that must be addressed in the new network. The need for security is obvious given the ubiquity of IP interconnects, and a number of solutions can be adopted, including firewalls and devices called *Session Border Controllers* or *Session Border Gateways*. The need for overload control is a side effect of the fact that NGN Call Agents may now offer service to a very large number of subscribers, and traditional blocks to signalling volume (such as the number of TDM-signalling circuits connected to the Call Agent) have been removed. This potentially allows call attempt volumes to reach levels that will cause the Call Agent processors to fail unless mechanisms are provided to throttle the load. This issue is complicated because in the PSTN, priority is required to be given to emergency calls and governmental calls and solutions must be found that permit priority calls to be identified while ensuring that the burden of analysis is not so heavy as to become comparable to the processing cost of actually carrying a call. A brief discussion on security is provided in Chapter 3 and overload control is considered in Chapter 8.

2.5 The Basic Call State Model

Because the BCSM is a key component of both PSTN switches and NGN Call Agents, it is worth exploring exactly what this does and how it does it. Probably the best example of a call control BCSM can be found in the ITU-T IN standards which defines the states a call moves through during set-up and identifies a set of trigger points within the call processing where value-added services can be invoked. The Intelligent Network (otherwise known as IN) standards defined a number of capability sets, which progressively introduced trigger points and expanded the call state model. Section 2.5.1 describes the originating call half of the BCSM as defined for IN capability set 2 [27] and Section 2.5.2 describes the terminating call half of the BCSM.

2.5.1 *The IN CS-2 Originating BCSM*

The originating call half is responsible for manipulating the call and invoking features from the perspective of the calling party and for controlling signalling and resources on the originating access or trunk interface. The states and triggers associated with the O-BCSM are shown in Figure 2.7.

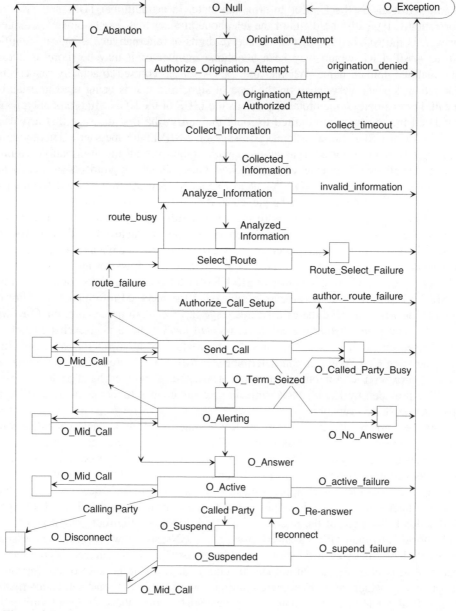

Figure 2.7   ITU-T CS-2 Originating BCSM (Reproduced with the kind permission of ITU)

When the call arrives at the Call Agent, it enters the Authorise_Origination_Attempt state. The Call Agent must determine whether the customer is entitled to make a call on the basis of his or her identity; typically this would be derived from the incoming physical line in a class 5 exchange; in an NGN network, this might use the incoming physical line for an H.248-based access or a logical identifier such as a MGCP domain name for an MGCP-based access or a Universal Resource Identifier (URI) for a SIP-based access.

If the call is authorised, then the Call Agent must collect sufficient information to route the call and it enters the Collect_Information state. In a traditional TDM network, this information is typically the digits of the telephone number that the customer has dialled; however, in the NGN, it is possible to receive digits or, alternatively, a logical identifier such as a SIP URI, which may be a telephone number or it may be more akin to a fully qualified domain name. Whether the Call Agent is required to actually request this information depends very much on the type of signalling that is being used to establish the call. For example, some protocols such as the ISUP or the NGN SIP protocol typically provide all the information required to route the call in the first message that arrives at the Call Agent (ISUP Initial Address Message or a SIP Invite message). This mode of working is typically referred to as *en-bloc* working because all the information required to route the call arrives together and unsolicited. Other signalling protocols such as H.248 and MGCP may require the Call Agent to explicitly request from the end terminal or access gateway that dialled digits be passed to it.

The Analyze_Information state is reached when sufficient information is received at the Call Agent to make an initial decision as to how to route the call. This routing decision is based on the digits received, if it is a traditional telephone number, and the subsequent analysis performed on those digits. The NGN allows for alternative formats such as fully qualified domain names in SIP URIs or telephone numbers contained within the SIP URI. In any case, the process of analysing the received information is followed to determine where to send the call. At this stage in PSTN call processing, the O-BCSM is able to determine whether the call is terminated locally (local inter-exchange call), is to be passed to a transit exchange via a trunk interface or is an international call. In a next-generation Call Agent, it can determine whether the call signalling will be passed to a SIP proxy server for onward routing or whether it will be handled locally with signalling provided by H.248, MGCP media gateway control functions or by a SIP User Agent Server (UAS) function.

When the Call Agent enters the Select_Route state, it selects a single candidate route from a list that it has identified as being available for the call. In a traditional network, these routes map to physical trunks and TDM circuits within the trunks. In the NGN, the route selection may map to a number of different next hop IP addresses or may be a null function. It is possible, for example, for the Call Agent to make a decision as to which of several interconnect points it wishes to use for an inter-network call. If the Call Agent has detailed knowledge of the underlying network topology, the route might potentially resolve to an individual MPLS tunnel; however, in NGNs, such low-level routing decisions are usually left to network-aware components such as bandwidth managers, as outlined in Chapter 8. It is possible that an individual route is unavailable because of congestion in the network, in which case, if there are multiple routes within the route set, an alternative is tried. If all routes within a route set are unavailable, then the Call Agent will return to the Analyse_Information state and try to identify an alternative route set for the called number. An example of this type of route processing is an emergency call, for example, 911, 112 or in the UK, 999. In this case, the dialled digits may resolve to a number of different destinations, which are the emergency call–handling centres. If all routes to a given centre are congested, then it is likely that the number analysis would be re-entered to select an alternative emergency call centre.

The Authorise_Call_Setup state is provided in the CS-2 call state model to give a final check to ensure that the selected route is compatible with the call type. If not, an error has occurred and exception processing is provided to terminate the call gracefully. It is unlikely that a next-generation Call Agent would require this call-processing state.

Having traversed the select route and, if applicable, the Authorise_Call_Setup state, the O-BCSM enters the Send_Call state. At this point, it passes a message internally to the T-BCSM informing it of the call details and requesting that the call be sent to the destination. This initiates call processing on the terminating leg. Having reached this point in the call state model, the O-BCSM may elect to send a signalling message back to the calling party to inform it that the call is progressing.

The O-BCSM enters the O_Alerting state when it is informed by the terminating call half that the called party is being alerted to the incoming call. In this state, a backward media path is required between the terminating exchange or user terminal and the originating exchange or user terminal. In the PSTN, the terminating end is responsible for playing an audible ringing tone to the calling party; because this tone is provided by the terminating exchange and not the originating exchange, inter-country calls will usually result in the calling party hearing the ringing tone of the terminating country. In some VoIP services, it is possible that the originating terminal will provide a ring tone to the calling party as this removes the need for connecting a media path from the far-end exchange. However, in reality, there are good reasons to provide a media path at this point in the call because it prevents clipping of speech caused by the customer starting to speak before the media path is fully switched through. This is possible if the backward path is not established before the called party is informed of the incoming call.

When the called party answers the phone, a signalling message is propagated back to the originating exchange and the terminating call half informs the O-BCSM that the call has been accepted and that billing can now start. The O-BCSM enters the O_Active state and two-way speech is possible between the calling and called parties. It is possible in this state for features to be invoked on the call, for example, the calling party might put the called party on hold to take an incoming call or initiate a new call, which could then be conferenced back into the first call, resulting in a three-party call. The call control was traditionally responsible for providing these features and manipulating the call state accordingly; however, the IN standards, by formalising the call model and identifying key trigger points within it, allowed for third-party control of services using stand-alone network entities called *Service Control Points*. This approach of distributed service processing has been carried forward in a more flexible way into the NGN and is discussed in detail in Chapter 6. The trigger points in the BCSM are shown in Figure 2.7 as small squares and have names such as "O_Called_Party_Busy" to allow the Service Control Points to arm themselves and uniquely identify which trigger has been hit.

The remaining state in the O-BCSM, the O_Suspended state, is entered if the call is placed on hold; such a call may return to the active state if required by the end users. The suspend state may also be entered for PSTN services should the called party hang up. In the PSTN, typically only the calling party can clear down a call and, hence, should the called party subsequently pick up the phone to try and initiate another call, they may find themselves connected back to the last call that they received. In reality, this rarely happens as the calling party has usually hung up and cleared the call. There are also specific services that might alter the clearing sequence for a call; for example, in the United Kingdom, there is a Malicious Call Identification service (MCID) which is invoked on the terminating line and prevents the calling party from clearing the call even if they hang up. This service was a pre-CLI (Calling Line Identity) solution to the problem of nuisance calls which allowed the victim of the crime to invoke the service, resulting in the line being held and allowing time for the call to be logged by the OSS and details passed to the relevant law enforcement agencies.

2.5.2 The IN CS-2 Terminating BCSM

The terminating call half is responsible for manipulating the call and invoking features from the perspective of the called party and for controlling signalling and resources on the terminating access or trunk interface. The T-BCSM state machine is shown in Figure 2.8.

The T-BCSM is inserted into the call chain when the O-BCSM reaches the Send_Call state and it passes the call details to the terminating side. The T-BCSM moves from the NULL state into the Authorise_Termination_Attempt state. In this state, the T-BCSM checks that the call is eligible to be carried over the terminating interface. A call might be rejected if the called party has a call-screening service in place, for example Anonymous Call Rejection (ACR), which would reject a call if the calling party has chosen to withhold its line identify from the called party.

Having determined that the call can be progressed towards the called party, the T-BCSM enters the Select_Facility state and checks (in a traditional TDM network) that a suitable bearer is selected to carry the call. A resource check is performed to ensure that the bearer is free and that sufficient resources exist within the Call Agent to handle the call. In an NGN where the call is terminating on an access gateway, the Call Agent may check the busy/free map it holds for the access gateway to determine whether the terminating subscriber is in a call; if so, then additional features such as call waiting (if active) may be applied to the call or the call may be rejected.

The T-BCSM then enters the Present_Call state and invokes the call-signalling facilities to inform the called party of the call. This signalling method varies, depending on the interface; on a trunk interface, an ISUP IAM (Initial Address Message) would be sent; if the call terminates on an access gateway, then an H.248 message would be sent to the gateway instructing it to alert the terminating subscriber and (if applicable) present the calling party's identity. When the T-BCSM receives confirmation from the signalling that the called party is being alerted, it enters the T_Alerting state and informs the O-BCSM that the called party is being alerted.

The T_Active state is entered when an indication is received from the signalling that the called party has answered the call, and the T-BCSM informs the O-BCSM that the call has been answered; in a traditional TDM network, this would be triggered by the reception of an ISUP Answer message; in an NGN using SIP signalling, this would typically be the 200 OK received for the original Invite.

During the call, the called party may go on hook which causes the T-BCSM to enter the T_Suspended state and inform the O-BCSM that the terminating call half has entered the

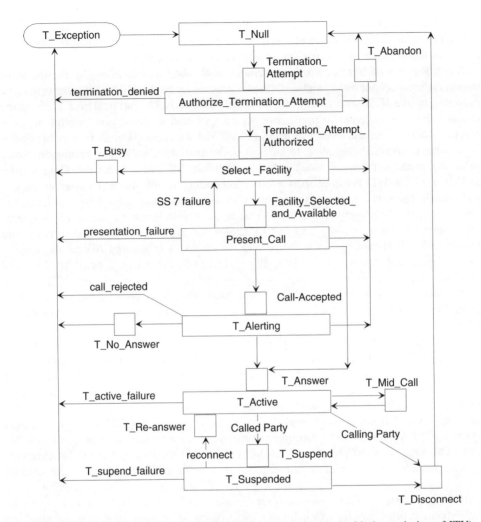

Figure 2.8   ITU-T CS-2 Terminating BCSM (Reproduced with the kind permission of ITU)

T_Suspended state. In this state, the resources allocated to the call remain in place while it is suspended and if the called party again goes off hook, the call will be reconnected. Depending on the interface, the T-BCSM may run a timer in this state to initiate call clearing in the event that the called party does not resume the call within a specified period.

2.6 Call Signalling in the NGN and the Role of SIP

As described previously, signalling is used in the NGN to set up calls between two end users and to pass information about the call state between the various Call Agents involved in handling the call. There are a number of signalling protocols that may be used, depending on the access type and whether the call requires interworking with the legacy PSTN. Typically, however, in the NGN, calls that originate or terminate on IP

phones and signalling between network Call Agents use the SIP protocol, as defined by
the IETF. In addition to SIP itself, SDP is used to communicate information about the
nature of the bearer that is being set up between the two parties.

The SIP protocol has become synonymous with VoIP and is today by far the most
important protocol for voice and multimedia call set-up in the NGN. SIP was originally
proposed by the IETF as a rival to the emerging ITU-T H.323 protocols for VoIP on the
grounds that it was easier to implement, more open and it would give control of voice
services back to the end user. From the outset, SIP was conceived to be flexible and it
is possible to use SIP signalling to support traditional voice calls, low bandwidth voice
calls and even peer-to-peer video calls. In addition, SIP used a text encoding similar
to that used for HTTP rather than a binary encoding which made it easier to encode
and decode for small start-up companies who were producing innovative products and
services for the VoIP market. For these reasons, it was selected by the 3GPP standards
organisation as the signalling protocol of choice for third-generation mobile voice and
multimedia call signalling. This choice had the effect of cementing SIP as the leading
protocol for VoIP, and today SIP is totally dominant in the industry, being used by end
terminals, Call Agents, media servers and even application servers. Because SIP is such
a ubiquitous protocol in NGNs, it is worth examining it in some detail to provide a
grounding for Chapter 6 which looks at how value-added services are implemented in
NGNs.

The primary function of SIP is to provide a method of signalling the call state between
two Call Agents or between user terminals and Call Agents. An additional role of SIP
is to provide a framework for identifying peer Call Agents and to locate the user within
the network. In order to achieve this, it uses the underlying capabilities of the packet
network to ensure that the signalling is transported between the relevant Call Agents, but
the SIP protocol itself provides routing functions to ensure that SIP messages are sent
to those Call Agents and user terminals that require information about the call state for
communication to occur. As part of the SIP call set-up, the two end terminals exchange
information about the type of communication they require; this is known as the *session*,
and information about it is exchanged using the IETF-defined SDP which is carried within
certain SIP messages. This forms an offer–answer mechanism whereby one party to the
call offers the other a range of media sessions that could be used for communication and
the other party replies with an acceptable session type.

In order to understand how the SIP protocol is used to establish a call, how it routes calls
through the packet network and how both ends of the call agree on mutually acceptable
codecs, this chapter considers in detail a simple end-to-end SIP call between two users
seeking to establish a voice telephony session. However, because SIP was developed in
isolation from the NGN architectures taking shape in the industry today, some of the
terms used may be confusing and so it is useful to examine the original SIP architecture
and describe how it maps to the NGN network elements that have been described in this
chapter.

2.6.1 A Brief Discussion of the SIP Architecture and Network Elements

When the SIP protocol was originally created, there was a considerable community in the
IETF that believed that the correct vision of future voice services was one of a core IP
network that primarily provided transport and, if necessary, call-routing functions but did

not attempt to replicate the feature-rich services of the PSTN. Instead, the SIP protocol was architected on the principle that the intelligence would move to the edge of the network into the hands of end users who would be free to innovate and implement their own services in SIP clients. This led to a very simple but powerful architecture around which SIP has been built. The key elements of this architecture are as follows:

- User Agents. (SIP UAs) which exist in the customer premises and implement services and features as required. A SIP User Agent may act as a User Agent Client (UAC) in which it initiates a call or it may act as a UAS in which case it responds to a request for a call. Simply put, the calling party SIP User Agent is the UAC and the called party SIP User Agent is the UAS.
- Proxy Servers. Proxy servers are network nodes that provide addressing and routing functions to get the SIP signalling through the network. A proxy server may be either stateful or stateless. A stateful proxy server maintains information about the state of the call throughout its duration and will usually insert itself into the call chain and ensure that it remains in the SIP-signalling path until either the call is cleared or a service that means it is no longer required is invoked. Stateless proxy servers do not maintain information about the call state but instead act only in response to a single SIP message such as the initial INVITE that initiates the call. Even if such a proxy server remains in the SIP-signalling chain throughout the call, it may only use the information available in a single SIP message to perform actions, as historical information about the call is not stored. The advantage of a stateful proxy server over a stateless proxy server is that it is able to invoke more services but its disadvantage is that it scales less well, as it has to hold the call state that ties up resource until the call clears.
- Back-to-Back User Agents. These are often abbreviated as B2BUA. These are typically nodes deployed within the network that behave on the side facing the calling party as a SIP UAS and on the side facing the called party as a SIP UAC. In effect, it terminates the call from the calling party and creates a new call on their behalf towards the called party. Such elements are able to provide useful functions, such as hiding the internal SIP topology of the network, but violate the original SIP vision of a network that is transparent to SIP signalling and may prevent the end user from being able to implement services locally.

The NGN vision that is emerging in the telecommunications industry clearly has a philosophy different from that of a transparent network with intelligence at the edge. One of the drivers of NGN is, after all, the ability to realise new value-added services for which revenue can be gained. However, it is to the credit of the SIP community that the protocol itself was easily adapted to such a model and SIP signalling is used between many of the elements in the NGN. Within the NGN SIP network, elements are deployed in the following ways.

- RGWs and IP phones are effectively SIP User Agents, existing in the customer premises and being used to initiate or receive SIP calls.
- Call Agents may map to stateful proxy servers or to Back-to-Back User Agents, depending on the functions they support and the network operator's preference. In general, however, a typical Call Agent can be considered to be a stateful SIP proxy server.

- SBCs are implemented as SIP Back-to-Back User Agents. Their major function is to provide security at the edge of the network and to hide the SIP routing (and, indeed, IP addressing) of the NGN from the end user.

In addition to these SIP elements, the protocol also envisages other types of nodes that can also be found in the NGN. These include SIP Registrars which allow users to register the location of their terminal(s) in the network, and SIP redirect servers that track the user's location (i.e. which one of their registered SIP terminals they are present at). The SIP redirect server provides a re-routing function to allow the end user to be reached from a single address of record, regardless of their physical location.

2.6.2 A Simple Call Set-up Using SIP Signalling

This section provides an example of how signalling is used in a VoIP network, where two customers are connected to the network using SIP terminals. The SIP signalling flow is shown in Figure 2.9 and the contents of the messages are explained in detail in the accompanying text and tables. A SIP terminal may be an IP phone (i.e. it looks like a traditional telephone but usually supports advanced features) or an application running on a PC. In this example, a customer Eric is using a PC to place a phone call to Amy who is using an IP phone. Because the end terminals are packet based and capable of SIP signalling, all the call set-up signalling can be based on SIP.

When Eric places the call, he selects Amy from his address book and instructs his SIP PC client to place the call. Because Eric has a high bandwidth Internet connection, his client is configured to request a 64 kbps voice channel but is prepared to accept a lower

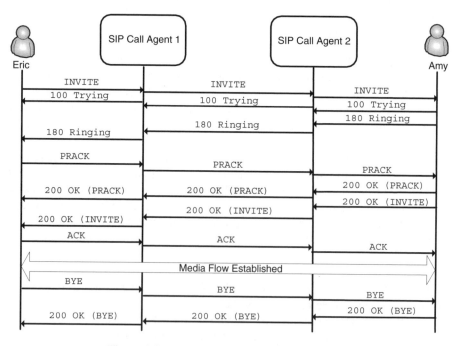

Figure 2.9   A simple call established using SIP

bandwidth codec if requested. Eric's client informs the network of his desire to place a call using the SIP-signalling protocol.

In order to use the SIP protocol to place a call, a SIP dialogue must be initiated. This is achieved by sending an INVITE message. The message is made up of a method (which is INVITE) and a series of header fields. This invite is shown in Table 2.1.

The Call Agent that serves Eric receives the INVITE and processes it as a new call request using its BCSM as described previously. As part of this process, the Call Agent must analyse the Request-URI header to determine the called party and perform a routing lookup to determine where this call is to be sent. This processing may require a DNS lookup; however, in this simple case the SIP URI can be parsed to identify the IP address. Having determined the next hop Call Agent that should be used for calls to Amy's network, the Call Agent forwards an INVITE onwards to that Call Agent. Because it acts in this way the Call Agent is seen by the SIP protocol as what is know as a stateful proxy server. This implies a certain type of SIP protocol behaviour and results in a modified SIP INVITE being sent onwards to the next hop Call Agent. The routing mechanisms for SIP are complex, and the routing process, how the next hop is determined and how the Request-URI differs at each hop in the call is described in Section 2.6.7 of this chapter.

The INVITE received by Eric's SIP Call Agent will be processed and as part of the SIP routing process, a new INVITE will be sent to the next Call Agent. This new INVITE copies many of the values from the original INVITE but some fields such as the Request-URI and the Record-route headers may differ and the Via header will be updated to reflect the identity of the SIP Call Agent that has forwarded the INVITE. Because Eric's Call Agent acts as a SIP Proxy Server, the other fields in the INVITE will remain mostly unchanged; however, Eric's Call Agent will add a Max-Forwards header to prevent loops (any subsequent Call Agents must decrement the value of this field by 1); because Eric's SIP User Agent did not include a Max-Forwards header, the default value of 70 is used by Eric's Call Agent. Note that the value of 70 is an arbitrary figure selected by the authors of the SIP RFC to be a good compromise between preventing loops and causing calls to fail because of an inability to transit SIP proxies. Eric's Call Agent is free to use other values if preferred by the network operator.

Eric's Call Agent may add additional headers, if it desires, to the INVITE before forwarding them and may also act in ways not strictly in conformance with the role of a SIP proxy as defined in the SIP protocol; for example, it might alter the SDP to constrain the codecs being requested by the user to a set that the network is happy to handle should it be necessary to interwork the call back to the TDM PSTN environment.

This processing of the INVITE is replicated at Amy's Call Agent, which is responsible for determining where Amy may be reached if she can be registered at a number of different locations. The SIP Request-URI in the INVITE sent to Amy will be addressed to the specific IP address that her Call Agent has determined to be used to reach her terminal; this may be based on location information derived from a SIP Registrar, see Section 2.6.6.

In this simple example, the INVITE reaches Amy's SIP User Agent (which is an IP phone) and immediately returns a provisional response, in this case a "100 Trying" message. The "100 Trying" message seen by Eric's SIP User Agent is shown in Table 2.2. The "100 Trying" message indicates that the SIP User Agent has received the INVITE and is engaged in call-processing activity.

Table 2.1 Invite

SIP method and headers	Comments
INVITE sip:9724411111@172.30.1.5; user = phone SIP/2.0	The INVITE keyword identifies the method as a SIP INVITE, the next part of the INVITE is the Request-URI, which identifies the entity that the request is addressed to, in this case the called party which is Amy. In this example, Amy is translated by Eric's address book as 97244111111@172.30.1.5; however, it could equally be a fully qualified domain name such as amy@bigtelconetwork.com. The user = phone parameter is discussed below and the version of the protocol is identified as version 2.0.
Via:SIP/2.0/UDP 172.17.2.29:5060;branch =z9hG4bKnashds8	The next header is the Via header. This top Via header value identifies the SIP node that last handled the request. In this case, it is Eric's SIP User Agent, which has the IP address 172.17.2.29 and is using port 5060. The UDP keyword identifies that SIP will use UDP over IP encapsulation for signalling.
	The branch parameter is used by SIP to provide a transaction Id that can be used to match responses to requests, to distinguish when a message arriving at a proxy server is caught in a loop and to assist in matching responses to requests in forking scenarios (described in the description of the From header below). The first seven characters of the branch are the so-called magic cookie and these are always "z9hG4bK". The magic cookie is required because SIP changed the use of Branch in RFC 3261 from its use in the previous version of SIP. Therefore, this magic cookie is used so that the SIP proxy or User Agent can determine which meaning of branch it should assume (since older implementations will not start the branch parameter with this magic cookie). The branch parameter is required to be unique across all transactions being handled by the SIP node and may be generated using, in part, a cryptographic hash of certain header fields. This approach is used if the SIP proxy uses the branch parameter to detect loops.
From:Eric Smith <sip:2143302105@172.17.2.29; user=phone>;tag=0E1D.8099	The From header identifies the address of record of the originating end point that is seeking to establish the dialogue, in this case it contains Eric's SIP URI; however, other address types are supported by SIP, depending on the application. It also contains a display name, "Eric Smith" which may be used by the called party's terminal equipment when the call is presented. The From header also contains a tag parameter which is used to identify the SIP dialogue

Table 2.1   *(continued)*

SIP method and headers	Comments
	as the combination of the Call-Id and the tags present in the To and the From headers uniquely identifies a dialogue. The tags are required in SIP because a single INVITE message to establish a dialogue can be "forked" by Call Agents within the network. This forking of requests results in multiple SIP INVITES being sent to different end points. However, one side effect of this is that the network attempts to establish multiple dialogues for a single call. The tag values can therefore be used to determine which dialogue a given message belongs to. Note that SIP nodes may also use the branch parameter to achieve the same result.
	The user = phone parameter is interesting because it is an illustration of one of the weaknesses of the SIP protocol which is its organic growth. There is no widespread understanding among implementers as to the meaning of this particular field; however, it is added because some SIP implementations may reject an INVITE if it is not present. This type of potential interoperability issue is one of the reasons that organisations such as the MSF provide IAs to define a subset of SIP headers to be used in a network scenario.

`To:`
`<sip:9724411111@172.30.1.5;`
`user=phone>`
	The To header contains the address of record of the entity that the request is addressed to. In this case, it is Amy's SIP URI. However, SIP also provides alternative addressing schemes, which may be used, depending on the application. The To header records the address to which the request was originally sent, the Request-URI is used for routing and so may be modified by intermediate SIP servers. Note that there is no tag present in the To header because a dialogue with the called party (Amy) has not been established yet. The tag will eventually be provided by Amy's SIP terminal. The user parameter field is largely arbitrary and defines the user type; in this case it is set to Phone.

`Call-ID:`
`0800.20CE.A152.3C5B.0E1D.`
`8099@172.17.2.29`
	The call-Id is used to uniquely identify the call that has been placed for future reference. In order to guarantee that the Call-Id is unique, its second part contains the address of the originating entity, in this case 172.17.2.29 and the first part is a hexadecimal identifier generated by the originating entity.

(continued overleaf)

Table 2.1   (*continued*)

SIP method and headers	Comments
CSeq: 21478 INVITE	Within a SIP dialogue, a number of transactions take place. A typical transaction would be a request from the calling party and a response from the called party. Because SIP is designed to run over unreliable Internet transport protocols such as UDP, a mechanism has to be provided to identify which transaction within a dialogue a given SIP message applies to, and the somewhat esoterically named CSeq header provides this information.
	The CSeq comprises two parts, a number and the name of a SIP method. The number is simply a count of requests sent by the originating entity and is incremented when a new request is sent. For a SIP request, the method field should be equal to the method used by the request, in this case INVITE.
Content-Length: 207	The Content Length header identifies the length of the payload contained within the SIP message.
Content-Type: application/sdp	The content type identifies the type of the payload that is being carried in the message. In this example (and by far the most common case in SIP), the content is identified as being SDP, which is used to describe the characteristics of the bearer that is being requested.
Contact: sip:2143302105@172.17.2.29; user=phone	The contact header provides a globally unique address, a SIP URI that can be used to reach the calling party at the equipment from which the dialogue is being established.
Session-Expires: 120	The Session-Expires header indicates that Eric's User Agent wishes to use a session timer of interval 120 seconds. Although Eric's User Agent has indicated a preference for the session timer by including this header because it is not included in the Require header, it is not mandatory for the called party's User Agent to support it. Eric's User Agent will not assume that the session timer is active unless a corresponding Session-Expires header is received in the final response to the INVITE.
	Session Timers are primarily used to provide a heartbeat/keep-alive function to ensure that a dialogue is still active.
Require: 100rel	The Require header informs the Call Agent that certain SIP methods must be supported in order to support the call. Because the SIP protocol has

Table 2.1   (*continued*)

SIP method and headers	Comments
	evolved, the mandatory core elements of SIP that will always be supported by SIP-based Call Agents are relatively small. Therefore, if a SIP terminal requires a particular method that is not part of this core set, it must explicitly signal it when it attempts to establish a dialogue.
	In this case, Eric's SIP terminal requires the network to support the 100rel method. This method is used to invoke reliable responses during call set-up, which it achieves by triggering the use of additional SIP messages (these messages will be seen as the call progresses).
	If a Call Agent is unable to support a method listed in the Require header, then it should reject the dialogue request with an error indicating that the required method is not supported. In most NGNs, this is likely to be the result of a configuration error, as Call Agents will implement all the capabilities required to support the network operator's services; however, in the more open world of Internet-based VoIP, the Require header provides a valuable function as if a dialogue were accepted but the Call Agent in question did not support the headers needed to support a service the service is likely to fail unpredictably if invoked.
`Allow: REFER, UPDATE, NOTIFY`	The Allow header lists the set of methods supported by the SIP User Agent. According to the SIP protocol, to be strictly compliant a User Agent or Call Agent should include all the methods that it supports in the Allow header if one is included (it is not required to be included). However, in reality most, if not all, SIP implementations include only those SIP methods that are not universal in their support. In this case, the User Agent informs the Call Agent that it supports the REFER, UPDATE and NOTIFY methods. This information may be used by the Call Agent to determine which signalling sequences to use when communicating the call state back to the User Agent.
`Supported: timer`	The Supported header contains information about optional SIP headers that are supported. In this case, the SIP terminal indicates that it supports the Timer header.

(*continued overleaf*)

Table 2.1 (*continued*)

SIP method and headers	Comments
`v=0`	The final element of the SIP INVITE is the body of the message. In this case, the body of the message contains an SDP message. The SDP is described in Section 2.7 and is used in NGN-signalling systems to describe the nature of the media (or bearer) that the user is requesting to establish.
`o=SIP-GW 7295 3647 IN IP4` `172.17.2.29`	SDP is hard to read at first sight and many of the conventions are historical and therefore potentially counter-intuitive. However, the key elements of the SDP are the c = line which identifies the originating media IP address of Eric's phone (172.17.2.31); the m = line which identifies the UDP port of the media, in this case 1000 and a list codecs which can be supported in preference order.
`s=SIP-GW SIP Call`	In this case, three codecs are identified using their static payload type. They are G.711 μ-law, G.723 and G.729. There is also a dynamic payload type (101) which is defined in the rtpmap attribute. This dynamic payload does not describe a codec as such but indicates that the end point supports the transfer of DTMF digits within the RTP protocol. The "telephone-event/8000" string has been defined by IANA as indicating support for RFC2833 DTMF in RTP. The a = fmtp line provides further attributes for codec 101, and in this case indicates that DTMF digits 0 to 15 can be signalled in the RTP. The use of RTP to carry DTMF will only be required if the two ends negotiate a low bit rate encoder that cannot carry the DTMF in-band because of the compression. For a G.711 codec, such mechanisms are not necessary because the DTMF tones can be carried transparently within the media.
`c=IN IP4 172.17.2.31`	The last a = line in this message describes the type of connection required; in this case, it is set to sendrecv which means that Eric's SIP phone will be able to send media into the network and receive media from it (in effect a bi-directional call).
`t=0 0`	It is interesting to note that in the SDP and also in other SIP fields, IP addresses and ports are present. This means that should the message encounter any NAT, special application-level processing will be required to translate the addresses in the message. This function is typically performed by SBCs (also known as SBGs) as described in Chapter 3.

Table 2.1 (*continued*)

SIP method and headers	Comments
m=audio 1000 RTP/AVP 0 4 18 101	
a=ptime:20	
a=rtpmap:101 telephone-event/8000	
a=fmtp:101 0-15	
a=sendrecv	

Table 2.2 100 Trying

SIP method and headers	Comments
SIP/2.0 100 Trying	Indicates that the message is a SIP 2.0 message and the method is provisional response, 100 Trying.
Via: SIP/2.0/UDP 172.17.2.29:5060; branch=z9hG4bKnashds8	The Via header in this case is copied from the incoming INVITE and must be returned unchanged. It points to Eric's Call Agent and the branch parameter matches that sent in the original via that Eric's Call Agent built and sent in the INVITE.
From: Eric Smith <sip:2143302105@172.17.2.29; user=phone>;tag=0E1D.8099	The From header is copied from the incoming From header and returned unaltered by Amy's SIP User Agent.
To: <sip:9724411111@172.30.1.5; user=phone>	The To header is returned unaltered.
Date: Thu, 27 Oct 2005 01:41:27 GMT	The Date header field contains the date and time at which the response was sent. There is no requirement for a User Agent to include a date in the response, and the field is of dubious utility; however, many SIP implementations do send the date.
Call-ID: 0800.20CE.A152.3C5B.0E1D.8099 @172.17.2.29	The Call-Id is copied from the original INVITE and returned unchanged in the response.
CSeq: 21478 INVITE	The CSeq header is also returned unchanged.
Content-Length: 0	The 100 Trying message has no body, just headers and so the content length is 0.

Although the 100 Trying message indicates that the INVITE has been received and therefore does not need to be re-sent, it does not in itself advance the call state and is not sent end to end but instead each SIP agent generates its own 100 Trying response to the upstream SIP Call Agent or User Agent. The only real impact of a 100 Trying message is to stop the retransmission timer running at the upstream node that would otherwise lead it to re-send the INVITE at some point in the future.

Because Amy's SIP User Agent is a relatively simple phone, with a single handset and no requirement to locate the end user, it is likely that it will immediately "alert" Amy to the incoming call by ringing. On taking this action, the Amy's SIP User Agent sends a 180 Ringing provisional response back through the network to Eric's SIP User Agent which receives the message shown in Table 2.3.

Because Amy's SIP User Agent generated a reliable provisional response, Eric's agent must explicitly confirm that it has received the response by sending a PRovisional ACKnowledgement (PRACK). Amy's provisional response contained an SDP with the answer to Eric's original session offer; therefore, Amy's User Agent will not finally accept the dialogue with a SIP 2xx message until it receives an explicit indication from Eric that his User Agent has received the provisional response. The content of the PRACK is shown in Table 2.4.

Having received the PRACK, Amy's SIP User Agent knows that the provisional response it sent to Eric has been received and acknowledges the PRACK with a 200 OK message which is received by Eric as shown in Table 2.5.

Having determined that Eric's SIP User Agent received Amy's offer and having acknowledged this, Amy's SIP User Agent now sends a final response to the session request contained in the original INVITE. For a description of the session timer, see Ref. 28. Note that the lack of a Session-Expires header in the 200 OK indicates that the session timer will not be active for this dialogue despite the fact that Eric's SIP User Agent requested it in the initial INVITE as seen in Table 2.6.

The 200 OK message means that the dialogue is no longer an early one but is now treated as a confirmed one. This has implications for how the dialogue is terminated. Eric's SIP User Agent completes the establishment of the dialogue by acknowledging the 200 OK (Table 2.7).

At this point in the call, the establishment is complete and a media session is flowing between Eric and Amy.

Within a confirmed dialogue, SIP allows the session to be modified; for example, the codec can be changed or the end point moved. Sessions can be modified using either an INVITE or the SIP UPDATE method. When an INVITE is sent within an established dialogue, it is referred to as a Re-INVITE; some implementations periodically send Re-INVITES to refresh session timers and to ensure that NAT bindings are kept in place within the network. If the session timer is active for a session and the SIP User Agent does not receive a Re-INVITE or an UPDATE before its session timer expires, then the User Agent will terminate the session. This provides protection against hanging SIP sessions in network Call Agents in the event that a failure causes the loss of the SIP messages terminating the dialogue.

Table 2.3 180 Ringing

SIP method and headers	Comments
SIP/2.0 180 Ringing	Indicates that the message is a SIP 2.0 message and the method is provisional response, 180 Ringing.
Via: SIP/2.0/UDP 172.17.2.29:5060; branch=z9hG4bKnashds8 From: Eric Smith <sip:2143302105@172.17.2.29; user=phone>;tag=0E1D.8099	The Via and From headers are copied from the original INVITE that was received from Eric's SIP User Agent.
To: <sip:9724411111@172.30.1.5; user=phone>;tag=5CEE58-2187	Amy's User Agent is required to append a tag to the received To header. The creation of this tag effectively results in a dialogue being established between Eric's and Amy's SIP User Agents. The combination of Call-Id plus the two tags uniquely identifies the dialogue. Because no final response has yet been sent (SIP 1xx responses are all provisional responses), the dialogue is known as an early dialogue.
Date: Thu, 27 Oct 2005 01:41:27 GMT Call-ID: 0800.20CE.A152.3C5B.0E1D.8099 @172.17.2.29 CSeq: 21478 INVITE	The Date, Call-Id and CSeq fields are the same as those sent in the 100 Trying response.
Require: 100rel	Amy's SIP User Agent inserts this header to indicate that it requires the support of reliable responses.
RSeq: 360	The RSeq header is included in order to ensure that this provisional response is sent reliably. This is required because the need for reliable provisional responses was specifically indicated by the previously received Require 100rel header added to the INVITE by Eric's SIP User Agent. The value of the RSeq chosen by Amy's SIP User Agent is random because it is the first reliably sent provisional response to the INVITE. The handling of reliable provisional responses and the RSeq header are described in Ref. 29.
Allow: UPDATE, REFER, NOTIFY	

(continued overleaf)

Table 2.3 (*continued*)

SIP method and headers	Comments

`Contact:`
`<sip:9724411111@172.30.1.5:5060;`
`user=phone>`
	The Contact header provides a globally unique address, a SIP URI that can be used to reach the Amy at the equipment from which the dialogue is being established.

`Content-Length: 204`	

`V=0`
`o=SIP-GW-UserAgent 1776 6232 IN IP4`
`172.30.1.5`
`s=SIP Call`
`c=IN IP4 172.30.1.5`
`t=0 0`
`m=audio 19374 RTP/AVP 0 101`
`a=rtpmap:0 PCMU/8000`
`a=rtpmap:101 telephone-event/8000`
`a=fmtp:101`
	The body of the message contains the SDP that describes the media session that Amy's terminal is prepared to accept. SIP uses an offer–answer model whereby the initiator of a session sends information about the type of media that he or she wishes to use for the session (in effect an offer of a session) and the responder returns information about the type of media that he or she is prepared to accept for the session (in effect the answer to the offer). The c = line contains the IP address of Amy's phone. The m = line contains the codec that Amy has selected for the call from Eric's list of supported codecs. In this case, RTP/AVP 0 indicates the selection of the G.711 codec with μ-law encoding. It is possible that the codec list offered to Amy was not the full set originally sent by Eric as the network itself may have altered the codec list to reflect preferences or to exclude codecs that it did not wish to support. Amy's end point also includes further information about the G.711 codec by explicitly defining RTP/AVP code 0 as meaning PCM μ-law with a clock rate of 8 kHz and it states that it supports the transport of DTMF in RTP; however, since G.711 is the selected codec (and is transparent to DTMF), this is not strictly required to support the call.

Table 2.4   Prack

SIP method and headers	Comments
PRACK sip:9724411111@172.30.1.5:5060; user=phone SIP/2.0	Indicates that his is a provisional response acknowledgement that is being sent to Amy's SIP User Agent. The PRACK is a new request within the existing dialogue as defined by the Call-Id, and the tags in the To/From fields. As such it must itself be acknowledged.
Via: SIP/2.0/UDP 172.17.2.29:5060; branch=z9hG4bK234567	Because this is a new request, a new branch parameter is calculated.
From: Eric Smith <sip:2143302105@172.17.2.29; user=phone>;tag=0E1D.8099	
To: <sip:9724411111@172.30.1.5; user=phone>;tag=5CEE58-2187	
Call-ID: 0800.20CE.A152.3C5B.0E1D.8099 @172.17.2.29	
CSeq: 21479 PRACK	Because the PRACK is a new request inside the dialogue, the CSeq value must be incremented by Eric's SIP User Agent.
Content-Length: 0	In this case, the PRACK does not contain any content, i.e. no SDP because the SIP offer and answer were carried in the initial INVITE and the first reliable response (the 180 ringing). However, in some modes of operation (e.g. network-initiated calls), it is possible that the INVITE might not contain an offer, in which case the first reliable response would contain SDP from Amy with her offer and an answer would be received as an SDP within the PRACK sent by the network to confirm the reception of the reliable response. This flexibility in when SIP passes session information is powerful and enables innovative services to be built but it often leads to interoperability problems if manufacturers make assumptions as to when session information will be available.
Contact: sip:2143302105@172.17.2.29; user=phone	

(*continued overleaf*)

Table 2.4 (*continued*)

SIP method and headers	Comments
RAck: 360 21478 INVITE	The contents of the RAck header allows Amy's SIP User Agent to determine that this is a valid acknowledgement. An acknowledgement is valid if all of the following applies:
	The dialogue matches that of the provisional response; the RAck field is coded such that the first number matches the RSeq sent in the reliable response; the next number matches the CSeq sequence number sent in the provisional response and the method matches that sent in the CSeq contained in the provisional response.

Table 2.5 200 Ok (Prack)

SIP method and headers	Comments
SIP/2.0 200 OK	Indicates that this is a final response to a request.
Via: SIP/2.0/UDP 172.17.2.29:5060; branch=z9hG4bK234567	
From: Eric Smith <sip:2143302105@172.17.2.29; user=phone>;tag=0E1D.8099	
To: <sip:9724411111@172.30.1.5; user=phone>;tag=5CEE58-2187	
Date: Thu, 27 Oct 2005 01:41:27 GMT	
Call-ID:	
0800.20CE.A152.3C5B.0E1D.8099 @172.17.2.29	
CSeq: 21479 PRACK	The CSeq field allows Eric's SIP User Agent to determine exactly which request this is a response to. It can determine from this field that this is a response to the PRACK request that was sent with CSeq value 21479 within this dialogue.
Content-Length: 0	

Table 2.6 200 Ok (Invite)

SIP method and headers	Comments
`SIP/2.0 200 OK`	Indicates that this is a final response to a request.

`Via: SIP/2.0/UDP`
`172.17.2.29:5060;`
`branch=z9hG4bKnashds8`	

`From: Eric Smith`
`<sip:2143302105@172.17.2.29;`
`user=phone>;tag=0E1D.8099`	

`To: <sip:9724411111@172.30.1.5;` `user=phone>;tag=5CEE58-2187`	
`Date: Thu, 27 Oct 2005 01:41:27` `GMT`	

`Call-ID:`
`0800.20CE.A152.3C5B.0E1D.8099`
`@172.17.2.29`	

`CSeq: 21478 INVITE`	The CSeq field allows Eric's SIP User Agent to determine exactly which request this is a response to. It can determine from this field that this is a response to the INVITE request that was sent with CSeq value 21478 within this dialogue; since this was the original INVITE, Eric's SIP User Agent can determine that this is the final response.
`Allow: UPDATE, REFER, NOTIFY`	

`Contact:`
`<sip:9724411111@172.30.1.5:5060;`
`user=phone>`	

`Content-Type: application/sdp`	
`Content-Length: 204`	

`V=0`
`o=SIP-GW-UserAgent 1776 6232 IN`
`IP4 172.30.1.5`
`s=SIP Call`
`c=IN IP4 172.30.1.5`
`t=0 0`
`m=audio 19374 RTP/AVP 0 101`
`a=rtpmap:0 PCMU/8000`
`a=rtpmap:101`
`telephone-event/8000`
`a=fmtp:101`
	The body of the message and confirmation of the session parameters.

Table 2.7   Ack

SIP method and headers	Comments
ACK sip:9724411111@172.30.1.5:5060; user=phone SIP/2.0	Indicates that this is an acknowledgement.
Via: SIP/2.0/UDP 172.17.2.29:5060;branch= z9hG4bKnashds8	
From: Eric Smith <sip:2143302105@172.17.2.29; user=phone>;tag=0E1D.8099	
To: <sip:9724411111@172.30.1.5; user=phone>;tag=5CEE58-2187	
Call-ID: 0800.20CE.A152.3C5B.0E1D.8099 @172.17.2.29	
CSeq: 21478 ACK	
Content-Length: 0	
Contact: sip:2143302105@172.17.2.29; user=phone	

2.6.3 Simple Call Clearing Using SIP Signalling

In the general case, once a confirmed SIP dialogue has been established, the call can be cleared by either end user (Eric or Amy in this case) by sending a SIP BYE message towards the other party. In this example, Eric decides to clear the call down which results in a BYE being sent as shown in Table 2.8.

Any node acting as a SIP User Agent or a SIP proxy server that receives the BYE will terminate the session and stop listening for or sending the media, regardless of whether the BYE is acknowledged by the far end. Eric's User Agent runs a timer and if no response to the BYE is received before the timer expires, then it will consider the dialogue terminated.

In this case, Amy's SIP User Agent receives the request and acknowledges by sending a final response to the BYE request (a SIP 200 OK message) as shown in Table 2.9.

SIP places restrictions on how a dialogue is terminated and provides an additional method – the CANCEL method – that may be used to terminate a dialogue that is still in the set-up phase. Specifically, SIP requires that for early dialogues, i.e. those for which the SIP User Agent has not received a final response, only the calling party may terminate a dialogue with a BYE. The called party may not send a BYE until it has received an ACK to a previously sent final response.

The CANCEL method allows the originator of a request to effectively abort the request. Typically sending a CANCEL results in the called party responding with a 200 OK to the CANCEL and sending a SIP 4xx final response to the INVITE that the CANCEL relates to. This response would typically be a 487 "request terminated" message.

Table 2.8 Bye

SIP method and headers	Comments
BYE sip:2143302105@172.17.2.29:5060; user=phone SIP/2.0	Indicates that this is a BYE request.
Via: SIP/2.0/UDP 172.30.1.5:5060; branch=z9hG4bKerib12345	
From: <sip:9724411111@172.30.1.5; user=phone>;tag=5CEE58-2187	The tags in the From and To headers and the Call-Id uniquely identify the dialogue. If Amy's SIP User Agent receives a BYE that does not correspond to a known dialogue, then it will be rejected with a SIP 481 message which means "call/transaction does not exist".
To: Eric Smith <sip:2143302105@172.17.2.29; User=phone>;tag=0E1D.8099	
Date: Thu, 27 Oct 2005 01:42:33 GMT	
Call-ID: 0800.20CE.A152.3C5B.0E1D.8099 @172.17.2.29	
Max-Forwards: 6	
CSeq: 21480 BYE	The BYE is a request within the dialogue and so the CSeq is incremented by one.
Content-Length: 0	

Table 2.9 200 Ok (Bye)

SIP method and headers	Comments
SIP/2.0 200 OK Via: SIP/2.0/UDP 172.30.1.5:5060; branch=z9hG4bKerib12345	Indicates that this is a final response to a request.
From: <sip:9724411111@172.30.1.5; user=phone>;tag=5CEE58-2187	
To: Eric Smith <sip:2143302105@172.17.2.29; user=phone>;tag=0E1D.8099	
Call-ID: 0800.20CE.A152.3C5B.0E1D.8099 @172.17.2.29	

Table 2.9 (*continued*)

SIP method and headers	Comments
`CSeq: 21480 BYE`	Indicates that this is the final response to the BYE request.
`Content-Length: 0`	
`Contact:` `sip:2143302105@172.17.2.29;`	
`user=phone`	

2.6.4 SIP Redirection Servers and SIP Forking

In order to support location services efficiently, SIP has defined a type of server known as a *redirection* server. Such a server holds location information for the subscriber and is able to provide a contact header for each location that the subscriber is at or may be resident at. The SIP redirection server effectively pushes the decision as to how to route a call back towards the originator of the call by returning a SIP 3xx "redirect" final response to any INVITE it receives for the subscriber. This 3xx response contains a contact header, which is populated with a list of possible locations for the subscriber. On receiving a 3xx final response, the originating SIP User Agent will attempt to route the call to the locations described within the Contact header. Note that the redirect server has no involvement in the future call set-up once it has returned the 3xx response.

An example of a call that has been redirected is shown in Figure 2.10. In this example, a call from SIP Call Agent 1 is passed to Call Agent 2 which forwards the INVITE to the location it currently holds for Amy. However, Amy has chosen to divert her calls to a voice mail box and so her SIP terminal returns a 3xx response, in this case 302 "moved temporarily". SIP Call agent 2 analyses the Contact header received in the 302 message and determines from it the location to which the call should be sent (in this case, Amy's voice mail). SIP Call Agent 2 sets up the call to the voice mail box but it includes a History header in the new INVITE to inform the next SIP Call Agent or User Agent that the call has been diverted.

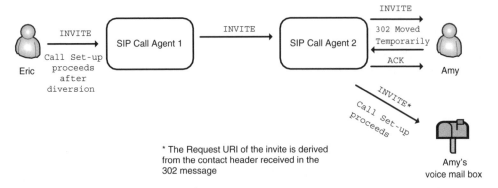

Figure 2.10   SIP redirection (Call forwarding)

A Call Agent may choose to forward an INVITE to multiple locations in parallel. An example of this scenario is a Call Agent that holds multiple possible locations for a given subscriber where they may be active and so forwards the INVITE to all locations simultaneously. This behaviour is known as SIP Forking and Call Agents that perform this action are often described within SIP as "Forking proxies".

Typically, a response would be received from one or more of the locations; it is possible that if multiple responses are received, the Call Agent that initiated the forking may determine which response will be used to establish a session. This could be based, for example, on a configured profile for a subscriber.

Each forked request requires separate tag values in the To and the From headers and forms a separate dialogue. The forking Call Agent can determine from the received tags which dialogue the response corresponds to. If the forking Call Agent no longer wishes to continue with a given dialogue, then it will send a BYE or a CANCEL to terminate the dialogue. An example of such a flow is shown in Figure 2.11.

SIP forking, although a simple concept, causes some significant issues for networks. The first issue is that it is possible for multiple INVITES to be sent out that in effect refer to a single media session. Any QoS mechanisms employed within the network have to be able to recognise that these INVITES are in some way linked and take this information into account when performing CAC for the session. Failure to identify the INVITES as being part of the same forked request may lead to CAC being too conservative in determining if the network is able to support the session. For example, if an INVITE is received and forked three times, but each of these INVITES are requesting a session over a common resource, CAC for that resource should be performed only once and not (as would normally be the case) three times.

Another area where SIP forking can cause a problem is where digest authentication is performed by the network (see Chapter 3). In this case, the forking Call Agent will effectively receive a challenge for each of the forked INVITES that it sent out. If it sends only one of these challenges back to the originating SIP User Agent, then only one of

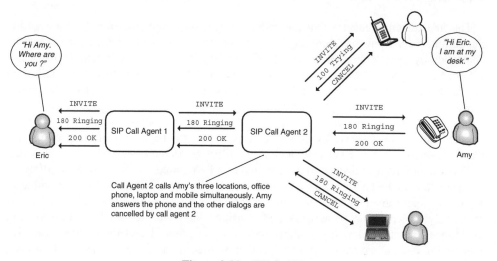

Figure 2.11   SIP forking

the forked destinations will receive a response to its challenge and, hence, only one of the forked INVITES can receive a response. In order to overcome this, SIP Call Agents that wish to fork INVITES must be prepared to send multiple challenges back to the originating SIP User Agent which in turn must be able to respond to each challenge. Note that the use of the tags in the To and From headers ensures that it is possible to determine which forked request the challenge/response applies to.

2.6.5 Privacy CLI and the SIP P-Asserted-Identity Header

Within many countries, a key property of the PSTN is the ability to identify the calling party with certainty from the call set-up signalling. In traditional telephone networks, a user's identity is determined and verified by his or her home network on the basis of his or her physical point of attachment. The local exchange on which the call originates can therefore determine the directory number associated with the subscriber and it uses this information within the ISUP call signalling, placing it in the calling party number field of the Initial Address Message (IAM). Because this number has been provided by a trusted network, any subsequent networks that carry the call can identify the subscriber on the basis of the call signalling and may present this number as a display number to the called party if the called party subscribes to a service such as CLI.

Within the United Kingdom, for a number of years the regulators responsible for telecommunications have maintained strong rules about supporting services such as CLI and they have published a CLI code of practice for network operators to adhere to [30 and 31]. At a very high level, this code of practice breaks down into two overriding principles:

- Only numbers that have been verified by a trusted network can be presented to the called party for display and the line identity should always represent the correct identity of the calling party.
- A subscriber may only have his or her number presented to the called party if the network operator providing the service to the called party can be sure that the calling party has been given the opportunity to restrict the display of their number.

The first principle ensures that the called party is protected from fraud by preventing the calling party from masquerading as somebody they are not (at least as far as can be controlled by the telephone network). It also ensures that law enforcement agencies and the emergency services can be sure of the originator of the call from the call signalling (in earlier days, the originator of a call had to be traced through the switch using a manual process). The second principle ensures that a customer who has subscribed to a calling number presentation restriction service and therefore has an expectation that his or her number will not be displayed will not be surprised by the behaviour of the network and can be assured that his or her confidentiality will be respected by the telephony provider.

These requirements have led to the creation of a P-Asserted-Identity header in SIP so as to ensure that the PSTN rules on CLI can be met. In principle, SIP poses no more problems for identifying a given subscriber than a mobile network. It is true that the subscriber is effectively hosted on an IP network and is using a computer terminal to place the call and hence cannot be identified reliably by his or her point of attachment; however, SIP registration and authentication procedures can be used by a network to establish the end user's credentials. However, because the SIP User Agent may be running on a device such as a PC and may be using an open source SIP stack, it is possible for the end user to modify the behaviour of his or her terminal. Therefore, the traditional SIP headers such as From and Contact cannot be used by intermediate networks and the called party to reliably determine the identity of the calling party because they are inserted by the calling party's SIP client. The P-Asserted-Identity header is inserted into the INVITE by the first network-based Call Agent that handles the SIP call and contains the identity of the caller. It is a requirement that the P-Asserted-Identity is only passed to trusted network elements and is never presented to a SIP User Agent; however, network-based SIP Call Agent that is serving the called party is able to use the contents of the P-Asserted-Identity header to create or overwrite the existing Display Name value in the SIP From header it receives, and this modified From header can be presented to the called party's SIP User Agent as a network-provided and verified address.

Like the PSTN, SIP must also support a mechanism to enable a subscriber to withhold their identity in the call set-up signalling. This is achieved in SIP by using the privacy mechanisms described in RFC 3323 [32] and RFC 3325 [33]. When a subscriber chooses to withhold his or her identity, the terminal signals this to the network by replacing the SIP URI in the From field with the value "sip:anonymous@anonymous.invalid" and using a display name of "anonymous". In addition, the subscriber's terminal requests a privacy service from the network using the Privacy header with a value of "Id" which informs the network that it must take all necessary steps to protect the identity of the user. This includes not just altering the SIP-signalling headers but also employing Back-to-Back User Agents (such as SBCs) to ensure that the IP addresses in the SDP and those used by the media stream itself do not betray the identity of the calling party.

As an example, if Eric had requested privacy when he called Amy, then the SIP signalling carried between Eric and Amy's network-based Call Agents would have had an INVITE as shown in Table 2.10.

The use of the anonymous URI and the Privacy header ensures that the networks involved know that the user requires their identity to be withheld from the calling party while the P-Asserted-Identity allows the networks themselves to know who originated the call so that emergency services and law enforcement agencies can be informed regarding the originator if necessary. Note that Amy's Call Agent will never receive the P-Asserted-Identity header and so the only information that she will receive will be the display name anonymous. It may also be that the originating network Id bigtelconetwork.com could be

Table 2.10   Invite with Privacy

```
INVITE sip:9724411111@172.30.1.5;user=phone SIP/2.0
Via: SIP/2.0/TCP bigtelconetwork.com;branch=z9hG4bK-124
To: <sip:9724411111@172.30.1.5;user=phone>
From: "Anonymous "<sip:anonymous@anonymous.invalid>;tag=9802748
Call-ID: 245780247857024504
CSeq: 2 INVITE
Contact: <sip:anonymous@bigtelconetwork.com>
Max-Forwards: 69
P-Asserted-Identity: "Eric Smith "<sip:2143302105@bigtelconetwork.com>
Privacy: id
```

removed from the Contact header and the Via header before passing it to Amy's User Agent so as to increase still further uncertainty as to the originator of the call.

2.6.6 SIP Registration Procedures

Because SIP was designed to operate over connectionless IP networks where users location may change as they move between network access points, SIP already supports a mechanism for dynamically registering the location of a user by providing a network element known as a *SIP registrar*. When a SIP terminal connects to the network, it registers by sending a SIP Register message to its chosen registrar. The choice of the registrar may be determined on the basis of the client configuration; it may be communicated by the network as part of a DHCP or a multicast register request may be sent. There is a convention defined in SIP that if the User Agent does not have a configured registrar, it should attempt to address a registrar using the host part of its address of record. For example, Eric's address of record might be "eric@bigtelconetwork.com" and therefore his registrations might be sent to "sip:bigtelconetwork.com."

Figure 2.12 shows a typical SIP register request and response and the content of the Register message sent by Eric's SIP User Agent is shown in Table 2.11.

Eric's registrar creates a binding for Eric that will ensure that any incoming INVITE for Eric arriving at his network-based SIP Call Agent will be resolved such that the INVITE will be forwarded to his SIP User Agent at his current location. In this example, the effect is to create a binding between "sip:9723542109" and the IP address 172.30.5.12. The SIP registrar confirms that it has created this binding by returning 200 OK as shown in Table 2.12. Note that the 200 OK message contains a list of all of the bindings that have been created for Eric by providing a contact header for each binding. In this example, Eric has one contact header and therefore one binding.

Each binding will expire if it is not refreshed by the User Agent before its expiry time has elapsed. The User Agent can refresh a binding by sending another Register request containing that binding.

In this simple example, no authentication of the user took place; however, in real networks, it is a requirement that the network ensures that it knows who has registered with it and that they have a valid set of credentials. There are a number of mechanisms to achieve this in SIP and these are discussed in Chapter 3.

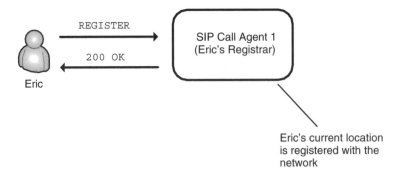

Figure 2.12   SIP registration

Table 2.11 Register

SIP method and headers	Comments
REGISTER sip:172.30.5.1; user=phone SIP/2.0	Indicates that this is a register request for the host 172.30.5.1.
From: 9723542109 <sip:9723542109@172.30.5.12; user=phone>;tag=02ffb006-d93ddf4f	The From field contains the address of record of the person responsible for the registration. In SIP, it is possible for a third-party registration to take place, whereby a binding is requested by one user on behalf of another. However, in this example, this is not the case.
To: 9723542109 <sip:9723542109 @172.30.5.12;user=phone>	The To field contains the address of record that is to be created.
Call-ID: 0eea65c0-630f598a-774af9dc@172.30.5.12	SIP recommends that the User Agent uses the same Call-Id for all register requests it sends.
Via: SIP/2.0/UDP 172.30.5.12	
CSeq: 439631084 REGISTER	
Supported: timer, 100rel	The Supported header allows the User Agent to inform the registrar of particular capabilities. It is also possible to use the Require header field, in which case if the network does not support any of the capabilities listed in the Require header field, it would reject the request.
Contact: 9723542109 <sip:9723542109@172.30.5.12; user=phone>	The Contact header field is optional in the request and may contain 0 or more address bindings. If multiple address bindings are present, the User Agent can prioritise them using the "q" parameter.
Content-Length: 0	

Table 2.12 200 Ok (Register)

SIP method and headers	Comments
SIP/2.0 200 OK	
From: 9723542109 <sip:9723542109@172.30.5.12; user=phone>;tag=02ffb006-d93ddf4f	
To: 9723542109<sip:9723542109 @172.30.5.12;user=phone>	

(continued overleaf)

Table 2.12 200 Ok (Register)

SIP method and headers	Comments
Call-ID: 0eea65c0-630f598a- 774af9dc@172.30.5.12	
CSeq: 439631084 REGISTER	Indicates that it is a 200 OK to the register sent with the given CSeq.
Contact: 9723542109 <sip:9723542109@172.30.5.12 ;user=phone;expires=3600>	The Contact header provides a list of bindings currently held and the time in seconds after which the binding expires.
Content-Length: 0	

2.6.7 Routing SIP Messages, Record-route, Route and Via Headers

As described previously, SIP supports the concept of a proxy server whose primary function is the routing of SIP messages towards their destination (as seen in Figure 2.13). SIP messages will traverse from the originating User Agent to the terminating User Agent via a number of network-based SIP Call Agents according to the SIP routing that is present within the network.

The decision as to which SIP Call Agent a call is sent to may be a purely routing-based consideration; for example, for this SIP domain my next hop SIP server is X or it may be based on a service decision, for example, for this service I must send the call via Service platform Y. The reasons for sending calls to service platforms such as service brokers and the mechanisms by which SIP achieves this are described in Chapter 6.

SIP has complex routing procedures and it is beyond the scope of the book to attempt to fully explain the SIP routing process in detail; however, it is useful to understand the basic procedures followed and further detailed information can be found in Refs 1 and 34. SIP has four headers that fundamentally determine how a call is routed (at the SIP level) in the network. These are the Request-URI, the Via header, the Route header and the Record-route header. It is important to note that the routing of a SIP request differs from that of the SIP response in that the request is routed on the basis of the contents of the Request-URI and, optionally, the Route header and the response is routed primarily on the basis of the contents of the Via header. The Record-route header is not directly used in routing the initial message but has a major impact on determining the routing used during a dialogue.

When Eric calls Amy, the initial INVITE sent by Eric's User Agent is passed to his network-based SIP Call Agent which is acting as a SIP proxy In Eric's case, his SIP Call Agent is determined by static configuration but it could also be determined by a dynamic mechanism such as DHCP. The Request-URI of the INVITE is set by Eric's SIP User Agent to the SIP URI that Eric understands is used to reach Amy and the Via header is set to point at Eric's SIP User Agent.

Eric's Call Agent acts as a SIP proxy server and in this role it handles an INVITE by examining the Request-URI contained within in it and resolving it to a set of target SIP URIs. It will send an INVITE to one or more of these SIP URIs either in sequence

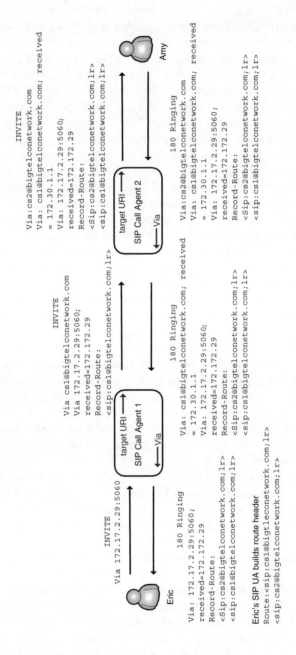

Figure 2.13 SIP routing mechanisms

or simultaneously, if it is acting as a forking proxy. The INVITE it sends will retain the To, From and Contact fields as set by Eric's SIP User Agent but it will modify the Request-URI, the Via and the Record-route header fields as part of SIP routing. It will generally copy any other header fields in the onward INVITE from the original message it received from Eric's SIP User Agent and it may also modify and add headers, depending on specific call-processing functionality.

In this case, it determines from the incoming INVITE Request-URI and its own call-routing data that it is not responsible for Amy's URI (i.e. it does not provide call handling services for Amy) and so the only action required is to forward the INVITE towards a SIP server that does. It therefore selects the incoming Request-URI as a single "target URI" to which it can forward the request. This target URI will become the Request-URI in the INVITE that it sends onwards towards Amy. Eric's network Call Agent also examines the Via received in Eric's invite and if it does not contain the same IP address as the originating IP address of the packet that contained it, then it adds a received parameter to the Via which is equal to the originating IP address of Eric's INVITE message. Eric's Call Agent must also add an additional Via header field value that contains a SIP URI that resolves to itself as the topmost via field in the message and if it wishes to ensure that it remains in the signalling path for the lifetime of the dialogue, it adds a Record-route header value which again contains a SIP URI that resolves to itself. Eric's Call Agent will forward the INVITE towards Amy and it may use DNS to resolve the actual location of the next hop SIP Call Agent for Amy's SIP URI.

In this example, the next SIP Call Agent that is reached is Amy's network SIP Call Agent. This Call Agent examines the incoming INVITE. It determines that it is responsible for handling calls to Amy's SIP URI (i.e. it provides call services on her behalf) and it again examines the topmost Via header, adding a received header if necessary. It resolves the Request-URI for Amy to match a number of destinations on the basis of the location service it is running; this could be based on bindings for Amy held in a SIP Registrar and could resolve to a SIP phone and a voice mail service. It selects each of these locations in the order of preference (as contained, for example, in the q parameter within a previously sent Register request). These locations are SIP URIs and form a list of target URIs to try, because the q parameters form a list in the order of preference, the Call Agent in this case decides to try them sequentially.

Before sending out the INVITE, Amy's network Call Agent must add a new Via header value that resolves to itself as the topmost Via header value (note that it does not remove the existing Via header values). Because it also wishes to remain in the call for the duration of the dialogue, it also adds a new Record-route header value as the topmost Record-route value; this again is a SIP URI that resolves to itself. It then forwards the URI to Amy's User Agent, which examines the incoming INVITE and determines that it is the destination.

Amy's User Agent sends a 100 Trying provisional response; however, since this is a special case in SIP in that it is sent only to the previous Call Agent, it is more illustrative to examine how it sends the subsequent 180 Ringing response which is sent end to end. The path followed by the 180 response is determined by the headers received in the incoming INVITE. Amy's SIP User Agent looks at the topmost Via header that was received in the INVITE and sends the response to the SIP URI contained within it or to the originating IP address of the IP packet that contained the INVITE, if it differs. The effect of this is to send the response to Amy's network-based SIP Call Agent since it had created the topmost Via header.

Amy's network-based SIP Call Agent receives the response and processes it, removes the topmost Via header (which happens to be itself), looks at the new topmost Via header, follows the same behaviour as Amy's SIP User Agent did and sends the 180 Ringing to the SIP URI or the IP address in the received parameter of the Via header, if present. This resolves to Eric's network-based Call Agent which processes the 180 Ringing message, removes the topmost Via header and determines the next hop from the new topmost Via header which resolves to Eric's SIP User Agent.

Eric's SIP User Agent removes the topmost Via header and determines that because no Via header values remain, the response is destined for it. It processes the 180 Ringing accordingly (generating ring tone locally to inform Eric that Amy's phone is ringing). Eric's SIP User Agent also identifies that the Response contained a Record-route header. This header contains a list of SIP URIs that resolve to each of the SIP nodes that identified itself as wishing to be informed about the call state as the INVITE passed through it; in this case, it is an ordered list of all the network-based SIP servers that the call passed through, that is, Eric's network-based Call Agent and Amy's network-based Call Agent. Eric's SIP User Agent takes this list and uses it to build a Route header. This Route header is a copy of the SIP URIs contained in the Record-route header. If Eric's SIP User Agent wishes to send a new Request, it must include this Route header in the Request and determine the target URI of the next hop from the Route header.

In this case, the next Request is the PRACK that is sent from Eric's SIP User Agent towards Amy. The Route header determines that the first hop will be Eric's network-based SIP Call Agent which behaves as for the INVITE except that it removes the topmost Route header and uses the SIP URI of the new top Route header as the target URI for the PRACK. This resolves to Amy's network-based Call Agent which behaves in the same way, that is, remove the topmost Route header which is the last one route the request to Amy not on the basis of the Route header but on the target URI that it has stored for the current dialogue; in this case, this is the same target URI that it used to send the call to Amy's SIP server.

Figure 2.13 summarises the use of the various routing related-headers at each leg of the call path between Eric and Amy.

One question that might be asked is why is the Record-route header needed when the backward signalling path is determined by the Via headers received by the called party and any new requests within the dialogue would use the target URIs stored at each intermediate node when determining how to forward the request. The answer is that SIP refreshes such as a Re-INVITE or an Update message may alter the target URI for the dialogue, for example, a subscriber might, mid-call, move from one point of attachment to another. In this case, the SIP routing would diverge at the first node with a different target URI, and nodes previously in the call-signalling path would drop out of it. However, if a route set has been built on the basis of the received Record-route headers, the request will contain a Route header based on the received Record-route header seen at the request's originator. This will ensure that nodes in the Route header are traversed in order and that the call-signalling path will diverge only at the last node that appears in the Route header. This example is shown in Figure 2.14.

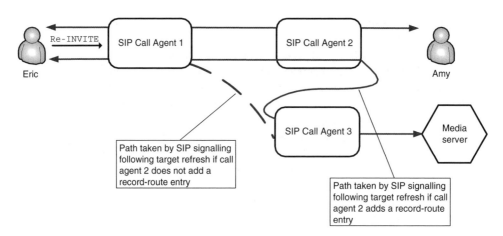

Figure 2.14   SIP-signalling and Record-route headers

As can be seen in this example, Eric is moved to a media server mid-call by a target refresh request; this will result in a more direct signalling path if SIP Call Agent 2 has not added a record-route entry because Call Agent 1 will resolve the URI to point at the new target, which is SIP Call Agent 3. However, if SIP Call Agent 2 set a record-route entry, then the Route header dictates that SIP signalling must always go to Call Agent 2 for the duration of the dialogue.

2.6.8 SIP Routing in Real Networks

The SIP behaviour described in Section 2.6.7 shows how SIP routes calls if the network-based Call Agents act as SIP Proxy nodes. However, it is not necessarily desirable in a carrier grade network to allow intimate details about the route taken by the SIP signalling to be seen at the customer's SIP User Agents; this is especially true if the network is diverting calls to a legal intercept platform. To avoid such problems, a network operator may deploy Call Agents that are directly accessible to the customer as Back-to-Back User Agents. In this case, SIP signalling is seen by the customer as always being sent to or received from the same SIP node. If Eric's network is configured in this way, it means that when Eric calls Amy, he will appear to be talking directly to Amy as though she was the network Call Agent. The network Call Agent then initiates a new SIP call on behalf of Eric towards Amy and strictly controls the propagation of headers between the two calls. It is usual in such cases to also force the media into the network rather than allow direct media connections between two customers. An example of this type of network-based Call Agent is the SBC which is discussed in Chapter 3.

2.6.9 The P-Charging-Vector Header

The P-Charging-Vector header provides a tool for network operators to pass billing infor-mation inside a SIP dialogue or transaction. Its primary use is to aid correlation between network entities and trusted networks. The P-Charging-Vector supports three pieces of information – the IMS Charging Identity (ICID), the address of the SIP Call Agent that created the value and up to two Inter-operator Identifiers (IOI). The ICID is a globally unique value and the IOI uniquely identifies a network on which the call originates or terminates. For further information about the use of P-Charging-Vector, see Ref. 35.

The information contained within the P-Charging-Vector header is meaningful to the billing systems of the networks handling the call; it can also be used to correlate related call legs. An example of the usefulness of P-Charging-Vectors can be seen in some network-initiated call applications which connect users together by initiating two separate SIP dialogues, one dialogue between the service platform and each user. Although these are separate SIP dialogues, they are related and will result in the set-up of a media flow directly between the two customers. The use of the same P-Charging-Vector in each dialogue allows the billing systems to determine that they are linked calls.

2.7 The SDP Protocol

SIP, in common with many other protocols, uses the IETF SDP [2] to signal the characteristics of the media stream that is being established. Like SIP, SDP is a text-based protocol and is extremely flexible and, therefore, can also pose interoperability problems when it is used between two network elements built by different equipment vendors. For this reason, the MSF has also created an Implementation Agreement for SDP[36] which seeks to narrow down the many options to a subset required for the support of peer-to-peer calls over an IP infrastructure.

In the NGN, the SIP protocol is used to set up a call between two or more end points and as part of this set-up process, it must identify a codec that is acceptable to both parties and also to the IP addresses and ports on which the media streams are to be sent and received by each end point. SIP achieves this by using a process known as *offer/answer* [37]. One party sends a block of SDP in a SIP message that contains an offer of a session and the other party considers the options and replies with an answer. The offer may contain a number of different codecs that can be used by the SIP end point (in preference order) but the answer dictates which codec will actually be used for the session.

For voice calls of PSTN quality over a packet network, for example, both end points would typically establish a session using the ITU-T G.711 codec, which supports 64 kbps voice circuits over a packet network. However, G.711 is typically inefficient because it transmits a constant stream of packets, regardless of whether the customer is actually talking or not. Other more efficient codecs can be used such as G.729, which can support silence suppression whereby packets are not sent in silent periods. This leads to a trade-off between bandwidth and quality, with a typical PSTN network using G.711 to preserve quality but VoIP services offered over low-speed access tails, such as cable networks, often use G.729 to maximise the bandwidth available on the link. It is possible in some circumstances that no compatible codecs can be agreed; for example, a PSTN subscriber places a call to a cable network–based subscriber who requires a low bit rate codec because of limited upstream bandwidth. The PSTN subscriber will offer G.711 in its SDP, and the Cable network subscriber will not be able to form an answer because it can only use G.729. In order to overcome this problem, the Call Agent may chose to intervene, altering the SDP being sent by the PSTN subscriber and inserting a transcoding network element (possibly at an IP–IP media gateway). In this case, the calling party would offer G.711 that would be answered with G.711 and the called party would see an offer of

G.729 that would be answered with G.729. If the Call Agent has passed a transcoding network element into the call path, the G.711 traffic will be transcoded into G.729 before it reaches the subscriber.

In order to assist in understanding what is being signalled with SIP and how the offer answer mechanism works, this section looks in detail at the SDP typically sent within SIP messages; it should be noted that there are other possible SDP lines and attribute values that are not considered here as they are not commonly used within SIP messages.

2.7.1 An Example Session Description

As an example, consider that in the simple call between Eric and Amy, Eric sent the following SDP block in the original INVITE.

```
v=0
o=SIP-GW 7295 3647 IN IP4 172.17.2.29
s=SIP-GW SIP Call
c=IN IP4 172.17.2.31
t=0 0
m=audio 1000 RTP/AVP 0 4 18 101
a=ptime:20
a=rtpmap:101 telephone-event/8000
a=fmtp:101 0-15
a=sendrecv
```

This block of SDP constitutes Eric's offer to Amy, so the eventual codec chosen for the session must be listed in this offer. The first lines, "v=, o= and s=" contain information to identify the session and the originator. They are largely included for historical reasons because legal SDP must have them even though, when used in SIP, much of the information they contain is redundant because it is already signalled with the SIP. Many implementations are likely to ignore the contents of these lines, although SDP parsers expect them to exist. The m = line describes the media stream (there may be multiple m = lines) and the a = lines describe the attributes of the media stream.

2.7.2 The v =, o =, s = and t = Lines

The v = line represents the start of the SDP and is the version of the SDP protocol; this is currently always set to 0. The o = line is the origin line and is made up of the following components: user name, session Id, version, network type, address type and address. The user name is a nebulous concept in SIP but here it is arbitrarily set to "SIP-GW" by Eric's User Agent. The session Id is a locally generated identifier so that the combination of user name, session Id, address type and address uniquely identifies the session; in this case, Eric's SIP User Agent sets it to "7295". The version refers to the version of the session announcement and is intended to be incremented each time the attributes of the session are modified; Eric's SIP User Agent sets it to an arbitrary value of 3647. The network type is always set to "IN" for Internet, the address type may be "IP4" for IPv4 networks or "IP6" for IPv6 networks; here it is set to "IP4" and the address is the IP address of Eric's SIP User Agent.

The s = line contains the subject of the session; this is not required for unicast sessions but may not be omitted because SDP requires it. Eric's Call Agent places "SIP-GW SIP

Call" in the line, the recommended value according to Ref. 37 is "-" but in any case, it will be ignored by the far end.

The c = line contains the connection data and is significant because it contains the IP address at which the SIP User Agent wishes to receive media; this line contains three parameters: network type, connection type and connection address. Eric's SIP User Agent sets the network type to "IN", the address type to "IPv4" to indicate that it wishes to receive media at address 172.17.2.31 (note that this is a different IP address from the one it uses for SIP signalling, which is not unusual as networks often use different IP subnets for signalling and media).

The t = line is another hangover from SDP, used within that protocol to define the start and stop times for the session. However, in SIP, the start and stop times are determined by the SIP control plane, and so each is set to 0.

2.7.3 The m = Line (Media Announcement)

The m = line is the media line and is used by SIP to signal all the possible codecs that can be supported by the end point in the order of preference. Below the m = line are a number of a = lines which are the attribute lines that relate to parts of the media line. SIP will negotiate which codec within a media line will be used but it is also possible to have multiple media lines because some services use multiple codecs running in parallel to achieve communication between end points. So multiple m = lines would indicate that multiple codecs should be active simultaneously (one codec from each media line) and this means that multiple media streams will be sent between the two end points. For example, video applications may use a video codec for pictures and separate audio codecs to carry the soundtrack. It is the responsibility of the application running in each end point to re-combine the media streams and play them out to the user correctly. In this example, a simple audio call is being set up which uses a single media stream and can be described with a single media line.

The m = line has a number of parameters, in order these are as follows: media type, transport port, transport protocol and media formats. The media type is one of "audio" for audio streams, "video" for video streams, "application" for data to be presented to a user on a PC, e.g. a whiteboard application and "data" for bulk data transfer. Typically, values of audio and video will mostly be seen, and in this case Eric's SIP User Agent sets this parameter to "audio". The transport port parameter is the port number on which Eric's SIP User Agent wishes to receive the media, set in this case to port 1000. Because media streams are typically sent using RTP, they should always use even port numbers (for reasons described later). Some codecs require that media be received on multiple ports simultaneously in which case, immediately following the transport parameter, the SDP will contain a "/" and the number of ports to use. These ports will be the first available ports starting with that specified in the transport parameter. If Eric's SIP User Agent had signalled 1000/2, it would mean that port 1000 would be used for the first RTP stream and port 1002 for the second (because RTP only uses even number ports). The transport type will be either "RTP/AVP" indicating the use of the IETF's RTP over UDP operating according to the RTP audio/video profile or "UDP" indicating that the stream is carried directly over UDP. This is set to "RTP/AVP" by Eric's SIP User Agent which is the usual setting for peer-to-peer calls in NGNs. The final parameters in the media line are all media formats; for the "RTP/AVP" transport type, these are defined as payload types, which usually (but not always) resolve to an individual codec. The IETF defines both static and dynamic payload types, which, in reality, define how they are interpreted by the SDP parser. Eric's SIP User Agent lists payload types "0 4 18 101" which are actually three statically defined payload types and one dynamic payload type.

2.7.4 Static and Dynamic RTP/AVP Payload Types

The IETF defined two different mechanisms for specifying the payload type within the RTP/AVP profile – a static definition and a dynamic definition. Statically defined payload types have a reserved number that always identifies the given payload type, for example, payload type 0 maps to G.711 (PCM) codec using μ-law encoding with a clock rate of 8 kHz while payload type 4 is a G.723 codec with a clock rate of 8 kHz. The list of defined static RTP/AVP payload types is defined in RFC 3551 [38]. The major audio codecs are summarised in Table 2.13.

From this table, it can be seen that Eric's Call Agent in its SDP offer has proposed in the order of preference G.711 μ-law, G.723 and G.729

Dynamic payload types use a reserved range of payload types of values between 96 and 127; these payload types are dynamic labels that allow further specification using the attribute (SDP a = line) rtpmap. The rtpmap attribute has the following format:

"a = rtpmap" payload type encoding name "/" clock rate "/" parameters.

The encoding name is the registered name of the payload format as held by IANA in Ref. 39 and usually defined in an RFC or an Internet draft. Most (but not all) of the important dynamic payload types are defined in RFC 3555 [40], the clock rate is defined for each codec in Ref. 39 as are the optional parameters. For example, the L16 audio codec is described as having the name L16, a required parameter rate (the number of samples per second) and optional parameters emphasis, channels (the number of interleaved audio streams), channel-order, ptime and maxptime. This codec would be defined in SDP as follows:

```
m=audio 1000 RTP/AVP 97
a=rtpmap 97 L16/11025/2
```

This defines dynamic payload type 97 as referring to the L16 codec with a rate of 11,025 Hz and two interleaved audio channels. In our simple example, Eric's SIP User Agent has included a dynamic payload in its list of codecs payload type 101 which is

Table 2.13 Some static payload-type meanings

RTP/AVP static payload type	Codec name
0	G.711 PCM μ-law
4	G.723
8	G.711 PCM A-law
9	G.722
13	Comfort noise
15	G.728
18	G.729

defined in the following rtpmap attribute "a = rtpmap:101 telephone-event/8000". Reference 39 informs us that telephone-event is defined in RFC 2833 [41] and is declaring Eric's SIP User Agent as supporting the capability to transfer DTMF tones over RTP.

2.7.5 SDP Attribute Lines

The attribute lines provide additional information about codecs and media handling for the session. Table 2.14 summarises the most important attributes defined for SDP when used in SIP.

2.7.6 Building an SDP Answer to and SDP Signalling Conventions in SIP

When the SDP offer arrives at the far end, it must respond with an answer. This may, in some cases, result in call failures if no acceptable codecs can be found in any of the offered media streams. In the simple example of Eric and Amy's call, Amy's SIP User Agent returns the following answer.

```
V=0
o=SIP-GW-UserAgent 1776 6232 IN IP4 172.30.1.5
s=SIP Call
c=IN IP4 172.30.1.5
t=0 0
m=audio 19374 RTP/AVP 0 101
a=rtpmap:0 PCMU/8000
a=rtpmap:101 telephone-event/8000
a=fmtp:101
```

Table 2.14 Common SDP attributes for SDP carried in SIP

Attribute name	Syntax and usage
a = fmtp	The syntax is codec specific but it starts with the defined payload type followed by codec-specific parameters. For example, for the telephony event, codec is defined in RFC 2833 which contains a number of defined code points for events and signals that can be supported by the end point; RFC 2833 explains how these are communicated in the a = fmtp SDP parameter. So if Eric's SIP User Agent's implementation of RFC 2833 can support digits 0–15, it may indicate this in the a = fmtpmap parameter as follows; "a = fmtp 101 0-15".
a = ptime	The ptime attribute is the packetisation time in milliseconds that the offering User Agent would like to be used for its media stream. This is defaulted to 20 ms in the absence of this attribute time. In this example Eric's SIP User Agent signals it explicitly with "a = ptime 20".
a = sendonly	Indicates that the SIP end point only wishes to send media and not receive it.
a = recvonly	Indicates that the SIP end point only wishes to receive media.
a = sendrecv	Indicates that the SIP end point wishes to receive and send media; this is the default setting.
a = inactive	Indicates that the media stream is inactive and the end point does not wish to send or receive media for this stream.

An examination of the SDP shows two items of interest. The first is that Amy's SIP User Agent has only returned a single codec RTP/AVP 0 which is G.711 μ-law. The second item of interest is that, although not necessary because RTP/AVP 0 is a static codec, the SIP User Agent has chosen to include an rtpmap attribute for it anyway which is recommended in RFC 3264. This is an example of how different implementations may generate slightly different SDP for the same set of codecs.

The rules for generating an answer in SIP state require that for every media line in the offer a corresponding media line must be generated in the answer and that the ordering is significant (i.e. the first m = line in the offer is answered by the first m = line in the answer). If the end point does not wish to support a media stream, then it should assign a port of 0 to it in the m = line (which is recommended); some implementations may chose to mark it as inactive using the a = inactive field.

For each media line for which an answer is being generated, the answering end point selects a list of codecs that it can support. It may express a preference by the order of these codecs in the same way that the offering end point did; however, it is recommended in RFC 3264 that the order of codecs remains the same in the answer as in the offer so as to ensure that both ends select the same codec. So, in the example, if Amy had chosen to accept either G.711 or G.729 even though the SIP end point preferred G.729, it should respond with an m = line of

```
m=audio 19374 RTP/AVP 0 18 101
```

which would result in the G.711 codec being selected.

There are a number of signalling conventions adopted by the offer–answer model which may seem counter-intuitive with reference to the removal of media streams from sessions and how media streams that have been put on hold are indicated in the SDP.

Because SIP requires each media line in an offer to get an explicit answer and because the ordering of the media lines is significant, the question arises how in mid-call bearer modification (which is achieved through another offer/answer exchange) should an individual media stream be disabled as it cannot be removed in a new offer because it would leave a mismatch in the number of media streams. The recommended approach is that a new offer be made with the port number for the removed media stream set to 0. However, some implementations may attempt to achieve the same by either setting the media stream to inactive "a=inactive" or providing a media-specific connection line with "c = IN IP4 0.0.0.0".

If a media stream is to be put on hold, rather than deleted, the fact that the far end is going to be put on hold is communicated by sending it a new offer with an altered media attribute. If the media stream was previously sendrecv, it is set to "a = sendonly" and if it was previously recvonly, it is set to "a = inactive". This means that each direction of the media stream is placed on hold independently, which has the side effect that the end placed on hold should not respond back with an answer containing the held SDP unless it wishes to also place the far end on hold (i.e. the answer attribute should remain "sendrecv" if it was a bi-directional call. Again, because this is a recent innovation, some applications place a call on hold by sending an offer with a connection line containing a null IP address; although deprecated by the IETF, this behaviour should still be supported by the end points to allow for backwards compatibility.

2.8 Media Transport Using RTP and RTCP

In packet networks, media sessions are transported using the IETF-defined Real-time Protocol [3]. RTP is typically run over UDP and provides additional services for real-time data transmission; in particular, it supports time-stamping and sequencing functions that allow the receiver to recover the data and play it out without requiring a synchronised and reliable transport network. The RTP allows for the definitions of payload types, that identify the media being carried, and profiles that extend the base RTP for a specific application. One example of a profile is given by RFC 3551 [38] which is the "RTP/AVP" profile that is used within NGNs for the transport of voice and multimedia data over RTP.

At the boundaries of the packet network, audio and video streams are packetised using the appropriate codec for the session and placed inside RTP frames for transmission. The interval at which this data is placed into the RTP frames is defined as the packetisation time. The longer the packetisation time, the more efficient the use of bandwidth in the network; as more media is encapsulated in the frame, however, the greater is the delay perceived by the end user. RTP specifies a default packetisation time of 20 ms, but for some applications, such as the PSTN, the regulator imposes strict delay budgets to control end-to-end quality and so it is not unusual for such networks to use a lower packetisation time, typically 10 ms. This adds significant overhead in terms of the bandwidth required to support media transport but keeps the delay within budget.

At the far end of the network, the packets are decoded and played out. In order to allow for the variation in delay that may be encountered in packet networks, the receiving end employs a jitter buffer that may expand and contract in size depending on the jitter encountered in the network and enables smooth playout of the media. Large values of jitter will again impact the delay, as significant buffering is required to ensure smooth playout, and so for PSTN, quality network design should ensure that the jitter remains within a very few milliseconds.

2.8.1 The RTP Header

Each RTP frame contains a fixed header, as defined in Ref. 3 and shown in Figure 2.15. The header fields are defined as follows:

- Version (V) identifies the version of the protocol, currently 2.

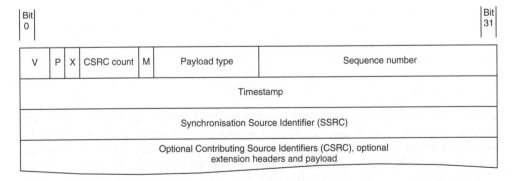

Figure 2.15   Real-time Protocol header format as defined in Ref. 3

- Padding bit (P) identifies the presence of padding fields which may be required for certain types of encrypted payload.
- Extension bit (X) identifies the presence of an extension header following the RTP header. Additional headers may be defined in RTP profiles; however, none are defined for profile RTP/AVP
- Marker bit (M). The Marker bit is a profile-specific field. For the RTP/AVP profile, it is used to support those applications that use silence suppression. Silence suppression increases the bandwidth efficiency by either not sending packets during periods of silence or injecting only occasional comfort noise into the media stream. In these applications, the marker bit is set to 1 in the first RTP packet that contains media after a period of silence; this is known as the beginning of a "talkspurt" and the use of the marker bit allows the receiver to detect the end of silence, making it easier to ensure a smooth playout of the media. At all other times, and for applications where silence suppression is not used, this bit is set to 0.
- Sequence Number allows the receiver to determine where in the stream the RTP packet belongs. The sequence number for a media stream starts at a random number (to prevent any encryption from being compromised) but increases by one with every RTP packet sent. Missing sequence numbers at the receiver indicate packet loss in the network.
- Timestamp. This represents the sampling time of the first octet of data in the RTP payload. Like the sequence number, the timestamp of the first octet of data in a media stream is set to a random value and increases with the sampling clock at the sender. The timestamp can be used by the receiver to re-assemble the media for playout and may also be used to determine the level of jitter being experienced in the network.
- Synchronisation source (SSRC). This identifies the synchronisation source of the packet. For a simple voice media session, each stream sent will have its own value of SSRC. For multimedia sessions where voice and video are sent in multiple separate media streams, each stream will be sent over RTP with its own SSRC and its own payload type. For this type of media, the SSRC is used primarily to identify the source that is being referred to in Real-time Control Protocol (RTCP) reports.
- Contributing Sources (CSRC). This identifies the list of contributing sources that make up the media stream. RTP-mixing devices typically take streams from multiple sources and mix them together (e.g. a multi-party conferencing bridge). In this case, the list of contributing sources is determined from the SSRCs seen in the incoming streams by the mixing application.
- For the RTP/AVP profile, no extension header is defined and the exact contents of the payload field depend on the payload type, which is again dependent on the codec used for the media. See Ref. 38 for information as to the encodings used for the various payload types supported in the RTP/AVP profile.

2.8.2 The RTCP Protocol

The RTCP is also defined in Ref. 3 and it was intended to support a number of functions, including the control of RTP sessions, support for quality monitoring and reporting and to assist synchronisation between multiple related RTP media streams. For voice services that utilise a single RTP media stream to carry the audio and where the set-up signalling is performed by SIP, H.248 or MGCP (depending on the application), RTCP is not strictly required and is used only for monitoring end-to-end connection quality. For video services that typically send video and audio over separate RTP media streams, RTCP is required to synchronise the playout of the individual streams at the receiver.

2.8.3 RTCP Reports

RTCP reports allow the sender of the stream to receive information about the packet loss, delay and jitter that were encountered by the RTP stream. Both the recipient and the sender periodically send reports and Ref. 3 defines the format of the SR (Sender Report) and the RR (Receiver Report). The RR is made up of a number of report blocks, one for each source that the recipient is aware of for the session. Each report block contains the following information:

- SSRC of the source of the RTP stream to which the report applies.
- The fraction of RTP packets lost from the source since the last RR was sent.
- The cumulative total of RTP packets lost since the beginning of reception from the source.
- Highest sequence number received at the time the RR was generated.
- Interarrival jitter. A measure of the jitter seen at the receiver for the RTP stream. The jitter value returned in the RR is calculated by the receiver on the basis of a formula defined in Ref. 3.
- Last SR timestamp. This identifies the Network Time Protocol (NTP) [42] timestamp or "wallclock" time contained in the last SR received.
- Delay since last SR. The time elapsed since the last SR was received.

The report block allows the sender to determine the packet loss, the jitter and the round trip delay experienced for communications with the receiver. For peer-to-peer voice and multimedia applications, this may be used to alert the application layers if the performance of the network means that the end user experience has fallen below acceptable levels. For voice gateways, this would be signalled to the Call Agent using the H.248 Quality Alert event defined in the network package (H.248.1 [15]) and the Quality Alert Ceasing event in the Quality Alert Ceasing package defined in H.248.13 [43].

The SR contains report blocks for each RTP source that it receives for the session but it also contains the following additional information sent by the sender of an RTP source (as defined by its SSRC) to assist the receivers of the source.

- SSRC. The SSRC used by the sender of the RTCP packet.
- NTP Timestamp. This contains what is known as the "wallclock time" as seen by the RTP source when the SR was sent. The wallclock time is a common time reference that is seen by the sending system that generates all the RTP streams required to support a given multimedia session (for voice this would be exactly one RTP stream but for video it would be at least two). Ideally, this would be an absolute time value as defined by NTP.
- RTP Timestamp. This contains the RTP timestamp as seen by the RTP source when the SR packet was sent. This corresponds to the value of the RTP timestamp that would have been placed in an RTP packet had it been sent at the instant the SR packet was sent.

- Senders Packet Count. This gives the total number of packets that have been sent for the session defined by the SSRC.
- Senders Octet Count. The total number of octets that have been sent for the session defined by the SSRC.

It is important to note the significance of the RTCP SR in the case where the session being supported is a video or a multimedia session. In the NGN, video sessions are supported by multiple RTP streams, for example, one for video and one for the soundtrack. The so-called wallclock time sent in the RTCP SR allows timing information that is common to both the video RTP stream and the audio RTP stream to be sent. The receiver can determine for each RTP stream the relationship of the wallclock time to the RTP timestamp values by examining the corresponding RTP timestamp value sent in the same SR packet, as this was generated at the same instant as the wallclock time. This allows the sender to understand where the two video session RTP streams' timestamps are with respect to the single wallclock time and, hence, it can align them on playout to ensure that (for example) lip synchronisation is maintained.

2.8.4 RTCP Extended Reports

In addition to the SR and RR defined in [3] the IETF has defined a set of RTCP extended reports (RTCP XR) defined in RFC 3611 [44]. RTCP (XR) extends the basic capabilities of RTP reports and defines additional report blocks that provide significant additional detail to be derived by the sender as to the performance of the RTP stream. The major benefits of these new report blocks are as follows:

- They allow additional information about lost, received and duplicate RTP packets.
- They allow the receiver to determine round trip delay by allowing it to send wallclock timestamps and receive information about the reception time of its reports as seen by the RTP sender.
- RTCP XR supports a summary report block that allows the receiver to summarise information about RTP losses, jitter, delay and IP-related information such as Time to live (TTL) information.
- A VoIP Metrics report block is defined that allows information specific to VoIP RTP streams to be passed to the receiver including information about the burstiness of the stream, the analogue side of the far-end receiver (e.g. the signal-to-noise ratio seen over the analogue line to the far-end subscriber, if the receiving system is an access gateway) and values of voice Mean Opinion Scores as calculated by the receiver.

RTCP XR is not required to support multimedia communication; however, it can provide network operators with significant additional information from which they can determine the actual performance of the network and whether they potentially fall within their Service-level Agreements (SLA) for any given customer. However, it is an issue for operators as to whether they are prepared to accept the additional RTCP traffic load on the network as well as the increased costs associated with buying gateways and clients that support the functions required to generate the extended reports.

Although RTCP XR adds significant capability to the basic RTCP reports, the migration of PSTN services onto packet networks have shown that even RTCP XR reports are not sufficient to provide information necessary to monitor certain services that must provide a very low loss rate or have very tight jitter requirements. To this end, work is underway in the industry to define further RTCP report blocks to fill this gap [45]. However, there is a significant danger that unless used sparingly by operators the quantity of RTCP data and the processing requirements it imposes in the end points may make it impossible for gateway vendors to hit the price/performance targets required by the NGN business cases.

2.8.5 *RTP Port Numbers and Symmetric RTP*

RTP does not have a well-known port assigned to it, instead the port numbers are signalled in the SDP that is exchanged as part of call set-up. It is a convention that is very widely used for RTP and RTCP that RTP uses only even port numbers and for any RTP stream the RTCP session that applies to it will have a port number equal to one greater than that of the RTP stream. Hence, it is not necessary to signal information in the SDP as to the RTCP port numbers that will apply for the session, as this convention is assumed to be in force. Explicit signalling of RTCP port numbers can be supported if, for whatever reason, an application needs to deviate from this norm.

Symmetric RTP is another convention that is very widely used and, in many cases, is essential to enable effective end-to-end communication. In many networks, the source of the media sits behind a device that effectively behaves as an integrated firewall and NAT device, for example, a residential customer's xDSL router. In order to overcome issues with NAT, solutions typically use an SBC; however, this does not help with traversing the local firewall functions at the customer's premises, which is the problem that symmetric RTP is designed to solve.

Symmetric RTP means that the customer's voice or multimedia application, for example a SIP phone, will send RTP and RTCP streams to its destination using as its source port and source IP address the same port and address values that it has indicated to the far end that it wishes to receive the media on. This means that as soon as the device sends RTP or RTCP traffic to its session peer, the firewall will log the source port and source address as being in use by an active session. When the media stream from the far-end peer subsequently arrives at the firewall with the same values in its destination address and port fields, the firewall identifies it as being related to a session initiated by the local user and allows it through. Otherwise, if no outgoing traffic had a source port and address that matched the incoming media stream's destination port and address, the firewall would have rejected the incoming media stream as being unrelated to any existing session and hence a security threat. Symmetric RTP therefore guarantees that provided the SIP phone or RGW sends RTP or RTCP traffic into the network first, it will be able to receive the relevant incoming RTP and RTCP flows when they arrive from the far end.

2.9 Addressing Issues

In the examples shown in this chapter, SIP messages have been sent using SIP URIs which have been used to identify the end user either as a name, for example sip:amy@bigtelconet work.com or as a number, for example sip:97244111111@172.30.1.5. However, today, the majority of people who communicate using the PSTN use traditional telephone numbers according to the ITU-T E.164 format. Even the new VoIP services may allocate telephone numbers to their customers to aide reachability from the PSTN and to allow black phones to be used as end user terminals. This creates two interesting issues, how should such numbers be represented in SIP signalling and how is it possible to translate between the computer world of SIP URIs and servers and the telephone numbers of the customers they support.

2.9.1 The SIP "tel-URI"

SIP defines a special URI format to allow telephone numbers to be used as addresses within the SIP domain [6]. This is known as a tel-URI and has a number of special parameters that are related to telephony. A SIP tel-URI is identified because it has the token "tel:" appended to the front of it. Global tel-URIs are identified by the use of the + symbol in front of the country code of the telephone number. Local tel-URIs consist of a telephone number and a separate phone-context parameter that defines the domain to which they apply. Three examples of tel-URIs are shown below.

tel:+44-20-79463426

which is a global number format tel-URI

tel:020-79463426; phone-context=+44
tel:3426; phone-context=bigtelco.com

which are two forms of a local number format tel-URI.

The tel-URI is defined in Ref. 46 and in addition to global and local numbers includes support for ISDN subaddresses and local extensions within a company's PBX.

It should be noted that there is no reason why a telephone number cannot be used as part of a normal SIP URI, (as opposed to the specialist tel-URI), for example sip:02079463426 @bigtelconetwork.com. In this case, the URI is treated as any other SIP URI and provided the telephone number is unique in its domain, this will be adequate to route the call.

2.9.2 Locating Telephone Numbers, ENUM

As networks start to converge and PSTN services migrate to the converged network, the issue of determining which SIP server to access in order to reach a given telephone network will become more pressing. The IETF, in collaboration with the ITU, has identified a solution that enables a simple lookup based on the mechanisms currently used by DNS. This is referred to by the industry as ENUM and is defined in IETF RFC 3761 [4]

ENUM allows a DNS NAPTR record to be created that relates an ITU-T E.164 telephone number to a SIP URI that can be used by the querying network element to route the call; to this end, the e.164.arpa domain has been reserved and is being populated with E.164 numbers. The SIP URI that is returned by the ENUM server is effectively the end user's address of record; the actual location of the subscriber may differ from the address of record but it will be resolved by the mechanisms defined in SIP for registration and redirection as discussed earlier in this chapter.

The current ENUM concept is based on individual end user registration which has a number of benefits but is not entirely suitable for large-scale carrier interconnect as carriers may just wish to be able to determine a SIP URI for call routing on the basis of a partial number. This might apply, for example, if all calls to a particular country were sent to the same network, in which case the network need only store information about a country code and not all numbers in the country. Some of the thinking in this regard may be driven by the traditional experience with number portability in different markets; for example, in the North American PSTN, an IN query is performed on a number to determine which network the subscriber is resident on, while in the United Kingdom, calls were traditionally routed to the default network for a given number range and may be routed onwards by that network if the customer has moved to another network. Work is currently underway in the IETF ENUM working group to analyse carrier requirements for IP interconnect using ENUM and to consider extending the solution.

It should be noted that the mechanisms defined by ENUM can be extended to provide support for routing E.164 numbers within an individual network operator's domain and that such solutions can be tailored to the needs of the network operator. It is therefore possible that a given network may deploy an "ENUM" server that is not a public ENUM server and does not handle requests for the e.164.arpa domain, but in all other respects, it behaves as an ENUM server.

2.10 Summary

This chapter has described how networks have evolved from the original completely separate TDM-based PSTN, and xDSL and cable broadband networks into a single converged network running IP that is capable of supporting both data services and PSTN voice services. The role of the Call Agent in the NGN has been described in detail, and it has been shown how each Call Agent has at its core a BCSM that is similar to those used for PSTN switches but that supports a separation of the call control service from the underlying packet network.

The technologies required to support voice and multimedia peer-to-peer services over an IP network have been examined in some depth, and it has been shown how the IETF-defined SIP is used to provide peer-to-peer call signalling between Call Agents. The contents of a typical SIP call set-up and tear down signalling trace have been described and the role of the SDP that is embedded within the SIP has been discussed. It has been shown that SDP is used by the end points in the call to convey information about the codecs that they wish to use for the call and to enable the far-end nodes to be configured to receive and send media when instructed to do so by call control. The role of the RTP in voice and video media transport over the packet network has been discussed and it has been shown that for multimedia video codecs the RTCP must also be used to allow the separate media streams for audio and video to be synchronised and re-combined by the end user's client.

Finally, some of the issues faced when deploying VoIP in real networks have also been looked at and the solutions adopted to overcome them described. These issues include the impact of firewalls on peer-to-peer multimedia services, which requires the use of symmetric RTP, and problems caused by addressing end users using traditional telephone numbers in an IP-based network, which can be addressed by using ENUM servers to provide mappings between E.164 telephone numbers and IP addresses.

However, there are other problems that must be overcome to enable the deployment of a peer-to-peer voice or multimedia service in the NGN. These are the issues of security, user authentication and NAT traversal. These problems and their solutions are considered in the next chapter.

References

1. J. Rosenberg, H. Schulzrinne, G. Camarillo, A. Johnston, J. Peterson, R. Sparks, M. Handley, E. Schooler. IETF RFC 3261 "SIP: Session Initiation Protocol" June 2002.
2. M. Handley, V. Jacobson. IETF RFC 2327 "SDP: Session Description Protocol" April 1998.
3. H. Schulzrinne, S. Casner, R. Frederick, V. Jacobson. IETF RFC 3550 (STD 0065) "RTP: A Transport Protocol for Real-Time Applications" July 2003.
4. P. Faltstrom, M. Mealling. IETF RFC 3761 "The E.164 to Uniform Resource Identifiers (URI) Dynamic Delegation Discovery System (DDDS) Application (ENUM)" April 2004.
5. K. Suda. "NTT's Network Vision and Approaches to Implement it". Presented at the MSF Carriers Panel in Nagoya October 18th 2005 – MSF Contribution msf2005.180.00. 2005.
6. P. Drew, C. Gallon. MSF-TR-ARCH-001-FINAL "Next-Generation VoIP Network Architecture" March 2003. Available at http://www.msforum.org.
7. S. Walker, P. Drew. MSF-ARCH-002.00-FINAL "MSF Release 2 Architecture" January 2005. Available at http://www.msforum.org.
8. ETS 300 347-1 "V Interfaces at the Digital Local Exchange (LE); V5.2 Interface for the Support of Access Network (AN); Part 1: V5.2 Interface Specification" 1999.
9. ITU-T Recommendation H.248.34 "Gateway Control Protocol: Stimulus Analogue Line Package" January 2005.
10. T. Taylor, S. Walker. MSF-ARCH-003.00-FINAL "MSF Release 3 Architecture" 2006. Available at http://www.msforum.org.
11. Integrated Services Digital Network (ISDN). ETSI 300 357 "Integrated Services Digital Network (ISDN); Completion of Calls to Busy Subscriber (CCBS) Supplementary Service; Service Description" October 1995.
12. ITU-T Recommendation Q.764 "Signalling System No. 7 ISDN User Part Signalling Procedures" December 1999.
13. ITU-T Recommendation Q.931 "ISDN User–Network Interface Layer 3 Specification for Basic Call Control" May 1998.
14. F. Andreasen, B. Foster. IETF RFC 3435 "Media Gateway Control Protocol (MGCP) Version 1.0" January 2003.
15. ITU-T H.248.1 "Gateway Control Protocol: Version 2" May 2002.
16. GR-303-CORE "Integrated Digital Loop Carrier System Generic Requirements, Objectives and Interface" Issue 4 December 2000.
17. ITU-T Q.1228 "Interface Recommendation for Intelligent Network Capability Set 2" September 1997.
18. S. Walker, P. Drew. MSF-IA-SIP.005-FINAL "Implementation Agreement for SIP Interface between Call Agent and Service Broker" September 2004.
19. ITU-T Q.1912.5 "Interworking between Session Initiation Protocol (SIP) and Bearer Independent Call Control protocol or ISDN User Part" March 2004.
20. L. Ong, I. Rytina, M. Garcia, H. Schwarzbauer, L. Coene, H. Lin, I. Juhasz, M. Holdrege, C. Sharp. IETF RFC 2719 "Framework Architecture for Signaling Transport" October 1999.
21. R. Stewart, Q. Xie, K. Morneault, C. Sharp, H. Schwarzbauer, T. Taylor, I. Rytina, M. Kalla, L. Zhang, V. Paxson. IETF RFC 2960 "Stream Control Transmission Protocol". J. Stone, R. Stewart, D. Otis. See also IETF RFC 3309 "Strem Control Transmission Protocol (SCTP) Checksum Change" September 2002.
22. K. Morneault, S. Rengasami, M. Kalla, G. Sidebottom. IETF RFC 4233 "Integrated Services Digital Network (ISDN) Q.921-User Adaptation Layer" January 2006.
23. K. Morneault, R. Dantu, G. Sidebottom, B. Bidulock, J. Heitz. IETF RFC 3331 "Signaling System 7 (SS7) Message Transfer Part 2 (MTP2)–User Adaptation Layer" September 2002.
24. G. Sidebottom, K. Morneault, J. Pastor-Balbas, (Editors). IETF RFC 3332 "Signaling System 7 (SS7) Message Transfer Part 3 (MTP3)–User Adaptation Layer (M3UA)" September 2002.
25. T. George, B. Bidulock, R. Dantu, H. Schwarzbauer, K. Morneault. IETF RFC 4165 "Signaling System 7 (SS7) Message Transfer Part 2 (MTP2)–User Peer-to-Peer Adaptation Layer (M2PA)" September 2005.

26. R. Mukundan, K. Morneault, M. Mangalpally. IETF RFC 4129 "Digital Private Network Signaling System (DPNSS)/Digital Access Signaling System 2 (DASS 2) Extensions to the IUA Protocol" August 2005.
27. ITU-T Q.1224 "Distributed Functional Plane for Intelligent Network Capability Set 2" September 1997.
28. S. Donovan, J. Rosenberg. IETF RFC 4028 "Session Timers in the Session Initiation Protocol (SIP)" April 2005.
29. IETF RFC 3262 "Reliability of Provisional Responses in the Session Initiation Protocol" June 2002.
30. Oftel (Office of Telecommunications) "Guidelines for the Provision of Calling Line Identification Facilities and other Related Services over Electronic Communications Networks. A statement issued by the Director General of Telecommunications", 28 August 2003. at the UK regulator (Ofcom) website http://www.ofcom.org.uk.
31. Oftel (Office of Telecommunications). Code of Practice for Network Operators In Relation to Customer Line Identification Display Services and Other Related Services, 3rd Edn., November 2001, available at the UK regulator (Ofcom) website http://www.ofcom.org.uk.
32. J. Peterson. IETF RFC 3323 "Privacy Mechanism for SIP" November 2002.
33. C. Jennings, J. Peterson, M. Watson. IETF RFC 3325 "Private Extensions to the Session Initiation Protocol (SIP) for Asserted Identity within Trusted Networks" November 2002.
34. J. Rosenberg, H. Schulzrinne. IETF RFC 3263 "Session Initiation Protocol (SIP):Locating SIP Servers" June 2002.
35. M. Garcia-Martin, E. Henrickson, D. Mills. IETF RFC 3455 "Private Header (P-Header) Extensions to the Session Initiation Protocol (SIP) for the 3rd-Generation Partnership Project (3GPP)" January 2003.
36. MSF-IA-SDP.001-FINAL "Implementation Agreement for SDP Usage & Codec Negotiation for GMI 2004" 2004.
37. J. Rosenberg, H. Schulzrinne. IETF RFC 3264 "An Offer Answer Model with the Session Description Protocol (SDP)" June 2002.
38. H. Schultzrinne, S. Casner. IETF RFC 3351 "RTP Profile for Audio and Video Conferences with Minimal Control" July 2003.
39. "IANA Assignments RTP Parameters". http://www.iana.org/assignments/rtp-parameters.
40. S. Casner, P. Hoschka. IETF RFC 3555 "MIME Type Registration of RTP Payload Formats" July 2003.
41. H. Schulzrinne, S. Petrack. RFC 2833 "RTP Payload for DTMF Digits Telephony Tones and Telephony Signals" May 2000.
42. D.L. Mills. IETF RFC 1305 "Network Time Protocol (Version 3) Specification, Implementation and Analysis" March 1992.
43. ITU-T H.248.13 "Gateway Control Protocol Quality Alert Ceasing Package" March 2002.
44. T. Friedman, R.Caceres, A. Clark, (Editors). IETF RFC 3611 "RTP Control Protocol Extended Reports (RTCP XR)" November 2003.
45. A. Clark, A. Pendleton, R. Kumar. IETF Draft "RTCP XR High Resolution VoIP Metrics Report Blocks" currently April 2006.
46. H. Schulzrinne. IETF RFC 3966 "The tel URI for Telephone Numbers" December 2004.

3

Securing the Network and the Role of Session Border Gateways

This chapter looks at security implications of running the Public Switched Telephone Network (PSTN), and its successor services over the packet network-based Next Generation Network (NGN). It looks at the additional dangers this implies for the network operator in terms of denial of service (DoS) attacks or theft of service attacks and it examines the solutions that are available to protect the NGN call control and related media devices from attack.

Central to the protection of the call control layer is the session border controller (SBC) or session border gateway, which also performs a number of other critical functions that are required by peer-to-peer voice and multimedia services. This chapter examines this network element in detail and looks at how it is continuing to evolve as the many diverse networks start to converge into the NGN.

This chapter does not provide a detailed analysis of all the security issues facing a converged multimedia network, as that is a huge topic in itself. Instead, it considers the general principles that any security solution must follow and looks at how some of the signalling interfaces in the NGN can be secured.

3.1 General Principles of Security and the NGN

Many of the issues confronted by a NGN are those that afflict existing widely deployed data networks. Some of the risks are specific to the NGN, of course and the nature of the NGN and its use may exclude some risks that other types of data networks are exposed to. But in essence the NGN should be secured as though it were a generic wide-area IP Network and any specialist solutions required to protect services such as PSTN voice should be layered on top of this foundation. The approach taken to solving this problem by the network operators and the threat models that they consider in doing so may well have a significant impact on how the call control layer is secured and the degree to which technologies such as encryption of signalling and media must be deployed.

Converged Multimedia Networks   Juliet Bates, Chris Gallon, Matthew Bocci, Stuart Walker and Tom Taylor
© 2006 John Wiley & Sons, Ltd

3.1.1 Security Assets

The first requirement for securing a network is to understand two, apparently simple, issues:

- What is being protected?
- What is it being protected from?

It is tempting to think of the network as being a collection of boxes in links. However, the security needs to encompass the services those boxes provide, and the information those boxes store.

For example, the service of making voice calls implies the security of many elements that work together to deliver the voice media stream only to the parties on the call, and to deliver billing only to authorised systems. Once the billing data has been stored, it needs to be protected from threats, which might copy it, alter it or destroy it.

A register therefore needs to be compiled which lists the services the network provides, and the data that is generated or required by those services in order to function. These services in turn require equipment, premises, people and other assets. Security technologies and processes need to address all of the identified items within the register.

Historically, telephone networks existed within physically secure buildings, which were exposed to users only through narrowband links. It was taken for granted that the employees of the telephone company were honest and the protocols that were in use within the network were obscure and the documentation hard to obtain. As remote management capability was added to telephone networks this was achieved using dedicated management networks that operated entirely separately from user plane traffic.

Today, the arrival of the converged NGN means that both of those assumptions are overturned. Customers now have broadband links to networks and they access the network with protocols and mechanisms that are widely understood. This situation is compounded by the well-documented problems with home user PC security which means that innocent people may be used as unwitting proxies by malicious individuals launching distributed DoS attacks. In some cases network operators are deploying management networks that operate in-band, that is, they share the same links over which customer traffic is carried. There are many techniques for separating traffic travelling over the same physical link, such as the use of MPLS BGP VPNS (see Chapter 7) or Provider Bridge VLANs[1] [1]; however this still increases the level of risk and must be mitigated against.

3.1.2 Risk Analysis

Once a list of the assets that are being protected and the threats to which they are exposed exists, a risk analysis can be conducted to consider the impact of those threats. An excellent methodology is described in Section 6 of BS7799-3:2006 [2]. In essence, a decision has to be made to:

- reduce the risk, by applying appropriate controls;

---

[1] VLAN stands for Virtual Local Area Network and is a technology used in Ethernet networks to keep customer traffic separated. The recent IEEE 802.1ad standard provides additional capabilities to scale VLANs so they can be used in carrier grade Ethernet aggregation networks.

- accept the risk, by agreeing that it is cost-effective or otherwise practical to deal with it;
- transfer the risk, by insuring against it or by using a more skilled outsource partner to handle part of the process;
- avoid the risk, by not providing the service in question.

At the end of this process, a (possibly zero) residual risk will remain. This must be accepted by the business in order to accept the analysis.

3.1.3 Common Pitfalls

Security assessments have a tendency to focus on external, technical threats. In reality the following are equally important threats:

- Insiders have more skills, more access and often more motivation to attack a network. The skilled, motivated network engineer with access to large parts of the network is a realistic opponent to consider.
- Outsiders will take a holistic approach, looking for weaknesses in physical and personnel security as well as obscure technical flaws. Never underestimate how much data can be extracted from an employee with a dinner, a fake job interview, the offer of money or threat of violence.

Threat models must consider these types of attack and the network operator must take steps to ensure that adequate protection is provided to the NGN beyond the more obvious measures taken to protect the network boundary.

3.2 The Problem of Secrets

3.2.1 Passwords

In a simple environment, such as a company LAN containing PCs and a server, users identify themselves with usernames and passwords. The identity of the computers is taken for granted by the users: they have no means of verifying the authenticity of the machine they are handing their username and passwords to. Although there are mechanisms within specific operating systems, which allow machines to prove their identities to other machines, in general, simple network attributes such as IP numbers or MAC addresses are used. The use of network monitors is often discounted, especially now that Ethernet switching has made their use hard.

This is clearly unacceptable in a wide-area network. People cannot be expected to supply secret material to equipment every time they are booted up. Equipment may be in locations where it is subject to examination by well-resourced attackers. Networks, which carry substantial amounts of traffic, may be available for monitoring to a suitably resourced opponent (consider the case of a Multi-service Access Node located in a street cabinet). This is complicated further because it is likely that different secrets will be required for each protocol, as the requirements (size, for example) will differ. Moreover, if a product is built from commodity software, consolidating the configuration so that secure material is read from a single source, it may not be straightforward.

3.2.2 Shared Secrets

At the simplest level, a secret is shared between communicating machines, and passed on clearly over the network. This is the model used by SNMPv1, for example, and it fails if someone is able to observe the network or obtain access to the equipment. Bearing in mind that this secret may be all that prevents an attacker from managing equipment to their own ends, this is clearly insufficient.

3.2.2.1 Shared Secrets and Hashes

This technique is used to secure the routing protocols OSPF and BGP, and the management protocol Simple Network Management Protocol (SNMPv3).[2] A secret is shared between communicating equipment, and is incorporated in a secure hash (such as MD5 or SHA1) calculated over messages and the shared secret. This provides security, within the limits of the security of the hash function, against an eavesdropper on the network. However, someone able to examine the equipment in detail may be able to extract the key. Specifically, it is likely that someone able to examine one piece of equipment will be able to attack its peers.

3.2.3 Public Key Infrastructure (PKI)

At the most complex level, public key cryptography can be used in the form of X.509 [3] certificates or other identification protocols; for further information consult Schneier's Applied Cryptography [4].

The most common public key algorithm is RSA (named for its inventor, Rivest, Shamir and Adelman). The mathematics are beyond the scope of this chapter, but the generation of an RSA key yields two components:

- The public part, which can be widely distributed,
- The private part, which must be kept secret.

You can think of these as two large prime numbers (the private part) and their product (the public part). It is computationally infeasible, for sufficiently large (2048 bit) keys, to recover the private key from the public.

Encryption is provided by applying the public part to the plain text. This allows anyone to send encrypted data, which can only be decrypted by the holder of the private part.

Authenticity is provided by applying the private part to the plain text (or a hash thereof). This allows the holders of the private key to prove to anyone who has the public key that they alone signed the data.

A certificate, which is not confidential, combines a public key with a digital signature made by a certification authority, by which means the certification authority warrants that the public key relates to a specific entity. There may be a chain of such certificates, by which means a specific certification authority traces its identity back to a 'root certificate' which all participants hold.

---

[2] SNMP stands for Simple Network Management Protocol and is widely used in the industry to set and get managed attributes on network elements. These attributes are defined for the network element by the set of MIBs (Management Information Base) that it supports. These MIBs may be private vendor MIBs or industry standard published MIBs.

In general, when X.509 certificates are used in e-business and similar applications, the process is as follows:

- An entity generates a private/public key pair.
- The entity submits the public key to a certification authority, together with proof of its identity.
- The certification authority returns a certificate containing the public key, signed by the certification authority.

This allows a machine to prove its identity to another without revealing the material required. However, Public Key Infrastructures (PKIs) are notoriously complex and the algorithms can be quite processor-intensive for use in high traffic environments. Someone wishing to attain this level of security would probably be better using IPsec to leverage the certificates into simpler systems.

3.3 IPsec

IPsec [5] provides a layer of security for the transport of arbitrary IP traffic. Over this can be run management, control or data protocols. This technology is commonly used to implement Virtual Private Networks. There are two different headers that can be used by IPsec and they provide two different modes of operation:

- IPsec AH (Authentication Header) [6]. Authentication Headers (AH) can be applied to IP packets, which provide assurances that the packets have been sent by their purported sender, and have not been altered in transit.
- IPsec ESP (Encapsulating Security Payload) [7]. Encapsulating Security Payload (ESP) encrypts the traffic itself, so that it cannot be observed by an attacker.

3.3.1 Key Management

IPsec has the same problem of key distribution, using shared secrets, key exchange protocols or X.509 certificates. The best practice is to use X.509 certificates to mutually prove identity, and then negotiate a temporary key for use for a short period (thus removing the process-intensive operations with public keys from the main flow of data). This process is repeated at intervals of the order of an hour. The protocol Internet Key Exchange (IKE) [8], supplied as part of an IPsec implementation, performs this task.

3.3.2 Key Distribution

The problem still remains of distributing keys to network elements. An X.509 certificate is accompanied by a private key, and an entity wishing to prove it is the holder of a particular certificate needs the private key. Were that key to be compromised, then the certificate can be misused.

However, the exposure of the private key of one network element does not compromise all of the network elements. The benefit of certificates is that an individual certificate can be installed on each network element; it can be confirmed as valid by anyone who has knowledge of the certification authority's certificate. Certification Revocation Lists provide a means of excluding certificates that have been compromised.

One approach would be to generate certificates at the point of manufacture, and load them into temper-resistant hardware. A full security analysis, focussing on risk analysis, must be carried out to determine the level of protection required.

3.4 Session Border Controllers and Session Border Gateways

The SBC performs a key role in the NGN and is an essential component of any network that provides real-time peer-to-peer voice and multimedia services to customers over an IP-based access interface. Its location at the edge of the network means that it acts as a gateway into the network for external packet flows, ensuring that only authorised media and signalling flows are permitted into the NGN. It is also the first call control function that packet flows entering the network will meet and therefore an obvious location for providing functions to protect against DoS attacks. The SBC has another important role, it provides a solution to the problem of NAT traversal for peer-to-peer voice and multimedia services and indeed this is probably its single most important function as an enabling technology for VoIP. Because of its key role in enabling packet voice services in the NGN and its role in securing the network it is worth looking at the functions of this product in some depth.

Traditional SBCs sit at the IP interconnect points of a VoIP operators network and provide the security and NAT traversal functions required to ensure that the VoIP service can be delivered and that call requests that have been rejected by the network's Call Agents are refused access to the network. To date, this has primarily been achieved by deploying SBCs that have Gigabit Ethernet interfaces for voice media and signalling traffic and have an integrated SIP (or Media Gateway Control Protocol [MGCP]) functionality that provides a subset of call control functions and signalling capabilities.

However, as the NGN evolves the SBC is adapting to become the session border gateway (SBG). Typically SBGs decompose into two parts, a data plane session border gateway (D-SBG) that handles the routing and switching functions required by the network and a signalling session border gateway (S-SBG) that provides the SIP signalling and call control intelligence functions. The S-SBG controls the configuration of the data plane in the D-SBG using a control interface that enables it to control how incoming packets are classified, policed, marked and forwarded.

Throughout this book references are made to both SBCs and SBGs. Where the term SBC is used it refers to a device with an integrated control and data plane, where the term SBG is used it refers to a device that has been decomposed into a data plane element (the data plane session border gateway (D-SBG)) and a signalling plane element (the signalling plane session border gateway (S-SBG)).

3.4.1 Functions of a Session Border Controller

The SBC performs three key functions:

1. Application Level Gateway for call setup signalling with topology hiding
2. NAT traversal of media flows by performing an address latching function
3. Policing and gating by acting as a cut down firewall.

An application level gateway for signalling is necessary because protocols such as SIP and MGCP violate the principles of layered protocols by embedding IP addresses and ports within their signalling, (in the case of SIP and MGCP in the embedded SDP). Since NAT in the network will take place at the IP layer without any reference to the call setup signalling it follows that the IP addresses contained within the SIP and MGCP will bear no relation to the actual IP addresses that the network is using to route the packets. Without a SBC to fix these IP addresses it would not be possible to support any VoIP services in today's IPv4 networks.

The same issues with NAT in the call setup signalling also apply to an extent to the IP addresses and ports used in the RTP streams. Although SIP and MGCP are used to instruct the User Agent or Residential Gateway which IP addresses and ports to use for the RTP stream, it is possible for the media stream to pass through any number of NAT devices before it reaches the service providers network. Therefore, the SBC has the job of fixing the media stream addresses so that traffic reaches the correct end-user. These functions can be seen in Figure 3.1 which shows how the SBC interacts with IP addresses in both the SIP signalling and the RTP (and RTCP) streams to perform NAT traversal.

The first point to note is that the SBC is acting as a Back-to-Back User Agent (B2BUA) in this example. Eric's SIP User Agent believes that it is talking to the Call Agent, but it is actually talking to the SBC and the Call Agent believes it is talking to Eric's SIP User Agent but it too is talking to the SBC. This creates two separate call legs, one between the SBC and the User Agent and one between the SBC and the Call Agent. The role of the SBC is to map the signalling between one leg and the other.

From a signalling perspective, the original INVITE from the SIP User Agent to the SBC contains within it the IP addresses that the SIP User Agent believes it has been allocated. However in this case the SIP User Agent is behind a home router that is performing NAT

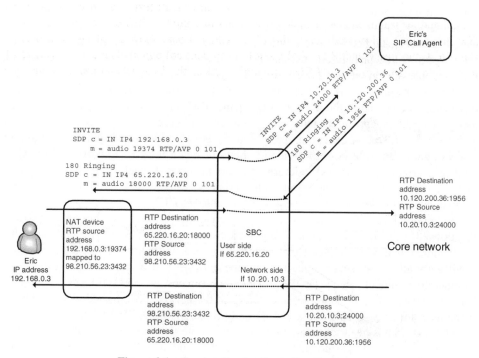

Figure 3.1   Session Border Controller functions

and so the IP address it is signalling in its SDP, in this case 192.168.0.3/24 is meaningless. The SBC, which has also been involved in the SIP Registration procedures, is aware that the SIP User Agent is a valid subscriber from their SIP URI and the SBC initiates a call setup request on their behalf towards the Call Agent, using the information it has received in the INVITE, but replacing the IP addresses and ports with values it has chosen from its core network interface. Similarly, when the 180 Ringing message is received from the network the SBC sends a 180 Ringing message back to the calling party's SIP User Agent, but it places IP addresses and ports in the SDP that it has chosen from suitable values on its access facing interface.

Because the SIP signalling has passed through any number of NAT devices to get to the SBC it sends the SIP 180 Ringing message to the originating IP address and originating port seen in the original SIP INVITE received from the SIP User Agent. The nature of NAT is such that this guarantees that the packet will get back to the correct SIP User Agents because at each intermediate NAT node a mapping is maintained between the source IP address and port used on the network side of the NAT and the equivalent values seen on the user side of the NAT. In effect if NAT has translated user traffic from 192.168.0.3 port 4002 to 180.65.220.9 port 1902 then traffic arriving on the network interface of the NAT device addressed to 180.65.220.9 port 1902 will be forwarded on the user side to 192.168.0.3 port 4002. These bindings are, however, highly dynamic and may time out in the absence of traffic. Because of this SIP and MGCP devices will carry out periodic registrations with the network to ensure that the SIP/MGCP NAT bindings are maintained and that they can receive incoming call signals.

Because the SBC has instructed the SIP User Agent to send media to its user side IP address, it must again translate the addresses in the RTP and RTCP to give the impression to the network, that they have appeared on the network side of the SBC. The SBC is able to recognise authorised media flows for a given call because it has instructed the SIP User Agent as to which IP address and port they are to be sent to and it can classify the incoming packets on these values. It sends the received RTP stream into the network with its chosen network side IP address and port as the origin of the session. When the far end media is received at the network side of the SBC interface, it remaps the IP address and port so that it appears to have originated from its user side and it sends it to the IP address and port so that the calling party's media stream is seen to originate from its user side interface. Because of NAT bindings this will reach the calling party's User Agent with the correct destination IP address and port in the RTP packets. It should be noted that this requires the end-user terminals to initiate sending of RTP traffic before they can receive any and relies on the convention of symmetric RTP.

The final function of the SBC of gating the network is a natural product of this process. It follows that any media stream arriving at the SBC from the user side which has not had an IP address and port reserved for on the network side will not be admitted and similarly it will only allow call setup signalling into the network if it originates from a registered and therefore authenticated user. When the call terminates, the SBC removes the bindings between the user side IP addresses and ports and the network side IP addresses and ports. If the user attempts to continue a call illegally by keeping the RTP stream running after the SIP session has been terminated the SBC will note that the RTP stream is being sent to an invalid IP address and port on its user side and it will reject the packets. This allocation of IP addresses and ports for RTP streams on the user side can be thought of as being a gate into the network and whether the gate is open or closed is entirely in the control of the SBC.

3.4.2 Session Border Gateways

Session border gateways perform the same functions as a SBC but they have been decomposed into the data plane functions and the signalling functions. The signalling is passed through the D-SBG up to the S-SBG which performs the same address translation functions as the SBC except that it instructs the data plane SBG what IP addresses and ports to expect traffic on and what they should be translated to (and on which interface) using a control protocol as shown in Figure 3.2.

This has the benefit of reducing the complexity required in the data plane and in some cases the data plane SBG does not even need to be a fully layer 3 aware device, it may for example be primarily a layer 2 device such as an Ethernet switching Digital Subscriber Line Access Multiplexer (DSLAM). However, the control plane signalling allows the specification of the following key properties associated with the expected streams:

1. The Destination IP address, port and network interface that the stream will have on the user side
2. The Source IP address, port and network interface to be used on the network side
3. The Destination IP address, port and network interface to be used on the network side
4. The permitted bandwidth for the stream (ideally expressed as a complete traffic descriptor to allow policing to be correctly performed)
5. Any QoS marking to be performed on the media stream packets (for example rewriting the Diffserv Code Point).

There are a number of control protocols that can be used for this interface, the ITU-T have suggested a Common Open Policy Service (COPS)-based protocol, the IETF midcom

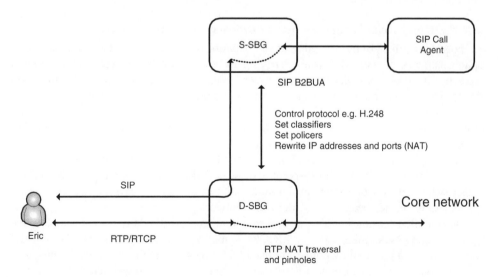

Figure 3.2 A decomposed session border gateway

working group proposed the use of SNMPv3, however, the leading candidates at the time of writing for real world deployment are the ETSI TISPAN Ia interface [9] which is based on the H.248 protocol or alternatively the ETSI TISPAN Gq' interface [10] which is based on the Diameter protocol.

3.4.3 Gates and Pinholes

When discussing SBGs the terms "gate" and "pinhole" are used. These terms are in practise interchangeable although there is an argument that the term "gate" implies a richer feature set. A gate (or pinhole) is defined in a D-SBG to allow a given media or signalling flow into the network. A gate at the D-SBG consists of a packet classifier, a policer and a forwarding behaviour. When a packet arrives at the D-SBG it is classified to see if it matches a defined gate. A typical classification may be to look for a given destination IP address and port number but may also check many other fields including source addresses and ports, the protocol Id and even the Diffserv Code Point (see Chapter 8). If a match is found then the packet will be policed according to the policer configured for the gate. If the packet is within the profile then it will be forwarded according to the forwarding behaviour defined for the gate. This forwarding behaviour may result in IP addresses being rewritten and the priority of the packet remarked. If the packet is out of profile then it will be dropped.

Packets that arrive at a D-SBG that do not match a defined gate are generally dropped, however, the network operator may for certain types of traffic choose to pass them and re-mark them as low priority traffic. The option chosen depends on the network deployment and the destination of the packets. For example, unauthorised packets would not be sent to a media gateway, or to a media server, or a point of PSTN interconnect. From the network operator's perspective, the effect of this is that packets arriving at a D-SBG that the Call Agent is not expecting (and therefore has not set up a gate) will be rejected. This provides a powerful solution to ensure that only authorised calls can access the network resources reserved for the voice service.

3.4.4 Preventing Denial of Service Attacks with Session Border Gateways

The position of the SBG at the edge of the network means that it is the first call control capable network element that traffic arriving in the operator's network will encounter. As such it is an obvious place for the network operator to plan to combat potential DoS attacks. These DoS attacks might include excessive registration attempts or excessive call setup attempts from SIP clients. The unfortunately lax nature of many home users security means that it is potentially possible for malicious individuals to stage co-ordinated SIP attacks on the NGN in the same way that they attack parts of the Internet today. The nature of such attacks is that it is almost impossible to prevent at least some disruption to service; however, the network operator must be able to take steps to ensure that any impact is limited in scope, brief in duration and critically is not able to disrupt the Call Agents sitting further back in the core of the network.

A decomposed session border gateway should be deployed on any packet access to the NGN that allows incoming SIP (or MGCP) signalling to the Call Agents, The D-SBG would be configured to allow only authorised RTP flows into the network but it must also allow signalling to pass through for subscribers who are either registered with or trying to register with a Call Agent or SIP registrar. This requires that a signalling gate must always be available through which, for example, SIP signalling will be passed from a user to Call Agent or registrar. The D-SBG might implement this by deploying a classifier that identifies SIP signalling destined for IP addresses that corresponds to a known S-SBG. Data flows that pass this classifier would then be policed to ensure that the quantity of signalling being sent to the S-SBG does not exceed a given limit; ideally this limit should be settable on a per individual user basis thus ensuring that a single user misbehaving could not consume more than their reasonable share of bandwidth. This approach is shown in Figure 3.3.

Because the total quantity of signalling that can arrive at the S-SBG is limited by the D-SBG policing it is possible to ensure that the S-SBG can be dimensioned such that it can handle the worst case load without an undue reduction in its steady-state call processing capacity (see Chapter 8 for a general discussion on processor overload control). This allows the S-SBG to parse the incoming SIP signalling accept or reject sessions, depending on a number of criteria including originating user, request type or call type (i.e. emergency calls from a trusted user would expect to gain priority) before passing them onto the Call Agents in the network. Again because the S-SBG can be configured to support a peak rate of calls to any Call Agent it is possible to dimension Call Agents to ensure that they do not become overloaded. It is of course possible that some event, such as a Call Agent hardware failure could mean that the S-SBGs are still sending a Call Agent more call attempts or register requests than it can handle. However, in this case as long as a signalling mechanism is provided to allow the Call Agent to request the S-SBGs to throttle back, the Call Agent will survive, as it can trust the S-SBGs to obey the request.

Because the combination of D-SBG, S-SBG and Call Agent overload controls ensures that the Call Agent can operate despite the attempted DoS attack, the network can continue

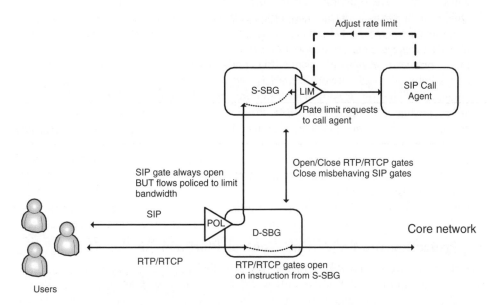

Figure 3.3   SBG architecture to prevent DoS attack

to function giving the networks operator's OSS time to identify the misbehaving users and bar them from the network.

It should be noted that while this mechanism is robust, allowing access to the call control layer (even via the SBG) to SIP clients on PCs is still less secure than allowing access via a black phone and an access gateway. Additionally, this approach does have limitations in that it is not easily possible to give priority to emergency or governmental callers at the SBG in all cases. For example, if the D-SBG is located at a point of network interconnect, rather than in the access domain (where users might have their own logical interface) it may not be possible to identify emergency calls before passing it upwards to the S-SBG because these fields in SIP signalling are not easily identified by the types of classifiers typically available in the routing layer. In addition, if encryption such as Transport Layer Security (TLS) is being used any SIP headers would be unreadable until they are decrypted at the S-SBG. This means that in such scenarios all calls would be policed identically regardless of any signalled priority or user Id.

3.4.5 Additional Functions of Session Border Gateways and Session Border Controllers

The position of the SBG/SBC at the edge of the network means that it may be used as a platform to support functions other than just straight NAT traversal and security functions. This is increasingly the case as the platform is split into the D-SBG and the S-SBG because it is possible to have a layer 2 switching or layer 3 routing platform dedicated to packet transport and a signalling platform optimised for network intelligence and call control functions, that is, a router and a high availability server farm. Some of the additional functions that such devices can support include the following:

1. Signalling interworking, for example to convert MGCP signalling from users residential gateways into SIP signalling towards the network SIP Call Agents.
2. Quality of Service, a simple solution to guaranteeing bandwidth for media flows, can be implemented at the SBG which may not scale to very large networks but is likely to be adequate for many networks, see Chapter 8.
3. Fixed mobile convergence; it is possible for the SBG to resolve signalling interworking issues between fixed network terminals and mobile networks when a fixed SIP User Agent is utilising services from an IMS core network. In particular, many SBG vendors are proposing to implement the IMS Proxy Call Session Control Function (P-CSCF) as part of their functionality (see Chapter 5).

Of course it follows that by breaking apart the architecture into D-SBG and S-SBG it is possible to add SBG control functions to Call Agents and provide an integrated Call Agent/S-SBG. As always, which solution is appropriate depends on the deployment scenario and the needs of the network operator.

3.5 Protecting the PSTN Call Control Platforms in the NGN

The closed nature of the PSTN meant that provided the network operator was able to protect access to the switches themselves, the management systems, and billing systems the network itself was largely secure. Customers were able to connect to the network either using analogue phones or ISDN terminals or PBXs which meant that only the next hop entity could be reached and that the signalling medium was either analogue or a very restricted telecommunications signalling protocol. The PSTN was vulnerable to DoS attacks on the inter-operator interconnect which allowed access to the SS7 signalling; however, this was a closely regulated interface, often using a restricted set of SS7 features and was open in the UK only to organisations that had been granted an SS7 license. Indeed, before an operator could even connect their switch into the SS7 network the hardware and software was required to have a special certification that it was safe to connect (the so-called blue stamp).

For the reasons described previously the NGN is a very different security environment and the level of threat is very much greater. It follows therefore that if the NGN can be compromised the attacker may now be able to bring down the critical infrastructure of a country (i.e. its basic phone service) in a way that was not possible before. To protect against these dangers, the following issues must be considered by the call control layer in the NGN:

- Prevention of DoS attacks on Call Agents by sending repeated bogus SIP INVITE messages or SIP REGISTER messages.
- Secure mechanisms for identifying the end-user, thus ensuring CLI services may be supported over the network.
- Prevention of theft of service attacks by masquerading as another user to place a call.
- Prevention of theft of service attacks by setting up unauthorised end-to-end media streams.
- Hiding the internal topology of the NGN to prevent end-users from analysing the signalling thus potentially allowing them to launch sophisticated DoS attacks or compromising legal intercept and security functions.

3.5.1 The Importance of Customer Access Type on Security

A major consideration when determining what security measures are necessary to secure the signalling and call control layer of the PSTN is the type of access the customer has to the network. A user might access the NGN in a number of ways including:

1. through a black phone connected to an access gateway;
2. using a SIP terminal or MGCP User Agent connected to a point-to-point access network such as x.DSL or Metro Ethernet;
3. using a SIP terminal or MGCP User Agent connected via an insecure radio interface or over the public Internet.

In the case of a black phone connected to an access gateway, the user has an analogue access to the network and their point of physical attachment at the access gateway can be used to identify the incoming line unambiguously. Internal network signalling is not visible to the end-user and so other than securing the management and H.248 control interfaces between the Call Agent and the access gateway and between the access gateway and its element manager (possibly using IPsec) there are minimal security implications (although the access gateway must implement overload controls to protect it from overload due to a mass call event). The internal network signalling between Call Agents may be SIP or

SIP-I; however, this is invisible to the end customer and takes place between trusted Call Agents within a secured network.

Where the end-user has a SIP terminal connected to a secure point-to-point access network then security is more complex but remains relatively straightforward. The point of attachment to the access network can be used to identify the end-user, although SIP registrars may wish to separately authenticate the user. In this case, however, the customer has an IP interface to the NGN and therefore a SBC (or SBG) should be used to protect the interface and it should be configured as a SIP B2BUA. The SBC should be dimensioned so as to be able to resist DoS attacks caused by excessive SIP messages from the end-user, filtering these messages to protect Call Agents deeper in the network. The use of traffic management and rate limiting of signalling in the access network elements, IP routers or the D-SBG may also be used to limit the volume of signalling that can be sent to the SBC or S-SBG.

Because the SBC or S-SBG is acting as a SIP Back-to-Back User Agent it follows that SIP signalling from the end-user must terminate on that device. This means that on the network side of the SBC (or S-SBG) all SIP signalling has originated from a trusted resource. This allows use of the P-Asserted-Identity header to verify the user Id and it also ensures that the SIP headers used for routing the call can be hidden from the end-user. By acting as an RTP proxy node (i.e. translating the IP addresses and ports used on the user side to different values on the network side) the SBC or SBG ensures that only authorised RTP flows are admitted to the network. Unauthorised RTP flows, that is, those for which there is no SIP signalling, can be blocked at this point.

Where the user is accessing the NGN using a SIP terminal over an unsecured interface, for example, a shared medium access the network, such as a WiFi hotspot, or over the public Internet, then additional security is required above and beyond that used for a secured access network. In this case, the point of attachment may not be known and the end-user devices' identity as signalled in the SIP may not be trusted. In addition, it may be possible for third parties to snoop the SIP signalling and potentially the session media as well. Therefore, some mechanism for authenticating the end-user and for encrypting the SIP signalling will be required in order to offer a secure service. Furthermore, it may be necessary to support a security solution such as IPsec or an alternative secure tunnel between the end-user and at least the SBC (D-SBG) through which the RTP flows may be passed in order to prevent eavesdropping on the conversation. Such mechanisms add expense and complexity to any network solution and should ideally be kept to the edges of the network and used only on those accesses that strictly require them.

3.5.2 SIP Security Mechanisms

SIP offers three tools for securing the network as described in [11], these are as follows:

1. TLS support for encrypted message transmission;
2. digest authentication to verify individual user identities;
3. Secure MIME (S/MIME) for encrypting the body of the SIP message to hide the SDP and also to allow SIP messages to be tunnelled end to end.

TLS is described in [12,13] and allows the setting up of a secure connection between SIP network elements over which SIP signalling would be sent. This requires an exchange of certificates between the two SIP network elements, however once established any SIP sent over the TLS connection will be encrypted and therefore the SIP call setup information will be hidden from any third party.

The desire to use TLS in SIP is indicated by the use of a special SIP Secure URI (written as a SIPS URI). For example, if Eric wishes to use TLS to communicate with Amy the INVITE would contain a SIPS URI and not a SIP URI as follows "INVITE sips:9724411111@172.30.1.5".

This would result in a secured TLS connection being used between Eric and his network-based Call Agent. TLS is a hop-by-hop security mechanism and so Eric's network-based Call Agent may determine not to use TLS for the next hop if it is to another trusted Call Agent over a secure network. For example, the call between Eric and Amy might be set up with just two TLS secured connections, one between Eric's SIP User Agent and his network Call Agent and one between Amy's network Call Agent and her SIP User Agent; the inter-call agent signalling may be sent in the clear. Typically, TLS would be used for SIP originating or terminating on an insecure access network and potentially also over any inter network operator interconnects to ensure that the Call Agents on each side of the interconnect can be certain that they are communicating with their chosen peer. TLS may also be used throughout the network if the security threat model requires it.

SIP also supports a digest authentication mechanism that is based on the established http digest authentication technology. This allows a SIP network element to challenge any SIP request received from an end-user and to obtain a response to the challenge from the end-user. The challenge/response mechanism relies on a shared secret between the two entities and uses the content of the message as part of the hash calculation in order to ensure that the message has not been tampered with by a third party, the so-called man in the middle attack. The http digest authentication mechanism is defined in [14] and rules on how to apply it to SIP are provided in [13]. The challenges are carried in SIP within the www-authenticate header and the responses carried in the authorisation header; this results in additional signalling as shown Figure 3.4, which outlines an example registration sequence with authentication.

The initial request is refused by the SIP Registrar by sending the 401 unauthorised messages containing a challenge back to the User Agent. The User Agent re-sends the SIP request (incrementing the CSeq by one as it is a new request) with the response contained within it.

Once a SIP User Agent has supplied credentials within a dialogue in response to a challenge it should include these credentials in all subsequent requests within the dialogue, these may be accepted by the network or alternatively it may issue a new challenge on receiving a request.

Note the digest authentication mechanism can be used in conjunction with TLS. For example TLS may be used between the User Agent and the Registrar, which ensures that the signalling is encrypted and secure, and that the User Agent is communicating with a legitimate SIP registrar and the digest authentication is used to authenticate the identity of the user.

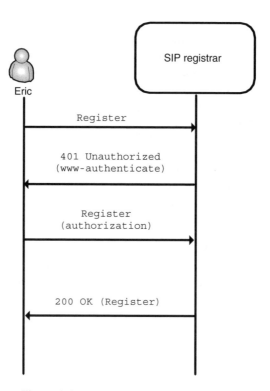

Figure 3.4   Registration with authentication

S/MIME is defined in [15] and can be used to encrypt the SDP that contains the session setup information or it may be used to encapsulate the original SIP message as seen by the sending SIP User Agent. In the NGN where SBCs or SBGs are deployed, the benefit of S/MIME is limited. If confidentiality is required then TLS provides a more complete solution, furthermore, the SBCs or SBGs must alter the SDP to fix any NAT issues in the access network and to translate the addresses to pin the media stream to their network facing interfaces. Therefore, it is unlikely that S/MIME would be used for security by a network operator. However the ability to use S/MIME to encrypt a SIP-signalling message may be of use to the end-users if they wish to pass information through a network between themselves in such a way that it cannot be altered by the network, for example they might wish to include the original INVITE as sent by the calling party complete with its original To and From headers. This would result in a body being added to the SIP message with the content type header set as follows "Content-Type:message/sip". A side effect of this is that SIP messages will increase in length and this may have an adverse impact on the network operators signalling throughput and capacity, and as such it is possible that the network operator might refuse to transport these message bodies.

3.5.3 The Impact of the Threat Model on Control Plane Security

As discussed earlier the threat model will have a significant impact on how the network operator addresses security and this has a particular impact on the control plane elements.

If the threat model is restricted to consider outside attacks on the network by malicious individuals then a sensible strategy would be to rely on the SBGs to keep the network secure and to allow unencrypted signalling to be used within the operators network (it would however be a sensible strategy to ensure that management interfaces are adequately protected).

However if the threat model is such that it is assumed that attacks may come from the network operators' technical staff, then there are additional considerations around issues of Legal Intercept that may require signalling to be encrypted even though it is being sent within a supposedly secure network. For SIP signalling this means the use of TLS. For the H.248 protocols that are used between the access and trunking gateways, and between the elements of the SBG then this implies IPsec with the ESP header. For the Diameter protocol, then either TLS or IPsec with the ESP header can be used. It is possible that in some network architectures it may be required to encrypt the media itself, in which case one solution would be to use IPsec with the ESP header on internal network interfaces at the D-SBG, the media servers, and the access and trunking gateways.

The encryption of signalling and media has both an operational and capital cost to the operator. From an operational perspective, there is some benefit in allowing staff to be able to snoop the signalling that is being carried internally in the network as it greatly assists in troubleshooting. It is also possible to buy dedicated network monitoring systems that can provide the network operator with information as to the health of the network based on information gleaned from the many signalling protocols in use. If encryption is in use in the network, then such platforms are seriously compromised. From a capital cost perspective the processing requirements of encrypting every message, and possibly media flow, adds a significant burden on network equipment. This means more equipment must be purchased to allow for the security overhead. In making decisions as to the need to deploy encryption internally within the network, the network operator must consider these cost implications.

3.6 Summary

This chapter has shown that the emergence of a truly converged network will have significant security implications, especially for those network operators looking to support national infrastructure such as the PSTN over such a network. This chapter has provided an overview of the security solutions that are available to the network operator considering the huge task of securing the NGN.

It has described how any security solution for the NGN must consider how best to secure the call setup signalling and potentially the media as well. The solution chosen will depend on the type of access being used by the end-users, the type of call setup signalling and the threat model that the network operator is using for their network.

SBCs and SBGs play a significant part in providing security by hiding network topology, providing a barrier to DoS attacks on the Call Agents and gating media flows to prevent unauthorised access to the network They also provide for the support of critical functions such as NAT traversal. This chapter has examined in detail the functions of SBCs and SBGs and has shown that the SBG is an evolution that will provide significant benefits in terms of scalability and new features.

This chapter has also considered how the signalling interfaces are protected from eavesdropping when carried over insecure networks or where the network threat model requires secure signalling transport. Where SIP is used for call signalling over an insecure access network TLS may be used to encrypt the signalling to the SBC (protecting the end-user from eavesdroppers) and digest authentication can be used to authenticate the individual user when they register with the network or when they set up a call. Other critical protocols such as H.248 or Diameter may use IPsec to provide a secure transport.

Above all a key point of this chapter is that these mechanisms for securing the call control layer of the NGN can work only as a part of a comprehensive security policy, which has analysed the assets that must be protected and the risks to which they are exposed.

References

 1. IEEE 802.1ad. "Virtual Bridged Local Area Networks – Amendment 4: Provider Bridges" Approved December 2005.
 2. BS 7799-3:2006. "Information Security Management Systems. Guidelines for Information Security Risk Management" Available from http://www.bsi-global.com. Published March 2006.
 3. ITU-T X.509. "Information Technology – Open Systems Interconnection – The Directory: Public-key and Attribute Certificate Frameworks" March 2000.
 4. B. Schneier. "Applied Cryptography: Protocols, Algorithms and Source Code in C" November 1995. ISBN: 0471117099.
 5. S. Kent, K. Seo. IETF RFC 4301 "Security Architecture for the Internet Protocol" December 2005.
 6. S. Kent. IETF RFC 4302 "IP Authentication Header" December 2005.
 7. S. Kent. IETF RFC 4303 "IP Encapsulating Security Payload (ESP)" December 2005.
 8. C. Kaufman (Editor). IETF RFC 4306 "Internet Key Exchange IKEv2 Protocol" March 2006.
 9. ETSI ES 283 018. "TISPAN NGN Release 1; RACS; H.248 Profile for the Ia Interface" March 2006.
10. ETSI TS 183 017. Telecommunications and Internet Converged Services and Protocols for Advanced Networking (TISPAN); Resource and Admission Control; DIAMETER Protocol for Session Based Policy Set-up Information Exchange Between the Application Function (AF) and the Service Policy Decision Function (SPDF); Protocol Specification January 1999
11. J. Rosenberg, H. Schulzrinne, G. Camarillo, A. Johnston, J. Peterson, R. Sparks, M. Handley, E. Schooler. IETF RFC 3261 "SIP: Session Initiation Protocol" June 2002.
12. T. Dierks, C. Allen. IETF RFC 2246 "The TLS Protocol version 1.0" June 2003.
13. D. Hopwood, J. Mikkelsen. IETF RFC 3546 "Transport Layer Security (TLS) Extensions" June 1999.
14. J. Franks, P-Hallam-Baker, J. Hostetler, S. Lawrence, P. Leach, A. Luotonen, L. Stewart. IETF RFC 2617 "HTTP Authentication: Basic and Digest Access Authentication" July 2004.
15. B. Ramsdell (Editor). IETF RFC 3851 "Secure/Multipurpose Internet Mail Extensions (S/MIME) Version 3.1 Message Specification" December 2005.

4

The NGN and the PSTN

The legacy telephone network, known formally as the Public Switched Telephone Network (PSTN), was introduced in Chapter 2. At the time of writing, the PSTN was still the dominant technology for telecommunications in most parts of the world. Continuing a long tradition of continuity in evolution, the designers of the NGN have accepted the necessity for the NGN and the PSTN to coexist at least for the next decade. This has required the development of an architecture and a suite of protocols to support interoperation between the two networks. Before going into these topics, however, it is worth reviewing the technology of the PSTN itself.

4.1 Circuits and What they Carry

In the PSTN, the communications path between any two parties in a call is built up from a series of *circuits* (See Figure 4.1). Some of the connections between these circuits, such as those at the cross connect in Figure 4.1, are semi-permanent, modified by network operators over a period of months or years as subscriptions and traffic patterns change. For our purposes, we will view the circuit segments thus linked as a single logical circuit. Other connections are made only for the duration of a telephone call. These connections are made under the control of *telephone exchanges*. The actual connecting equipment may be the switching fabric of the exchange itself or that of a remote peripheral controlled by the exchange.[1]

Circuits may use analogue or digital transmission technology. Standard analogue voice circuits were designed to fit into a 4000 Hz frequency band, although only 3400 Hz is available to the user. Dedicated *special service circuits* for purposes ranging from low-bandwidth alarm telemetry to high-fidelity transmission of radio programmes between studios and broadcasting stations were designed for requirements outside this range.

Restriction of data transmission to special circuits was unsatisfactory. To allow dial-up data transmission service over voice circuits, various modem[2] schemes were devised.

[1] In North America, telephone exchanges are more commonly known as *offices* or, more generically, as *telephone switches*. The logical circuits between offices are known as *trunks*.

[2] Modem is short for "modulator-demodulator". Most modems use either phase-shift keying (PSK) or frequency-shift keying (FSK).

Converged Multimedia Networks   Juliet Bates, Chris Gallon, Matthew Bocci, Stuart Walker and Tom Taylor
© 2006 John Wiley & Sons, Ltd

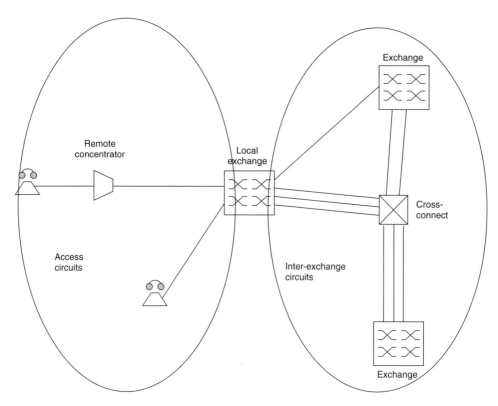

Figure 4.1   PSTN circuits

These schemes allow bits to be encoded as analogue signals using frequencies within the operating range of a voice circuit. Similarly, facsimile services were also devised to use voice circuit frequencies.

Beginning in the 1960s in North America and later elsewhere, the analogue circuits between exchanges began to be replaced by digital technology. The standard digital circuit is based on the standard voice circuit just described. The voice signal is sampled 8000 times a second, as required by the Nyquist sampling theorem to faithfully reproduce the entire 4000 Hz of bandwidth. Each signed sample amplitude is encoded into 8 bits, giving a total bit rate of 64,000 bits/second. However, to reduce the impact of quantisation error on perceived voice quality, a logarithmic transformation is first applied to the sample amplitude[3] before the result is encoded. This encoding and the inverse decoding process at the receiving end are standardised in ITU-T Recommendation G.711[4] [1].

Unfortunately, the specific logarithmic transformation used varies between regions. North America uses *μ-law* companding. The rest of the world uses *A-law*.

---

[3] This process is known as *companding*, a word coined from "compressing–expanding". Companding effectively results in using more bits to represent a given absolute change in amplitude when it occurs on top of a quieter base signal than when it occurs on top of a louder one. This provides a better match to the perceptive capabilities of the human ear than a straight linear encoding.

[4] G.711 is an example of a *codec*, an abbreviation for "coder–decoder".

The *access circuit* connects the subscriber to the local 'exchange.[5] It is also known as the *local loop* or simply as a *line*. For economic reasons, it took longer to digitise access circuits than inter-exchange circuits. Beginning in the late 1970s, however, a massive standardisation effort was undertaken under the umbrella of the Integrated Services Digital Network (ISDN), to allow digital technology to move into the access network. A major aim of this work was to carry a broader range of services over standard circuits, thus reducing the expense of designing and maintaining special service circuits. One of the services that eventually became easier to implement was video conferencing, based on the H.320 standard [2].

Summing up, the NGN must interwork with a PSTN that operates using a mixture of analogue and digital circuits. Either technology may be used to carry voice and also data, facsimile and multimedia services. When voice is carried on digital circuits, it is encoded using the G.711 codec, but the specific details of encoding vary between North America and the rest of the world.

4.2 Signalling and Supervision

Call signalling is the process whereby the caller's request for service is transmitted through the network. Call signalling is carried in numerous protocols, with major differences between those used on the access circuit and those used between exchanges. Earlier protocols simply carried requests forward until they were either successful or met with some form of failure. ISDN signalling permits the caller's terminal to negotiate the service received, both with the network and with the called party's terminal at the far end.

Call signalling on analogue circuits may be carried by tones or by changes of electrical state. Pre-ISDN call signalling on digital circuits is carried by tones. ISDN access signalling is carried on a separate signalling channel (the D-channel) physically associated with the user channels (B-channels). Between exchanges, the ISDN uses *common channel signalling*, with messages carried over a logically separate signalling network.[6]

Supervision is the process of monitoring a circuit for changes of state relevant to call processing. The simplest circuits have two states: busy and idle. More complex circuits go through a number of states, marking different phases of call signalling as well as indicating when the circuit is unavailable for calls. For analogue circuits, circuit states may be indicated by electrical states or by tones. For digital circuits, circuit states are indicated by bits associated with the circuit.[7] The interplay of call signalling and supervision will be described in greater detail in the next two sections.

---

[5] In fact, the subscriber may be served by an access network, which in turn connects to the local exchange. Whether the subscriber is connected directly or through an access network, the initial circuit leaving the subscriber's premises is dedicated to that subscriber.

[6] In Europe, the individual links of the signalling network tend to be carried in the same transmission facilities as the digital circuits they control. In North America, the signalling more commonly takes a path independent of the voice circuits, but statically determined for a given origin and destination.

[7] Digital circuits are multiplexed together in hierarchical groups. The smallest grouping of the hierarchy in Europe is the El, carrying 32 digital channels at 64 kbit/s each. Circuit states are indicated for E1 channels using octets transmitted in channel 16. In North America, the smallest grouping of the digital hierarchy is the T1, carrying 24 digital channels at 64 kbit/s each. Circuit states are indicated using bits robbed from the low-order bits of the respective voice signals. For channels using such *robbed-bit signalling*, only 56 kbit/s rather than 64 kbit/s is available for user data, even though the supervision bits are stolen only from every sixth frame.

4.2.1 Signalling and Supervision on the Access Link

The simplest example of call signalling and supervision is the ordinary analogue telephone line. The supervisory states consist of on-hook and off-hook, indicated by electrical conditions (e.g. open or closed circuit, respectively). However, the interpretation of the state transitions depends on the call context. An initial transition to off-hook at the calling end indicates a call initiation. An off-hook state at the called end when the call arrives indicates "line busy". However, a transition to off-hook at the called end during alerting indicates call answer. Finally, a transition from off-hook to on-hook at either end indicates end of call.

Just to complicate matters, transitions between hook states are subject to further interpretations if the line has specific services enabled. One example is dial pulse signalling, an older method of passing dialled digits to the local exchange. Dial pulse signalling consists of a series of rapid on-hook, off-hook transitions (at a typical rate of 10 cycles/s). Another example is the hook-flash, a transition from off-hook to on-hook and back again, with a duration within specified limits. If the subscriber has the necessary features enabled, the local exchange interprets hook-flash as a request to put the current call on hold (and typically, as a request to make a second call while the first is held). The hook-flash is ignored if the subscriber has not subscribed to the features that use it.

Signalling on an analogue line proceeds using tones and announcements. The network signals that it is ready to receive dialled digits by returning the dial tone after off-hook is detected. The subscriber indicates what service is wanted by dialling a set of digits. Typically, the digits will be signalled to the local exchange using DTMF (Dual Tone Multiple Frequency) tones.[8]

DTMF tones have other uses. At the beginning of a call, they may be used to enter additional information such as telephone credit card numbers. They may also be used in mid-call to signal far-end subscriber equipment such as voicemail servers or, very commonly these days, integrated voice-response (IVR) systems. Finally, the network may interpret certain DTMF signals in mid-call as additional call requests. One example is when the caller is allowed to use a long-duration "#" to indicate that he or she wishes to make another call without hanging up, for example, to avoid having to re-enter a credit card number.

A description of analogue line signalling would be incomplete without the mention of alerting. At the called party end, alerting is carried out by applying an intermittent alternating current to the called line. This causes the called telephone to ring. The called party's local exchange also returns an in-band tone, "ringing tone" (also known as *audible ringback*) to the calling party. When the called party answers, the called party's local exchange detects the off-hook condition on the called line. It stops the application of the ringing current, stops sending a ringing tone to the caller and provides a two-way speech path connection between the called party and the outgoing circuit to the caller.

Figure 4.2 summarises the signals exchanged on the access links for a basic analogue line call.

Supervision for some types of access links is far more complex than the simple on-hook and off-hook of the simple analogue line. The telephone network still supports a

---

[8] The DTMF tone system is specified in ITU-T Recommendation Q.23 [3], which makes quaint reading these days when DTMF dialling is so widely deployed. There are 16 signals in all, each generated using a pair of tones. One is selected from a set of four lower frequencies, the other from a set of four higher frequencies.

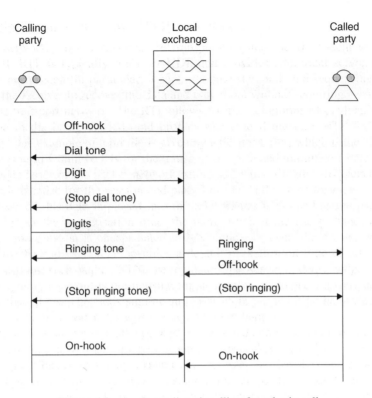

Figure 4.2   Analogue line signalling for a basic call

variety of proprietary analogue private branch exchange (PBX) supervision schemes, not to mention different types of payphone interface. The repertoire of electrical signals can include open and grounded circuits, positive and negative applied voltages and pulses, all applied in specific sequences.

The most straightforward network architecture on the access side is one where the access link extends all the way to the local exchange. As Figure 4.3 shows, it is also possible that the access link is served by a remote peripheral which responds to signals from the telephone switch central control. In support of such an architecture, European standards bodies created the ETSI protocol standards V5.1 [4] and V5.2 [5]. The V5.x protocol model places all signalling relating to a given access link within a signalling session, begun and ended by specific messages. The remote peripheral controlled by V5.x can report specific conditions which the subscriber equipment has applied to the access link and can in turn be asked to send specific electrical signals to that equipment. The repertoire of commands and reports available in V5.x signalling correlates directly with the repertoire of electrical conditions which can be generated and received on the access link.

The primary difference between V5.1 and V5.2 is that V5.1 assumes a fixed association between each access link served by the remote peripheral and a circuit from the remote peripheral to the local exchange. V5.2 provides additional commands to allow the central control in the local exchange to order the remote peripheral to connect the access link to a

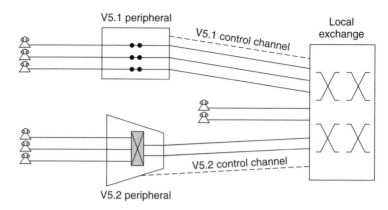

Figure 4.3   Direct, V5.1, and V5.2 access connections

specific circuit joining it to the local exchange. This allows the network operator to provide fewer circuits between the remote peripheral and the local exchange than the number of access links served, using statistical calculations to ensure that the probability that all circuits are busy when an additional one is needed does not exceed a specified value.

We mentioned the Integrated Digital Services Network (ISDN) above. The ITU-T created a binary, packet-based protocol for signalling over ISDN access links. This protocol is defined by the ITU-T Recommendation Q.931 [6]. The ISDN model is quite different from the analogue line model discussed so far. Gone are the electrical line states. Many of the in-band tones have also been replaced by indications generated within the subscriber devices themselves. Call signalling is logically separate from the "bearer services" which may be requested for a given call.[9] These bearer services may range from connections suitable for speech alone, through connections suitable for analogue modem or facsimile traffic, through connections offering a full 64 kbit/s (or multiples thereof) of digital transmission to be used in whatever way the user equipment is capable of using it.

Q.931 signalling allows for negotiation of services – in the first instance, between the calling terminal and the network(s) between it and the called party terminal and also between the calling and called terminals themselves. For instance, the calling terminal can propose a 64 kbit/s unrestricted digital connection (e.g. as a first step to setting up an H.320 [2] or H.324 [7] video conferencing session) and determine whether the intervening networks and the far end can support this service. The calling terminal can also indicate to the called end what higher-layer protocols it proposes to use on the connection.

Some services are invoked when subscribers press particular keys on their telephone sets. Q.931 allows the signalling of these services in two different ways, reflecting the two sides of an old controversy. The first approach supported is that of "stimulus signalling". The user equipment reports which stimulus (e.g. key press) occurred and leaves it up to the network to interpret what service is being invoked. The second approach is that of "functional signalling" – the user equipment decides what service is being invoked and identifies it to the network.

---

[9] It is possible to have ISDN calls consisting solely of signalling, for example, to invoke a feature such as "Do Not Disturb".

Figure 4.4 shows a basic call flow using Q.931 signalling. Compare this figure with Figure 4.2. Notice that the user interface – provision of dial tone (if any), ringing and ringing tone (if any) – is now the responsibility of the user set. In fact, the tones may be replaced or supplemented by displayed status information. This idea was carried over into Session Initiation Protocol (SIP).

Some services do not involve bearers directly. One example of a signalling associated service is "User-to-User Data". Subject to agreement with the network operator, the subscriber equipment can include messages of its own, directed at the far-end subscriber equipment, but encapsulated within the user-to-network signalling. A typical use of this capability is for peer-to-peer feature invocations between networked PBXs.[10]

The NGN needs to be able to interwork with the types of equipment described here at the signalling and bearer level. We will see again later in the chapter what has just been described, as we consider how information is transferred from the electrical signals and tones of the analogue network and the more sophisticated signalling of the ISDN to the

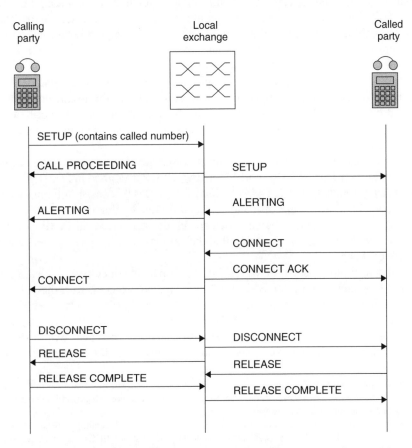

Figure 4.4 Q.931 messages exchanged for a basic call

---

[10] See, for example, the specification of QSIG [8,9].

fundamental protocols, SIP, Session Description Protocol (SDP) and RTP, on which the NGN is built.

4.2.2 Inter-exchange Signalling and Supervision

Inter-exchange signalling must satisfy a number of requirements beyond those served by access signalling. In the first place, it must carry the information needed to account for and bill the call. Secondly, it must carry information that the network requires to provide the services to which the caller has subscribed. On top of these, inter-exchange signalling must carry information needed to satisfy regulatory requirements. As a result, the protocols used for inter-exchange signalling have always been more complex than those used for signalling on the access link.

As with access links, several generations of inter-exchange call-signalling protocols are still in use. The analogue variants use tones instead of electrical signals to provide supervision.[11] They also use multifrequency tones to carry digits and other information. Signalling sequences are specified, typically at a national level, defining the order in which the called number, calling number and other information are sent across the link. Analogue signalling is carried on the same link that will subsequently be used for speech. It is not appropriate to describe any analogue inter-exchange signalling systems in detail here, but Ref. 10 provides an introduction to a number of them.

As indicated in the introduction to this topic, some of the analogue-signalling systems have variants for use on digital transmission systems. Supervision is carried on bits from a dedicated channel (in Europe) or stolen from the speech channel (in North America). The same tone combinations used on analogue circuits are also used on digital circuits to carry digits and other information.

The form of inter-exchange signalling most commonly in use now, however, is Integrated Services User Part (ISUP), short for "ISUP". ISUP is defined in ITU-T Recommendations Q.761 through Q.764 [11–14].[12] Like Q.931, ISUP consists of binary-encoded messages, containing a mixture of mandatory and optional parameters, depending on the message type and the services being provided. Figure 4.5 shows an ISUP call flow for the same call as in Figure 4.4, assuming that the calling and called subscribers are served by separate exchanges.

The complete protocol stack of which ISUP is a part is collectively called *Signalling System No. 7*, or more familiarly, SS7.[13] SS7 is a layered protocol, with the bottom three layers providing a common transport service for the rest. Collectively, they are known as the *Message Transfer Part (MTP)*. MTP 1 describes the framing of messages on the wire. MTP 2 provides the procedures for reliable transfer of messages over a single link. MTP 3 is concerned with network aspects and contains procedures for message routing

---

[11] Dial pulse signalling was at one time in use in inter-exchange circuits. The pulse rate was in the order of 20 pulses/s, double the rate used on access circuits. Access circuits exist in a more electrically noisy environment, so signalling had to operate more slowly for reliability.

[12] Beginning in the late 1990s, ISUP was re-engineered to separate the parts relating to bearers (e.g. circuits) from those relating to call signalling. This Bearer Independent Call Control (BICC) was first devised to allow control of ATM connections and subsequently extended to control of RTP media streams over IP. 3GPP has standardised on BICC as the call control protocol for cellular networks providing voice services, as opposed to the multimedia services provided by the IMS core network using SIP signalling.

[13] For a general description of SS7 and its parts, see ITU-T Recommendation Q.700 [15].

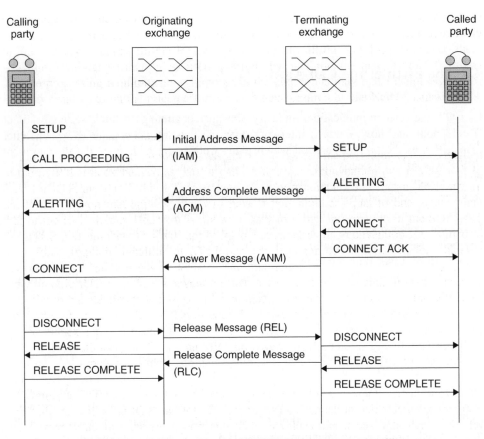

Figure 4.5 Q.931 combined with ISUP inter-exchange signalling

and for the management of message routing within the SS7 network. The management procedures include means for warning upstream nodes when a given node experiences congestion, so that they can route around it if possible.

The SS7 network is designed to be highly reliable. Traditionally, it has been argued that SS7 is also highly secure, both because it runs on a physically segregated network and because a different protocol is used between exchanges from that used on access links. Internet experts are inclined to view these claims with some skepticism. Nevertheless, high reliability and a high degree of security[14] are requirements for any IP-based signalling transport mechanism intended to be used in the NGN.

We will note one final feature of SS7: the presence of a transaction capability separate from call signalling within the protocol suite. The Transaction Capabilities Application Part (TCAP) allows an exchange to query a database when this is necessary in the course of providing services. One example of this is the Freephone service: the caller dials a 0800 number, which is then translated to an ordinary telephone number through a database lookup using a TCAP query. In the transition to the future NGN, an operator may need

---

[14] More precisely, signalling transport must offer the services of integrity, confidentiality and message source authentication.

to either support or interwork with a TCAP interface, from either the exchange or the database side.

4.3 The Birth of the Call Agent

4.3.1 History of an Idea

In 1997, the telecommunications industry was just beginning to think seriously about IP telephony and how it would interwork with the existing network. H.323 [16] rather than SIP, was the protocol around which plans were being laid, since the development of SIP was still in its early stages. H.323 postulated a network architecture consisting of H.323 end points, Gatekeepers and Gateways. The H.323 Gatekeeper provides various services, including admission control, registration and, in theory, bandwidth management. The H.323 Gateway has a simple definition: it is the logical equivalent of one or more H.323 end points operating back to back with terminals using another protocol suite. The key point was that both signalling and media flowed through one monolithic entity.

In April of 1998, Bellcore (now Telcordia) made public a protocol built around a new architectural proposal. This proposal was to decompose the gateway into a call control element called a Call Agent and media-processing elements simply called *gateways*. Between the Call Agent and the gateway would run a new protocol called the *Simple Gateway Control Protocol (SGCP)*. We will examine the characteristics of SGCP as reflected in its offspring shortly.

The Bellcore announcement caused a furor in some standards bodies: ITU-T Study Group 16, where H.323 was being developed, and the ETSI TIPHON project, where a new network architecture was being designed around H.323. The concern was that the new architecture could be used to bypass H.323 signalling while using IP to carry voice traffic. In fact, SGCP's primary author, Christian Huitema, revealed a few years later that this was Bellcore's precise intention, because H.323 at that stage of development took far too long to set up calls, particularly if it had to be interworked with ISUP. Bellcore was interested in a PSTN-bridging application. The media streams would use the IP network but the Call Agents would signal to one another using ISUP rather than interworking ISUP to H.323 in the middle. We will return to the theme of PSTN bridging later.

The Bellcore announcement also had another effect: it spurred a parallel protocol development project sponsored by the carrier Level 3. This group accepted the Call Agent–gateway decomposition and worked on their own gateway control protocol called *IPDC* (short for IP Device Control). IPDC had a short life. Some of its ideas were taken up into SGCP, and the result was renamed the Media Gateway Control Protocol (MGCP). MGCP persists to this day [17]. The IPDC project had a broader effect than this, however, because the participants put together a serious effort to promote work on the protocol in various standards bodies: the Internet Engineering Task Force (IETF), ITU-T Study Group 16 and the ETSI TIPHON project. By the end of the summer, the furor had died down and all three of these bodies were prepared to consider the new architecture.

The IETF decided that the decomposition did not go quite far enough. Their conclusion, reflected in RFC 2719 [18], was that a complete gateway between the PSTN and the Internet had three functional elements: the media gateway (MGW) the media gateway controller (MGC), and the signalling gateway (SGW). The MGC is at the centre of this ensemble, receiving and processing call signalling and doing any interworking needed between the PSTN and Internet call–signalling protocols. It controls the MG, to set up and take down the bearer connections needed for the call. The SG is actually a transport converter, exchanging the transport layers under the call signalling on the PSTN side for a new protocol stack on the Internet side, but leaving the actual call signalling unchanged.

Figure 4.6 illustrates the IETF architecture in the context of a broader network. The shaded box encloses the complete gateway. ISUP signalling is just one example; the signalling could be Q.931 from an ISDN terminal or a PBX. In the latter case particularly, the SG has to be co-located physically with the MG, because the Q.931 signalling enters the gateway on the same digital link as the voice circuits it controls. In the non-ISDN case, when signalling is associated with the voice circuit, the SG is absent but the MG has to relay the signalling to and from the MGC. The IETF decomposition continues to be valid; the MG, MGC and SG functional entities appear in the NGN architectures currently being standardised.

Leaving aside the Bellcore team's original motivation, decomposition offers several advantages: flexibility in engineering the network, flexibility in evolving the design of the individual components and, to the extent to which adherence to standards creates interoperability, increased competition among vendors with the resulting advantages to network operators. The decomposition is actually reminiscent of the structure of PSTN exchanges, which, aside from the switching fabric, characteristically consist of a central controller and a variety of specialised peripherals, though with proprietary rather than standardised protocols running between them.

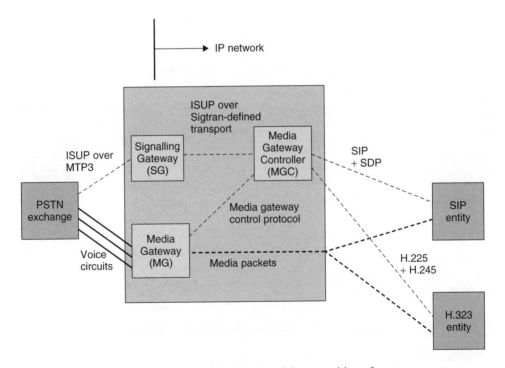

Figure 4.6   IETF view of the functional decomposition of a gateway

The IETF created two Working Groups to formulate the protocols needed by the decomposed architecture. The Megaco Working Group had the task of developing the protocol between MGC and the MG. The Sigtran Working Group was responsible for the protocol stack needed to carry PSTN call signalling between the MGC and the SG. We will look at the output of both groups in detail.[15]

4.3.2 Applying the Architecture
4.3.2.1 PSTN Bridging

PSTN bridging is the use of the IP network to carry media and, possibly, signalling between two PSTN entities. Ironically, this first potential application of the Call Agent concept may grow in importance as the replacement of circuits by IP networks begins to carve the circuit network into islands. PSTN bridging can be applied in a number of ways. Most commonly, the IP "bridge" lies on the path between two PSTN exchanges. Parts of RFC 2833 [20] anticipated the use of an IP bridge to replace part of the access link between a telephone set and the local exchange, although it is unclear whether this was ever implemented. Finally, in some sense the transmission of signals such as DTMF or data or facsimile modem tones between two PSTN end points can be thought of as PSTN bridging, since the content being carried was designed for transmission on PSTN circuits. We will confine ourselves to a look at a few inter-exchange bridging cases here, followed by a brief discussion of alternatives for carrying DTMF signalling.

The first bridging case we consider is that envisioned by Bellcore when they first designed SGCP, where the IP network is used as a substitute for voice circuits in an ISUP-controlled network. Figure 4.7 shows this case, with the assumption that the ISUP is carried over the Sigtran protocol stack across the IP network.

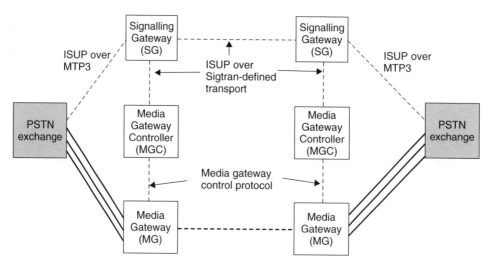

Figure 4.7   PSTN bridging using ISUP signalling only

---

[15] SGCP was published as an Internet Draft which may still be found on some archival web sites. The last version published was draft draft-huitema-sgcp-v1-02.txt [19].

This case presents two challenges for signalling between the two MGCs: how to signal packet-based codecs and how to signal the addresses and ports to which the packet streams should be directed. ISUP as originally defined in ITU-T Recommendation Q.761–Q.764 [11–14] cannot carry this information, though it could easily be extended to do so. A more important consideration is that ISUP signalling is closely tied to the circuit which the upstream exchange has allocated to the call.

The ISUP problem was solved when the ITU-T re-engineered the protocol to make it independent of the bearer type through which the connection would be provided. This Bearer Independent Call Control protocol (BICC) is documented in ITU-T Recommendations Q.1902.1 through Q.1902.4 [21–24]. A simplified view of the complete signalling architecture is shown in Figure 4.8 [25, 26]. It starts with the idea of a Bearer Control Function which is logically and possibly physically separate from the Call Service Function which processes the BICC signalling. When these two functions are physically separated, a "vertical" bearer control protocol (ITU-T Rec. Q.1950 [27]) passes between them. This protocol is effectively a profile of Megaco/H.248 (see below).

In the BICC architecture, it is the Bearer Control Functions at the two ends of a connection that signal codec details and addresses and ports to each other. To do this, they use the IP Bearer Control Protocol (ITU-T Rec. Q.1970 [28]), which is a profile of the SDP. The IP Bearer Control Protocol is tunnelled through the same transport stream that carries the BICC messages. BICC signalling was adopted by 3GPP for use in circuit-based backbone networks carrying mobile voice calls, but is basically a dead end given the increasing volume of multimedia communications.

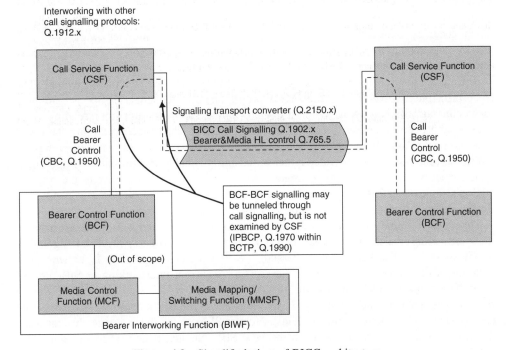

Figure 4.8 Simplified view of BICC architecture

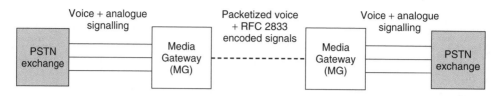

Figure 4.9   Bridging analogue inter-exchange signalling using RFC 2833

This model can be thought of as a further decomposition of the IETF MGC. The MG is logically separate from and subordinate to the Bearer Control Function. The advantage of the decomposition is that it allows the call processor, which has to process the binary BICC protocol anyway, to avoid dealing with the text-based SDP too.

The second bridging case we consider is one where analogue inter-exchange signalling and the accompanying voice traffic are carried across an IP network. Figure 4.9 shows this situation. It is not clear whether such systems are widely deployed, particularly with the replacement of analogue signalling with ISUP in most regions, but they have definitely been implemented. The role of the MGs in Figure 4.9 is to convert the analogue signalling as well as the voice traffic into packets. The signalling continues to travel along the same path as the voice traffic. At the far end, the packets are converted back to analogue signals and voice on the circuits leading to the destination exchange. In this case, no MGCs are necessary. The MGs can be configured to do their job without the need to signal between them.

An early motivation for using IP transport was to save on bandwidth by using low-bandwidth codecs whenever possible. However, codecs such as G.723.1 [29] or G.729 [30] are optimised for voice and do not reproduce the simple tones used for signalling with sufficient accuracy to meet the requirements of existing exchange equipment. RFC 2833 [20] was initially created to solve this problem for DTMF signals, but was expanded to include codepoints for the more common analogue-signalling systems.

RFC 2833 defines a specialised type of RTP [31] payload, "telephony event". The basic principle involved is that the MG is programmed to recognise the tones belonging to specific signalling systems. When it does recognise such a tone, it stops sending voice packets and instead sends a packet of the RFC 2833 payload type containing the event code corresponding to the tone it has detected. It will continue to send event reports while the tone lasts. This provides concise and fairly reliable transmission of signalling, which when received at the downstream gateway can be converted back to the original analogue tones. Real-life implementations of RFC 2833 must be prepared to handle the timing issues introduced by congestion and jitter in the IP network.

The third example of PSTN bridging is one that is most likely to be seen in NGN networks. This uses the mechanism of ISUP encapsulation to carry the ISUP signalling in SIP messages. The ISUP is interworked to SIP at the entrance to the IP network, but continues to be carried across the network. The concurrent interworking to SIP and use of SIP for transport make it possible to take advantage of SIP's routing capabilities to find the best path for the signalling. It also allows for the SIP network to provide some call services, which may result in changes to the SIP part of the message. In that case, the ISUP has to be updated to match the new SIP contents. If and when the signalling reaches a gateway to the PSTN, the ISUP is retrieved from the SIP and sent on its way.

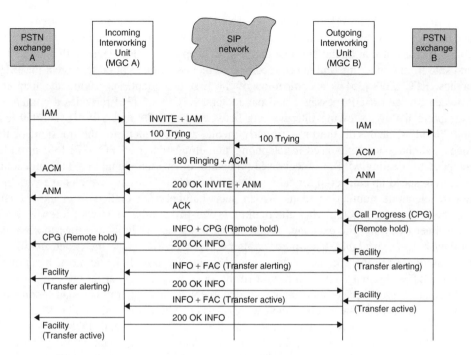

Figure 4.10 Basic call and call hold leading to call transfer with SIP-I

The idea for this procedure dates back to the late 1990s, when people were first thinking seriously about using SIP signalling. To make it possible, it was necessary to standardise a way to carry ISUP in SIP messages. This was achieved with the definition of the MIME encoding of ISUP (RFC 3204 [32]) so that it can be carried in the SIP message body. A second requirement arose because of the mismatch between ISUP and SIP call models. Under some circumstances, ISUP requires that a message be sent, but no SIP message is required at that point in the call. The SIP INFO method (RFC 2976 [33]) was defined to take care of this situation. Figure 4.10 shows a typical combined SIP-ISUP call flow for an ISUP call transiting the IP network. In addition to the basic call setup and takedown, the sequence shows three mid-call messages (Call Progress (CPG), Facility (FAC), Facility) sent when the called party in the PSTN performs a call transfer. Since the transfer does not modify any of the states relevant to SIP, these messages have to be carried using INFO requests.

4.3.2.2 Interworking between SIP and ISUP

To a protocol purist, interworking standards are unnecessary. An interworking device has two (or more) interfaces, as well defined as their respective protocols. What happens in passing between one and the other is a matter of implementation. In actual fact, there is room for variation in the details. This is well illustrated by the case of interworking between SIP and ISUP.

Industry interest in the interworking between these two protocols was sufficiently great that both the IETF and the ITU-T created standards for the task. The IETF finished its work first. RFC 3372 [34] introduced the term SIP-T (SIP for telephony) to apply to the use of ISUP encapsulation and showed the different ways it could be used. The security considerations of that document note the need to keep the ISUP content both confidential and safe from tampering, and hence recommend the use of encryption to protect the ISUP bodies. RFC 3398 [35] is a companion of the first document, specifying the mapping between SIP and ISUP messages and parameters. RFC 3666 [36] provides a number of example call flows showing interworking between the PSTN and SIP. RFCs 3959 [37] and 3960 [38] deal with the topic of early media, a requirement for the transport of the tones and announcements used in telephony signalling. Finally, RFC 3578 [39] provides methods for dealing with the specialised procedure of overlap signalling. This procedure is used to speed up call setup for calls in countries such as Germany which have variable-length telephone numbering plans. When such numbers are dialled from an ordinary telephone, it is hard to tell when the dialling is finished. Rather than wait a few seconds for a timer to expire, the exchange sends out an initial set of digits and then more digits as they become available. This process fits poorly with SIP, since the called number is typically placed in the To: header field as well as the Request-URI, and more digits mean new INVITEs that may not even be routed the same way as the earlier ones.

The ITU-T preferred to build their own document from scratch rather than modify the RFCs just described. The result of their work was ITU-T Recommendation Q.1912.5 [40]. This document covers the same territory as RFCs 3372 3398, and 3578 combined. It defines three profiles of usage: A, B and C. Profiles A and B are pure interworking profiles: SIP on one side, ISUP on the other. The difference between them is that Profile A was defined specifically to meet the requirements of the 3GPP mobile networks, while Profile B is more general. In particular, Profile B specifies the handling of overlap signalling. Overlap is unnecessary in a mobile network because the numbering plan is of fixed length and cell phones provide explicit indication of the end of dialling without the need to wait for a timeout. Profile C is also called *SIP-I* (SIP plus ISUP), because it describes the use of ISUP encapsulation.

The ITU-T procedures rely on the concept of trust between signalling entities. Encapsulated ISUP must be sent only to entities that understand it and can be trusted to handle it appropriately. This in itself is a choice of one of several architectural alternatives offered in RFC 3372. Going beyond that point, Q.1912.5 makes use of the P-Asserted-Identity header field to represent the calling party number. This usage imposes trust requirements that are documented in RFC 3325 [41]. It is the assumption of a trusted environment that most distinguishes Q.1912.5 from the IETF work. In fact, there is not much of a difference between the mappings of Q.1912.5 and those of RFC 3398. The treatment of overlap signalling in Q.1912.5, although worked out independently and somewhat painfully, turned out to use one of the three approaches described in RFC 3578.

One point of difference between RFC 3398 and Q.1912.5 may be illustrative of the differing mindset between the two bodies. SIP has status codes to indicate the reason for the rejection of a request. The analogue to these in ISUP are ISUP's cause codes. The IETF and, at first, the ITU-T tried to define a mapping between status codes and cause codes that matched their semantics. In the end, however, the ITU-T greatly simplified its mapping, for two reasons. One was that the semantic mapping implicitly treats the complete call path as a single system, whereas the operators represented at the ITU-T felt that representing SIP network failures as if they were ISUP failures would make the cause codes useless as diagnostic indicators and vice versa for SIP status codes. The second point was an awareness that SIP status codes are likely to be propagated to the end user. The operators considered that it was unnecessary to pass details of network problems to subscribers. Thus, aside from obvious mappings like "Busy", many of the ISUP to SIP mappings became either 480 Temporarily Unavailable, meaning: "Please try again in a little while" or 500 Server Internal Error, meaning: "Please don't try again". In practice, vendors make the mappings configurable, since each operator has its own preferences.

4.3.2.3 PSTN Emulation and PSTN Simulation

ISUP still supports many features that SIP does not. A telephone network operator evolving towards the NGN would not want to lose features and the corresponding revenues while doing so. In some cases, SIP can provide equivalent features, but they operate differently. The operator is reluctant to risk subscriber irritation as a result of the differences. Finally, some features may be required by regulation, but do not as yet have adequate support in SIP. How to bridge the gap until SIP catches up?

One way to do it is by encapsulating ISUP in SIP as previously described, but processing the ISUP in the SIP network rather than bridging it to another part of the PSTN. This is the basic idea of "PSTN emulation". The thought is that the encapsulated ISUP is sent to specialised application servers that run both SIP and ISDN call models. As a result, they can provide the services requested at both protocol levels. This architecture and its details are still being debated at the time of writing, although it was anticipated by some of alternatives in RFC 3372. At the protocol level, PSTN emulation is equivalent to Q.1912.5 Profile C.

"PSTN simulation" is a term used to describe the addition of SIP extensions to the point where SIP can duplicate PSTN features. Partly, this effort is necessary to support features covered by regulatory requirements. Typically, regulators define how particular services must be performed independently of the technology used to provide them. PSTN simulation is, within limits, a valid response to the problem of making the transition from the PSTN to the NGN. If it forces SIP to be used in unnatural ways, however, it places into question the whole rationale for transition to SIP as the common protocol of the NGN. It is thus likely that PSTN simulation will be confined to the most commonly used features, easing the adjustment to the new network while allowing new services to take advantage of the capabilities of SIP in a natural way.

4.3.2.4 Transport of DTMF Signalling

In the late 1990s, as the voice on IP industry began to take shape, the way to carry DTMF signalling across an IP network became a controversial topic. The reader will recall that DTMF has more general uses than simply signalling the called number to the local exchange. Within the network, one example of DTMF usage is to enter telephone credit card numbers. Another is to signal the local exchange that the current call is finished but the caller wishes to make another one without hanging up (and having to re-enter a credit card number, for example).[16] Off-net applications range from the control of voicemail systems through telephone banking.

---

[16] Typically, in North America this is signalled using a long-duration "#".

A close look at these different applications reveals that different transport mechanisms are needed in different cases. Off-net applications will operate satisfactorily if the DTMF is transported along the voice path. An application such as the new call signal needs to be sent to a call-processing entity, to be translated into a BYE followed by a new INVITE after digits have been collected. The third possibility is that a network entity not normally in the call path needs to know about digits entered by the caller, for example, a conference control unit.

At first, the only mechanism available for DTMF transport was the use of call-signalling messages. H.323 signalling, in particular, has fields for the transport of DTMF information. But this has drawbacks: it adds extra burden to entities all along the signalling path, while requiring that the information somehow be passed from signalling entities to end points or other servers in two of the three cases just described. In fact, if unchecked by the network, rapidly and repeatedly hitting a DTMF key by a caller could overload every signalling entity along the call path, making them unavailable to other subscribers.

The answer, in part, is to send DTMF along the voice path instead. The problem was that low bitrate codecs, particularly G.723.1, could distort the DTMF signals to the point where they were misinterpreted at the receiving end. This problem led to the development of RFC 2833 [20], as described in Section 4.3.2.1.

There remains the problem of accommodating the need of some servers to know about DTMF signalling without being in the call signalling path. This requirement was met when SIP acquired event notification capability, with SUBSCRIBE and NOTIFY (RFC 3265 [42]). The server subscribes to the SIP events and receives notification when they occur, but is otherwise not required to process signalling associated with the call. draft-ietf-sipping-app-interaction-framework-05.txt [43], approved but not yet published as an RFC, provides an overall view of this process. The approved draft-ietf-sipping-kpml-07.txt [44] gives the detailed protocol means to carry it out.

4.3.2.5 Summing up

The last few sections are focussed on a higher-level view of how the NGN and PSTN will coexist. The remaining sections of this chapter provide a detailed look at the protocols that emerged from the decomposed gateway model. We begin with a description of MGs and the media gateway control protocols and Megaco/H.248.

4.4 Media Gateways

MGs actually emerged as a number of specialised types. One type is the residential gateway (also known as an *analogue terminal adapter* (*ATA*) or integrated access device (IAD)), which sits in the home at the end of a high-speed Internet connection and is controlled by an MGC operated by the Internet access provider (e.g. the telephone or cable company). The residential gateway is small in scale, supporting one to four lines. For historical reasons, residential gateways tend to be controlled by MGCP.

Another variant is the access gateway, which sits in the service provider's premises and connects to hundreds or thousands of access circuits. A third primary variant is the trunking gateway, which connects to inter-exchange circuits rather than subscriber lines. Again, the scale may be in the thousands. More specialised gateways provide tones and announcements, packet-to-packet transcoding and a number of other services. The predominant protocol for access and trunking gateways is Megaco/H.248 [45,46]. This protocol was used to control the specialised gateways, but is now being supplanted by other protocols operating in combination with SIP.

Figure 4.11 shows a schematic view of the interior of a hypothetical MG.

In this example device, circuit and packet transmission media terminate on circuit packs that handle the sending and receiving of signals at the physical level. As commands come in from the MGC, the control processor processes them and dispatches its own commands to subordinate digital signal processors (DSPs). As a result, the DSPs load the required codec algorithms into memory, begin to acquire signals from the bus on one side, process them according to the selected algorithms and output them to the bus on the other side. Sometimes, a command may require the DSP to load an additional algorithm to generate a tone to be inserted into the media stream. At other times, or possibly throughout the session, the DSP may be programmed to detect particular tones if they appear in the stream. Other commands may be directed towards the circuit packs terminating the transmission media, directing them to report supervisory events such as on- or off-hook. Alternatively, the circuit pack may be directed to pass out-of-band signals outwards, one example being ringing current on an analogue line.

This brief look inside a MG is provided to make a couple of points. One is that the MG needs a certain amount of configuration data per circuit terminating on it. The circuit must carry an identity that the MGC can refer to. Information must be provided on the

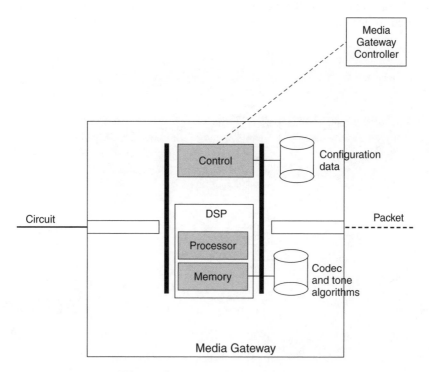

Figure 4.11   Inside a media gateway

particular characteristics of the circuit: whether it is analogue or digital, to start with, and also details such as the cadence of the ringing signal that gets applied to it if it is an analogue line.

A second point is that the MG is designed with a certain set of capabilities, but its ability to exercise them will vary over time. The DSP has algorithms available for a particular set of codecs and tones. But at any time, the memory may be too full to hold another algorithm, so even though the DSP could support a given codec or a given tone, it is unable to do so for the moment. The bottleneck could be DSP processor time or even control processor time. What this means for a media gateway control protocol is that the protocol must allow the MG to inform the controller when a request has failed because of lack of resources.

Occasionally, it is necessary to make a circuit-to-circuit connection through the MG, when a call involves two lines served by it. Some MGs are able to make the connection internally. Others are not designed with this ability. As a result, the media flow from one circuit has to be sent out as a packet stream and reflected back to another port at the nearest router. Only then can it be passed to the other circuit. The MGC has to be aware of this limitation, and if it is not, the media gateway control protocol has to be able to carry diagnostic information indicating why a particular connection cannot be made.

Let us now step back and have an extended look at the functions a media gateway control protocol must support. Figure 4.12 shows a moderately complex example that is nevertheless realistic. The MG serves analogue lines and also supports connections over an Asynchronous Transfer Mode (ATM) transmission link to a PSTN exchange. The media gateway controller (MGC) controls the MG and an associated audio server which can be used to generate announcements and special tones. The MGC also receives and

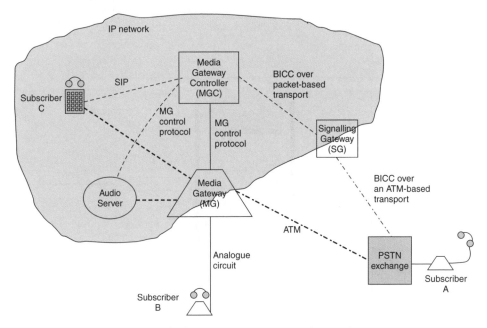

Figure 4.12 Media gateway control example

generates call signalling to the PSTN exchange and to SIP end points. The signalling to and from the PSTN exchange uses BICC rather than ISUP, consistently with the use of ATM transmission. BICC was described in Section 4.3.2.1.

Consider a calling scenario where subscriber A at the PSTN exchange first calls subscriber B. The phone rings a few times, but B does not answer. Subscriber B has the "Call Forwarding No Reply" feature[17] activated, so after a predetermined time elapses, the MGC forwards the call to subscriber C. Before doing so, the MGC causes an announcement to play out, informing subscriber A of the change in routing. Then subscriber C is alerted, answers her phone and conversation ensues between A and C. Eventually subscriber A hangs up, subscriber C hangs up and the call is cleared.

Having looked at this from the point of view of the subscribers, let us now consider the signalling involved at each step. We are primarily interested in the interactions between the MGC and the MG, but the other signalling received and sent by the MGC is described in order to set the context.

1. *Initial Address Message (IAM) received from the PSTN exchange.* The Initial Address Message (IAM) is received in association with bearer setup information, indicating the type of connection to create over the ATM link and the codec, typically G.711, to be used for the call. The IAM identifies subscriber B as the called party.
   In processing the IAM, the MGC sends the MG commands to do the following:
   - establish the connection on the ATM link and report the details back;
   - connect that bearer path to subscriber B's line;
   - prepare to transcode between subscribers A and B if the codec to subscriber A is not G.711;
   - begin ringing subscriber B's line;
   - begin sending the ringing tone back to subscriber A;
   - report when subscriber B goes off-hook. If B is already off-hook, report immediately. In the case of a collision like this, typically the called party has priority: the call from subscriber A will be cleared down and subscriber B will be able to proceed with his call.
   The MGC in this case is aware that subscriber B has the Call Forwarding No Reply feature activated, so the MGC sets an internal timer to revisit the call if nothing else happens before the timer expires. Without the operation of this feature, the MGC would typically set a much longer timer, in the order of minutes.
   To keep the PSTN exchange aware of progress and tell it what its bearer control function needs to know about the ATM connection set up for the call, the MGC sends back an Address Complete Message (ACM).
2. *Ringing timer expires.* Subscriber B fails to answer before the expiry of the timer the MGC has set to monitor ringing. The MGC sends the MG commands to do the following:
   - stop ringing on subscriber B's line;
   - disconnect subscriber B's line from the bearer connection over the ATM link to the PSTN exchange, while leaving the bearer connection intact;
   - stop sending the ringing tone to subscriber A;

---

[17] This feature is documented in ITU-T Recommendation Q.732.2 [47].

- allocate an IP address and port for receiving a media flow from the audio server and prepare to direct that flow to the bearer connection over the ATM link towards subscriber A. The MG returns the allocated address and port.
- continue watching for off-hook on subscriber B's line (normal idle line treatment). The MGC sends commands to the audio server to play out a call forwarding announcement once only to the address and port returned from the MG.[18] It asks the audio server to report back when the announcement is finished. The MGC also sends a CPG message back to the PSTN exchange to let it know that an announcement is in progress.

3. *Audio server reports announcement completion.* The MGC sends commands to the audio server to deallocate any resources set up to play out the announcement. It sends a request to the MG to indicate what codecs it can currently support between the IP address port and the ATM-based bearer connection. Upon receiving the reply, it sends a SIP INVITE to subscriber C's phone, offering in the SDP body a session description based on the information received from the MG.

4. *SIP phone sends a 180 Ringing provisional response.* The MGC sends another CPG message back to the PSTN exchange, indicating that the call has been forwarded and the new party is being alerted. It sends a request to the MG to resume sending the ringing tone back to subscriber A.

5. *The SIP phone returns a 200 OK INVITE final response with an SDP answer.* The MGC returns an ACK to the SIP phone. It then sends commands to the MG to request the following:
   - stop sending the ringing tone to subscriber A;
   - set up a two-way media flow between the bearer connection over the ATM link and subscriber C's phone using the local address and port originally allocated for the announcement, the remote address and port returned in the answering session description in the 200 OK INVITE and the codec selection also returned in that answer.

   The MGC sends an Answer Message (ANM) to the PSTN exchange. This is important to allow billing at that exchange to proceed. The MGC will typically keep its own billing records.

6. *The MGC receives a Release (REL) message from the PSTN exchange.* This message indicates that subscriber A has gone on-hook. The bearer connection over the ATM link is no longer needed. The MGC sends commands to the MG to request the following:
   - disconnect the flow between the IP side and the bearer connection over the ATM link;
   - deallocate the local address and port on the IP side;
   - tear down and deallocate the bearer over the ATM link to the PSTN exchange.

   The MGC returns a Release Complete (RLC) message to the PSTN exchange and sends a BYE to the SIP phone. When the SIP phone returns 200 OK BYE, the call is complete and the MGC deallocates all resources associated with it.

---

[18] The use of such an announcement for Call Forwarding No Reply is not specified in Rec. Q.732.2. It is included here for purposes of illustration because it is a realistic possibility for other features. For simplicity, the procedure described above ignores the setting of the codec to be used between the audio server and the MG and the establishment of an address at the audio server to which the MG can send RTCP receiver reports.

This extended example illustrates most of the requirements that have to be satisfied by a media gateway control protocol:

1. It must support the allocation and deallocation of reusable transmission resources such as User Datagram Protocol (UDP) ports and ATM virtual circuits.
2. It must carry requests to direct media flows between specific bearers (IP, circuit, ATM). These flows may be one- or two-way and at times they may have to be deactivated or looped back to the sender for testing. The example did not show another common requirement: connecting more than two parties in a conference. A further possible requirement is to support multimedia flows – text, voice and video all at once.
3. It must support codec negotiation and assignment. In many cases, the MG will be requested to reserve more than one codec in the same session and be prepared to switch between them (e.g. between G.711 for voice and T.38 [48] for facsimile) depending on the nature of the media content.
4. It must support requests for the MG to detect particular events. These events may be in-band (e.g. DTMF tones), out-of-band (e.g. on-hook and off-hook) or internal to the MG. It must also provide the means for the MG to report events when it has detected them. In our example, the requested events were
   - subscriber B off-hook;
   - completion of the announcement playout by the audio server.
5. It must support requests for the MG to apply signals. Again, these could be in-band or out-of-band. In our example, the signals were
   - ringing on subscriber B's line (out-of-band);
   - ringing tone to subscriber A (in-band);
   - announcement from the audio server to the MG and thence to subscriber A (in-band).

One final set of requirements on a practical gateway control protocol relates to housekeeping. The protocol may need to provide for startup and takedown of control relationships and should also provide the possibility of auditing the gateway state. Auditing allows for recovery if something causes the MG and MGC to get out of step with each other.[19]

4.5 A Look at Media Gateway Control Protocols

This section provides a detailed examination of two competing media gateway control protocols, MGCP and Megaco/H.248. Both are in common use.

4.5.1 SGCP and MGCP

SGCP [19] introduced a number of design ideas that were carried forward in MGCP [17]. The first was a careful integration with IETF created protocols instead of those created by the ITU-T. MGCP is text based rather than binary. This allows it to use SDP [50] for negotiation of media sessions. Since the SIP also uses SDP, it is easy for the MGC to transfer codec information between the two protocols when negotiating the media session with its remote peer.

---

[19] The Megaco Working Group compiled an extensive set of requirements for the protocol with the help of other standards bodies. These requirements are to be found in RFC 2805 [49].

MGCP messages consist of a series of lines of text. The syntax is keyword based. Each line begins with a token indicating the content of the line, and subsequent parameters are generally specified in the form of keyword, separator and value. The initial line of a request begins with the token defining the request type. The initial line of a response begins with a three-digit response code. SDP session descriptions are preceded by a blank line and conform to RFC 2327 [50]. As it developed, MGCP added the idea of "bundling" of messages within the same UDP packet. Succeeding messages are separated by lines containing only a period ("."), and are intended to be processed in the order in which they appear. This is similar, as we shall see, to the Megaco/H.248 idea of a transaction.

MGCP is designed around a connection model that features two constructs: the end point and the connection. End points are sources or sinks of data and can be physical (e.g. circuits) or virtual (e.g. a specific tone from an audio server). A connection is an association between two end points to allow data to flow between them. This description needs some qualification: it takes two connections to achieve a flow, one for each end point. This is easy to see if the end points are in separate gateways, but it is also true if they are both in the same gateway. (MGCP has an optimised command syntax variation to cover this special case.) A given end point can take part in multiple connections simultaneously. The Call Agent can control the routing of media flows when multiple connections are present.

To simplify handoff from one Call Agent to another in the case of failure or overload, the protocol tries to minimise the state associated with the relationship between the Call Agent and the gateway. The gateway is configured with the identity of the Call Agent, preferably as a domain name, with the idea that it should look up the address via DNS (and refresh it occasionally in case the address changes). The transport for the protocol is specified as UDP rather than a session-oriented protocol such as TCP. Further robustness measures in the protocol are the inclusion of the gateway domain name as part of the end point identifier (so that the call agent does not have to maintain a signalling session identifier) and the association of a "NotifiedEntity" with each end point that can be reset in most commands to change the controlling Call Agent on the fly.

MGCP has nine commands:

- CreateConnection (CRCX), sent by the Call Agent to create a new connection;
- ModifyConnection (MDCX), sent by a Call Agent to change the parameters of a connection (e.g. to provide the address and session description applicable to the remote end point);
- DeleteConnection (DLCX), sent by a Call Agent to terminate an existing connection or by a gateway to indicate that it was unable to sustain a connection. Upon executing DLCX, the gateway returns some basic statistics associated with the flows carried by the connection.
- NotifyRequest (RQNT), which allows a Call Agent to indicate events the gateway is asked to report and signals that the gateway is to apply if the events occur. The events and signals may apply to end points named within the request or to specific connections. CRCX, MDCX and DLCX may contain the same parameters as would appear in a stand-alone RQNT, but the stand-alone version is needed in the protocol to cover the case when the end point is not involved with a connection.
- Notify (NTFY), which is sent by a gateway to indicate that an event has occurred.

- EndpointConfiguration (EPCF) allows the Call Agent to specify bearer characteristics for the end point. One example is to specify whether the end point uses A-law or μ-law companding. Normally, one would expect this information to be statically provisioned in the gateway, but the EPCF command allows it to vary if necessary. As with RQNT, CRCX, MDCX and DLCX may contain the same parameters as would appear in a stand-alone EPCF command.
- AuditEndpoint (AUEP) allows the Call Agent to find out specified information about the state of an end point.
- AuditConnection (AUCX) allows the Call Agent to find out specified information about the state of a connection.
- RestartInProgress (RSIP) is sent by the gateway to indicate that an end point or group of end points is being put into service or being taken out of service.

MGCP picked up the idea of "packages" from IPDC. A package contains a specific list of events and signals. It can also define other characteristics of an end point. The intent is that an end point, by provisioning, supports a specific default package. It is also possible for the end point to support other packages. The default package can be thought of as defining the end point's basic character (e.g., access line, digital inter-exchange circuit). Since the same event or signal may be used by more than one type of end point, MGCP allows events and signals to appear in multiple packages. MGCP also picked up from IPDC the idea of adding parameters to events and signals, including a core set specified in the base protocol.

The task of collecting dialled digits is potentially burdensome and even dangerous to both the gateway and the Call Agent. If the gateway reports each digit as it detects it, repeatedly hitting a key sufficiently rapidly could cause a messaging overload. MGCP therefore includes in the base protocol a means for controlling the reporting of digits. When the Call Agent asks the gateway to start watching for digits from an end point, it can specify a dialling plan containing one or more patterns to be matched. The gateway waits until the set of digits collected either matches one of the specified patterns or the gateway determines that no match is possible (or a timeout occurs). The gateway then reports the set of digits collected up to that point.

Upon consideration, it was realised that digits must often be collected in multiple stages. Thus the Call Agent will specify an initial digit map, wait for the result and then specify a new map to handle subsequent digits. During the time required for the second map to arrive, however, new digits may be dialled and lost. So the Call Agent needs to be able to specify with the initial NotificationRequest that any digits coming in after the digit map has been satisfied or not matched should be saved in a buffer. When the Call Agent requests subsequent digits for the called number, it needs to be able to indicate that the buffer should be processed before processing new digits. Moreover, once the number is complete, the buffering should stop and excess digits should be discarded. Considerations such as these resulted in considerable elaboration in MGCP both of the event-processing model and the controls the Call Agent could use to direct that processing.

MGCP is a transaction-oriented protocol, consisting of requests and responses. Each command carries a transaction identifier. The first line of the response to each command contains a response code and the transaction identifier of the corresponding request. A set of response codes was specified, with two success codes (200 for normal execution, 250 for successful connection deletion). MGCP also has interim response codes to acknowledge commands that require substantial time to execute and has the ability to acknowledge responses in a three-way handshake.

The original Internet Draft describing SGCP was 54 pages long. MGCP has been through two major design iterations since then. The latest version, in RFC 3435 [17], does not include package specifications, but has 210 pages even without them. To give an idea of the flavour of the protocol without getting into too much detail, a basic call flow involving MGCP control is shown in Figure 4.13. This covers the same call as Figure 4.2, but assumes that the called party is a SIP end point. The Call Agent uses some database to convert the called digits to a SIP URI. One possibility for that database is DNS, using the ENUM capability defined in RFC 3761 [51].

Figure 4.13(a) shows the initial stages of the call, up to the point where digits have been collected. The first RQNT would be sent at the time the line goes idle, typically well before the new call begins. The second RQNT provides a digit map and an indication of which digits are to be reported. The effect of the digit map is to cause digits to be collected until a pattern in the digit map is matched, it becomes clear that no pattern will be matched or dialling times out. At that point, all the digits collected so far are reported in the same message. MGCP has a rule that any signal that is being applied when an event is detected will immediately stop unless the event parameters say otherwise. Thus,

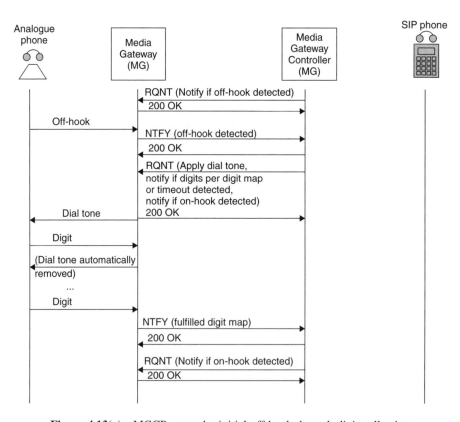

Figure 4.13(a)   MGCP example, initial off-hook through digit collection

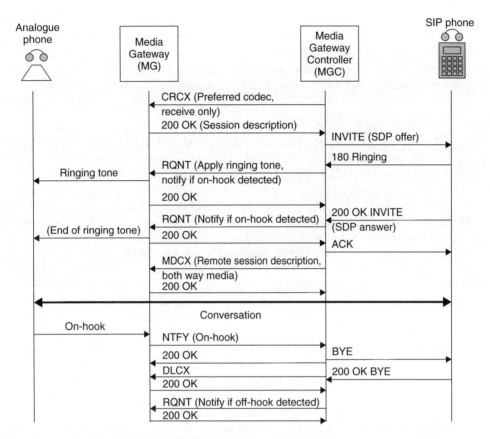

Figure 4.13(b)   MGCP example, connection, conversation and cleardown

the dial tone stops as soon as the first digit is detected, without an explicit instruction from the MGC. The final RQNT asks that the MG report if the line goes back on-hook. The need to monitor for on-hook persists for the duration of the call.

In Figure 4.13(b), the CRCX asks the MG to allocate a codec and packet-side ports for the receipt of media and RTCP reports. The MG returns an SDP session description indicating the allocated codec, RTP payload type value and packet-side ports. The MGC places this session description into a SIP INVITE as an SDP offer and sends it to the SIP phone that it has found to be the destination on the basis of the dialled digits. The SIP phone returns a 180 Ringing provisional response, indicating that the subscriber is free. In reaction to this, the MGC sends a RQNT asking the MG to transmit the ringing tone to the caller.

When the MGC receives the 200 OK INVITE from the SIP phone indicating that the called party has answered, it sends another RQNT asking only to monitor for on-hook. By default, the omission of the ringing tone signal from this request terminates the application of the ringing tone by the MG. The MGC also sends an MDCX to pass on the SDP answer received in the 200 OK INVITE to the MG. This answer contains the remote address and ports for the sending of media and RTCP reports. The MDCX thus enables two-way media flow, allowing conversation to take place. At the end of the conversation, Figure 4.13(b) assumes that the caller hangs up first. An on-hook event is reported to the MGC, which terminates the call on the SIP side. The DLCX tells the MG to deallocate the codec and packet-side resources that were allocated to the call. The final RQNT is the same as the one at the beginning of Figure 4.13(a), readying the line for the start of the next call.

4.5.2 The Megaco/H.248 Protocol

Section 4.3.1 took us up to the point where IPDC and SGCP had merged into MGCP and the IETF had created the Megaco Working Group. However, the IPDC team had also managed to interest ITU-T Study Group 16 in the task of defining a gateway control protocol. Megaco began its work in the usual IETF manner, by arguing out requirements, at its initial meeting and by E-mail.[20] Study Group 16 began to tackle the protocol itself. An early commitment to MGCP as a base was rejected by both groups. The MGCP connection model was the focus of a debate at a Study Group 16 meeting in February 1999 and was also rejected. At an IETF meeting the next month, the antagonists sat down in a hotel room and argued their way to a breakthrough: a new connection model inspired neither by MGCP nor by IPDC, but satisfying the requirements for generality cited in Section 4.4.

Before looking at that connection model, let us complete the history of the development of the Megaco/H.248 protocol. Behind the scenes, ITU-T and IETF management came to the agreement that they would work jointly on the protocol and both organisations would have to consent to it before it was finally approved. However, the ITU-T, rather than the IETF, would own the resulting standard. Work on the initial draft of the protocol was essentially complete by February 2000, with steady progress in the IETF alternating with marathon bursts of effort at Study Group 16 meetings. The IETF Last Call process was completed in the next few months, some corrections were made as a result, and the document was finally approved as ITU-T Recommendation H.248 [45] in June 2000. The identical specification was published by the IETF as RFC 3015 [46].

In 2002, the ITU-T issued a second version of the protocol [52], incorporating a large number of corrections and clarifications as well as some new capabilities. At this point, a number of package specifications had also been published as annexes to the base specification, but these were essentially independent of the version of the base protocol. Rather than reissue everything because of the change in the base specification, the ITU-T chose to give independent numbering to the base protocol document and the additional package specifications. Thus, the base protocol specification is officially ITU-T Recommendation H.248.1 (now up to its third version [53]), and packages have been standardised in Recommendations H.248.2 through H.248.37 with more on the way.[21] This history should help alleviate confusion over the relationship of Megaco to H.248 to H.248.1 – they are all the same. For simplicity, the protocol will be called H.248 from here onwards. Successive versions of the protocol will be called out by their ITU-T designations: H.248.1 (2002) for the second version and H.248.1 (2005) for the third version.

---

[20] At its peak, the Megaco E-mail list had 1200 subscribers. (It still has 800.) At least 10,000 messages were sent on the list in the 14 months leading up to February 2000.

[21] The ITU-T is not the only body creating package specifications for H.248. Additional publicly specified packages have been registered by ETSI and 3GPP, and private packages by vendors LM Ericsson and Nokia. The complete list of packages is maintained by IANA at http://www.iana.org/numbers.html and also appears in an ITU-T H-series Supplement.

4.5.2.1 A Brief Digression: H.248 Syntax

Syntax is one of the less-important aspects of a protocol, although for H.248 it was a continuing item of contention. Nevertheless, a brief description of the syntax will be given here, to prepare the reader for the examples in succeeding sections.

The people developing H.248 were divided into two camps: those, typically with an ITU-T protocol background, who wanted binary encoding to keep message sizes down and those who wanted a protocol in the spirit of HTTP, SIP and SDP for easy debugging and extensibility. Allied to this question was how media sessions would be specified: using SDP for compatibility with SIP or using H.245 [54] for compatibility with H.323. After six months of debate, the Megaco Chair announced that it was clear that no one had a compelling argument either way, and the sensible thing was to toss a coin and move on.

The coin toss came out "text", but the next ITU-T meeting came up with an interesting counter-proposal. Syntax specification should be in ASN.1 (ITU-T Recommendation X.680 [55]), but could be encoded as text using a proposed set of "Text Encoding Rules". ASN.1 is usually encoded in binary form, and this would also be permitted. Neither side was happy with the coding efficiency of this approach, because of the constraints it imposed. In the end, the decision was to provide both ASN.1 and ABNF (Augmented Backus-Naur Form, RFC 4234 [56]) descriptions of the syntax, with the editors responsible for keeping the two syntax descriptions synchronised. The binary form of the protocol is used with BICC (which was described in Section 4.3.2.1), but the text form is by far the most widely implemented and will be used in the illustrations that follow. The binary form of the Protocol uses the Basic Encoding Rules (BER) as specified in ITU-T Recommendation X.690 [57]

The general form of constructs in the text encoding (with some exceptions) is:

$$<keyword>[= \text{identifier}]\{ \text{ associated content } \}$$

This structure applies recursively to the transaction, action, command, descriptor and individual signal and event levels, all of which will be explained in the sections which follow. The identifier is present at the transaction level to identify the transaction, at the action level to identify the context, at the command level to identify the termination to which the command applies, in the Events descriptor to identify the request and in the Stream descriptor to identify the stream. The associated content is a set of zero or more entities at the next lower level, separated by commas. At the lowest level, parameters and properties in the narrow sense (see Section 4.5.2.4) have the simpler form: name = value.

The protocol is case insensitive, except for literals. White space, including end of line and comments, is allowed on either side of the equals signs ("="), braces ("{","}") and commas separating terms within the braces. In general, the protocol does not constrain the order in which terms at the same syntactical level are presented. The protocol uses SDP for media descriptions. SDP expressions follow the syntactical rules of SDP rather than the rules just stated. Hence, the SDP expressions are case sensitive and consist of an ordered series of lines each beginning with a single-character keyword. Outside of the SDP, the protocol defines both a long and a short form for most keywords. The examples below will use the long forms for readability.

4.5.2.2 Protocol Structure

The H.248 protocol is transaction based. Each transaction consists of a request, possible interim replies and a final reply. The interim replies carry no information other than identification of the transaction and serve only to prevent timeout at the requestor on the rare occasion when a command takes a long time to execute. The reply may contain information asked for in the request and other information showing the results of commands invoked by the request.

H.248 messages may contain multiple transaction requests, replies and interim replies. The protocol does not specify the order in which multiple transactions in the same message are processed, but it is recommended that replies are processed before requests. One constraint on mixing requests and replies in the same message is that H.248 requires requests to be sent to the address negotiated during control association startup, but replies are always sent to the source address from which the original request was sent.[22] If these addresses differ, requests and replies must be in separate messages.

The contents of an individual transaction are guaranteed to be executed in the order in which they are presented, up to the point of failure, if any. In contrast to the database concept of a transaction, H.248 does not offer all-or-none execution of a transaction. Clean-up in case of failure is limited to the failing command, to the extent possible.[23]

A transaction is organised as a sequence of *actions*, each of which identifies an H.248 context to which the contents of the action apply. These contents consist of context property settings, context property audit requests and commands. Actions and commands are illustrated in Section 4.5.2.3.

Each command applies to a specific termination. The complete list of H.248 commands, all of which are described below, is as follows: Add, Subtract, Move, Modify, AuditCapability, AuditValue, Notify and ServiceChange. ServiceChange can be used with individual terminations as well as with ROOT, to indicate their coming into or out of service or to cause them to be reset.

Commands enclose descriptors, which organise the properties (in the broad sense) being set on or reported from a termination. Table 4.1 in Section 4.5.2.4 tabulates the different H.248 descriptors and the commands that can contain them. A descriptor may be empty or contain a number of individual descriptor parameters, many of which will have been defined in packages.

4.5.2.3 The H.248 Connection Model

In H.248, connections are made by adding *terminations* to a *context*. The context is the easy part of this statement to deal with, since it has no MGCP equivalent. Conceptually, it is the equivalent of a conference bridge, interconnecting all the terminations it contains. The term "termination" covers both more and less than the MGCP term "end point". "Physical terminations" such as circuits have the same scope as MGCP's "physical end points" – they are persistent entities representing hardware of some sort (or a shared piece of it, like a digital circuit multiplexed with others on the same optical fibre). However,

---

[22] This helps pass replies through firewalls and Network Address Translators (NATs) lying between the MG and MGC.

[23] The protocol offers the ability to mark commands as optional, meaning that if they fail, the receiver continues to execute subsequent commands in the same transaction.

H.248 also has "ephemeral terminations", which give a separate existence and identity to the packet side of the MGCP connection. An ephemeral termination is created by adding it to a context and destroyed by removing it from any context. Any termination can belong to at most one context at any one time. A context from which the last termination is removed ceases to exist.

MGCP has the idea of "virtual end points", with a particular tone from an audio source as one example. The names of such virtual end points are meant to be recognised by the gateway. H.248 lacks this concept. From the H.248 point of view, any audio source external to the gateway is represented by a termination, typically ephemeral, which must be added to the context when it is needed and then taken away again. We will see an illustration of this sort of operation shortly. If the gateway has internal audio sources, they are invoked by applying signals defined by packages such as the Call Progress Tones Generator (cg) package in Recommendation H.248.1 Annex E.

Three of H.248's eight commands deal with the movement of terminations into and out of contexts:

- The Add command places a termination into a context. The context may be an existing one or a new one created as a side effect of the Add command, depending on what the MGC specifies. Similarly, the termination may be a physical termination or an ephemeral one created by the Add command.
- The Subtract command removes a termination from a context. If the termination is physical, it continues to exist, but resides in the *NULL context*[24] in the reinitialised state. An ephemeral termination is destroyed when it is subtracted. If this was the last termination in the context, the context itself is destroyed. Similarly to the DLCX in MGCP, the Subtract returns statistics on the media sessions supported by the termination while it was in the context.
- The Move command moves a termination from one context to another. Move is useful, for example, to put a termination on hold by putting it into a context by itself. Move is not the same as Subtract followed by Add because it preserves the state of the termination (and does not destroy ephemeral terminations).

Figure 4.14 shows an example of the use of these commands. A and B are talking. C calls A, who has the Call Waiting feature. Upon being alerted to the waiting call, A puts B on hold and switches to C. B gets tired of waiting and hangs up. It is assumed that A is served directly by the controlled MG (and is therefore represented by a physical termination). B and C are served by other equipment and are thus represented by ephemeral (IP) terminations.

Figure 4.14(a) illustrates several points. The upper part of the figure shows the context connecting A and B while they are in conversation. The context has a numerical identifier, normally assigned by the MG. Each termination also has an identifier, which may be a mixture of letters, numbers, slashes and underscores and may also include a domain name following an "@" sign. The identifier for the termination serving A shows a typical convention for identifying an analogue line. The three levels of the hierarchy might be frame, shelf and port, for instance. This identifier would be configured at the MG and MGC. The identifier for the termination serving B shows a convention that might be used for an IP ephemeral termination: leading characters identifying the transport, followed by a numerical value typically assigned by the MG.

---

[24] The NULL context is not really a context, but a logical holding point for idle terminations.

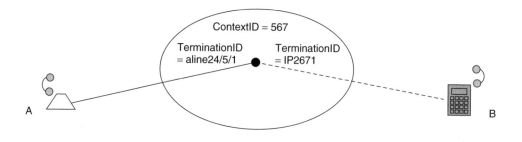

Request: Context = ${Add = ${description of connection to C}}
Reply: Context = 572 {Add = IP2675{local media description}}

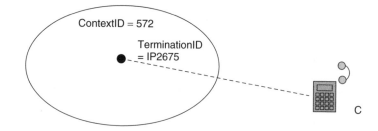

Figure 4.14(a)   A and B in conversation, call from C arrives

The middle part of Figure 4.14(a) shows the request that would be used to set up a termination representing caller C and the reply from the MG. Creating a termination is necessary so that caller C can receive the ringing tone while waiting for A to answer. Creating a context is also necessary, since an ephemeral termination cannot exist outside a context. The "$" signs in the request are CHOOSE wildcards. They ask the MG to create instances of a context and an ephemeral termination matching the contained description. The response tells the MGC what values were assigned. The result is shown in the lower part of Figure 4.14(a): another context co-existing with the first one.

Figure 4.14(b) shows the use of the Move command to disconnect A from B and connect A to C instead. The request identifies the termination to be moved and the context to which it is to be moved. This time both these identifiers are already defined, so they are stated explicitly. The reply merely confirms the request. It would also be possible to move a termination to a new context, using "$" as the context identifier.

The top of Figure 4.14(c) shows the situation immediately after B has lost patience and gone on-hook. The middle of the figure shows the issuance of a Subtract command to delete the connection to B. By default, a Subtract returns statistics for the session just terminated. The bottom part of the figure shows the conversation between A and C continuing, but the context which originally held A and B was deleted when the last termination it contained was subtracted. In physical terms, the resources allocated to the context and termination, including their identifiers, were released for reuse.

Summing up, where MGCP was optimised for circuit-to-packet connections, the H.248 connection model provides an elegant solution to the problem of making more general connections. The H.248 model offers an additional, less-obvious benefit. It is possible in

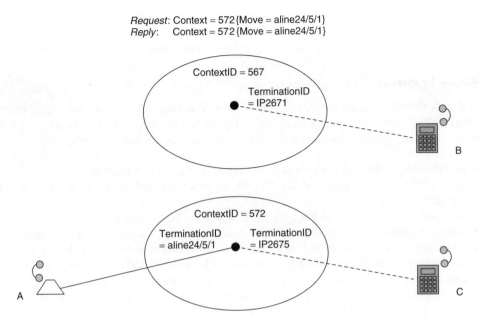

Request: Context = 572{Move = aline24/5/1}
Reply: Context = 572{Move = aline24/5/1}

ContextID = 567

TerminationID = IP2671

B

ContextID = 572

TerminationID = aline24/5/1   TerminationID = IP2675

A   C

Figure 4.14(b)   A puts B on hold and switches to C

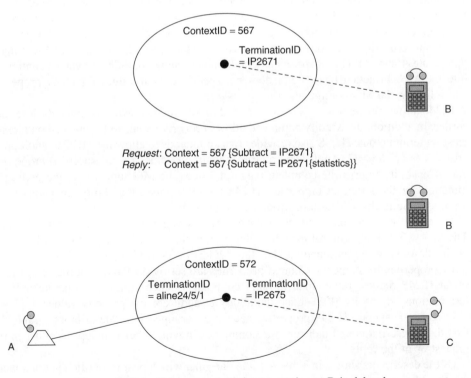

ContextID = 567

TerminationID = IP2671

B

Request: Context = 567 {Subtract = IP2671}
Reply: Context = 567 {Subtract = IP2671{statistics}}

B

ContextID = 572

TerminationID = aline24/5/1   TerminationID = IP2675

A   C

Figure 4.14(c)   B hangs up and the connection to B is deleted

H.248 to observe events and play out signals on any termination, which means that the packet side can be manipulated independently of the circuit side. In contrast, MGCP must work either with the end point (i.e. the circuit) or with the connection as a whole.

4.5.2.4 Properties

H.248 deals with a large amount of state. The H.248 specification uses the term "properties" in its broadest sense to mean the complete set of state values associated with a termination or context. The state defined for a termination includes the following:

- descriptions of the media flows sent and received by the termination;
- descriptions of the events that the MG has to watch for and report if they happen on the termination;
- descriptions of signals that the MG has to apply to the termination;
- control parameters that modify the operation of the protocol upon the termination;
- other parameters defining the characteristics of the termination as a whole or the characteristics of individual media streams such as their direction of flow which are independent of the actual codecs used. These other parameters and the control parameters of the previous bullet are called *properties* in the narrower sense used by the H.248 specification. The reader should be aware of the double usage of this term.

In theory, H.248 properties (speaking from this point on in the broad sense) all have specified default values. These defaults apply to all physical terminations when a MG first boots up or when the terminations are subtracted from a context. They also apply to ephemeral terminations when they are first created, although some of the property values are typically overridden immediately by values provided in the ADD command that creates them. All H.248-specified default values can be modified by values configured into the MG. In actual fact, H.248 does not specify default values for many properties and relies on the MG configuration to provide them.

H.248 allows property values to be specified in the Add and Move commands described earlier. In addition, the Modify command allows property values to be changed on already existing terminations. H.248 also provides two commands to allow the MGC to audit property values on the MG. AuditValue returns the current values for the specified properties. AuditCapability returns the complete range of values the MG supports on the given termination for the specified properties. H.248.1 (2002) added the ability for audits to be more specific in the selection of properties to be audited.

The basic properties of a context are the emergency flag, a priority level and its *topology*. The first two of these control resource allocation within the MG, while the third controls media flows between terminations. H.248.1 (2005) introduced the possibility of defining other properties for contexts. Context properties are set by declaration at the action level of an H.248 request rather than through commands, since commands operate only on terminations. H.248.1 (2002) added the ability to audit context property values.

In terms of syntax, H.248 properties have been grouped into "descriptors". Table 4.1 lists these descriptors and indicates the commands in which they can appear, either in the request or in the reply.

Of the descriptors shown in Table 4.1, the ones that will appear most often in commands are the Media, Signals and Events descriptors. The Media descriptor actually encloses

Table 4.1 H.248 descriptors and the commands that can contain them[25]

Descriptor	Add Move Modify	Subtract	Audit (Value or Capability)	Notify	Service Change
Media	Y	Reply[a]	Reply[a]	N	N
Events	Y	Reply[a]	Reply[a]	N	N
EventBuffer	Y	Reply[a]	Reply[a]	N	N
Signals	Y	Reply[a]	Reply[a]	N	N
DigitMap	Y[a]	Reply[a]	Reply[a]	N	N
Packages	Reply[a]	Reply[a]	Reply[a]	N	N
ObservedEvents	Reply[a]	Reply[a]	Reply[a]	Request	N
Statistics[b]	Y[a,c]	Reply[d]	Reply[a]	N	N
Audit	Request	Request	Request	N	N
ServiceChange	N	N	N	N	Y
Topology **ContextAttribute**	Applicable at the context level only				

General note: Except for the ServiceChange descriptor, all descriptors are optional in any specific instance of a command.
Note a: These descriptors are present in the reply only if requested within an Audit descriptor in the request.
Note b: The Statistics descriptor may occur at the termination and/or individual stream level. Stream level statistics were introduced in H.248.1 (2005).
Note c: The Statistics descriptor is present in the request only to activate the collection of specific statistics out of the set supported on the termination. Activation/deactivation was introduced in H.248.1 (2005).
Note d: Statistics are always reported on Subtract unless specifically disabled in the Audit descriptor.

other descriptors: the TerminationState descriptor and one Stream descriptor per stream supported by the termination. In turn, the Stream descriptor encloses the LocalControl, Local, Remote and Statistics descriptors. This is summarised in Figure 4.15.

As a syntactic optimisation, if the terminations of a context support just one stream (see the next section), the Stream descriptor level of the syntax may be ignored. In this case the syntax becomes:

Media { LocalControl {...}, etc. }, rather than

Media { Stream = <stream number>{ LocalControl {...}, etc. }}

4.5.2.5 Media Flows

MGCP can handle multimedia flows, but contains no special design features for that purpose. H.248 designers took multimedia as a basic design requirement and created a design specifically addressed to the management of multiple simultaneous media flows. As we shall see, H.248 also created its own model for the negotiation of media sessions

---

[25] The Modem and Mux descriptors are omitted from this table, the former because it has been deprecated and the latter because of its rarity of application

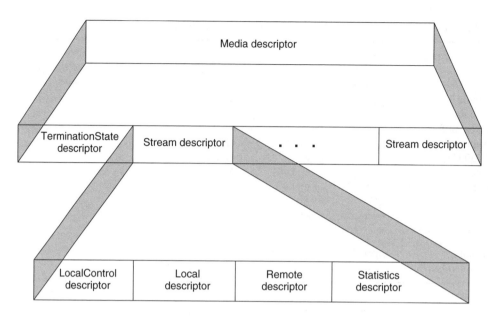

Figure 4.15 Components of the Media descriptor

between the MGC and the MG. An unfortunate side effect of this design is that H.248 media negotiation fits poorly with the SIP offer–answer model [3] (which it predates), and SDP manipulation is required at the MGC when interworking between H.248 locally and SIP across the network.

The basic mechanism that H.248 uses to specify media flows is the *stream*. A stream is a single bidirectional flow of media. In H.248, each direction can be specified independently, although the normal practice is to use symmetric flows (same codec in each direction). Each stream has an identifying number, given as part of the Stream descriptor. H.248 specifies that each stream is mixed independently across the context. A given termination is not required to support every stream. As an example, Figure 4.16 illustrates a context with three terminations T1, T2 and T3. T1 and T2 can handle both audio and video. T3 can only handle audio. The MGC has specified for each termination that the audio will flow in stream 0 and the video in stream 1. Accordingly, all three terminations support stream 0, but only T1 and T2 support stream 1.

For each stream, the MG must perform four tasks: enabling and disabling of flows, transcoding, mixing and mediation between transmission media. H.248 provides the means to control at least the first three of these tasks explicitly.

Enabling and disabling of flows can be accomplished at two levels: at the termination level for each stream and at the context level. At the termination level, the Mode parameter within the LocalControl descriptor for each stream determines its direction of flow: one way into or out of the context, both ways, inactive or looped back for testing purposes. At the context level, the Topology descriptor specifies the direction of flow between specific pairs of terminations. In the absence of a Topology descriptor (the normal case), all flows are two way. H.248.1 (2005) added the ability to specify the topology at the individual stream level.

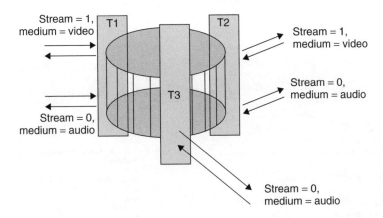

Figure 4.16 Media streams in a context

Transcoding and mediation between transport modes (e.g., packet to circuit) are done implicitly within the context. For each stream at each termination, the media encoding for the outgoing flow and the incoming flow are specified separately, in the Remote and Local descriptors respectively.[26] As described in more detail in the next section these two descriptors each contain one or more SDP session descriptions, with the restriction that each session description has only one media ("m=") line. This restriction was imposed to be consistent with the idea that a stream supports only one media type. Note the impact on the MGC when negotiating with a SIP end point at the other end: multiple session descriptions, one or more for each stream, sent between the MGC and the MG, must be rearranged into a single session description with an ordered sequence of media descriptions to send to the SIP peer and vice versa. In any event, transcoding is required when, for a given stream, there is a mismatch between the codec(s) in the Local descriptor for one termination and those in the Remote descriptor for another termination in the same context. Figure 4.17 illustrates an example of such a situation, where for illustration both terminations use IP transport.

The need for transcoding for the flow in the reverse direction in Figure 4.17 is similarly determined, by comparing the Local descriptor for T2 with the Remote descriptor for T1. Note that if more than two terminations are in the context, the MG may have to transcode all audio flows to a base encoding (i.e. linear pulse coded modulation) so it can mix them before re-encoding them according to the requirements of each termination.

The "mixing" applied by the context is obviously media specific and is often a matter of implementation: audio mixing generally means addition of inputs, but video can be "mixed" in a number of ways, with display of each input in a separate window as one example. Recommendation H.248.19 [58] provides a number of packages that can be used to specify mixing behaviour in a standard manner.

The termination transport type is determined by the address family used on the "c=" line in the session descriptions in the Local and Remote descriptors. "IN" denotes "Internet". RFC 3108 [59] provides conventions for describing ATM bearers in SDP, including the

---

[26] For circuit terminations, the media encoding is typically specified through provisioning and never appears explicitly in H.248 commands.

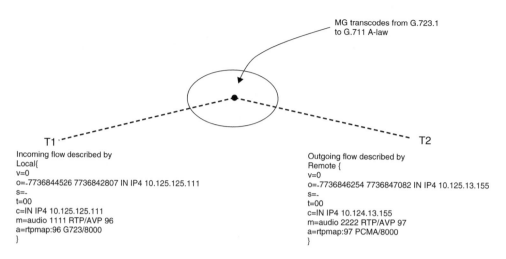

MG transcodes from G.723.1
to G.711 A-law

T1
Incoming flow described by
Local{
v=0
o=-7736844526 7736842807 IN IP4 10.125.125.111
s=-.
t=00
c=IN IP4 10.125.125.111
m=audio 1111 RTP/AVP 96
a=rtpmap:96 G723/8000
}

T2
Outgoing flow described by
Remote {
v=0
o=-7736846254 7736847082 IN IP4 10.125.13.155
s=-.
t=00
c=IN IP4 10.124.13.155
m=audio 2222 RTP/AVP 97
a=rtpmap:97 PCMA/8000
}

Figure 4.17   H.248 example of a situation where transcoding is required

use of "ATM" to identify the address family. Circuit bearers are normally identified implicitly through pre-provisioning on the MG.

4.5.2.6 H. 248 Media Negotiation

Consider a typical media negotiation sequence involving a MGC and MG at the originating end of the call, a SIP device at the other end and an IP network in between. Note that the media path is represented by an ephemeral termination in the MG. The complete connection through the MG requires the creation of a context that will contain this ephemeral termination. We are not concerned particularly with the other side of the connection, but it will be represented by another termination in the same context.

1. Call signalling received at the MGC provides some description of what media flows the connection should support. For instance, the description might have come via the Bearer Capability element of a Q.931 (ISDN) Setup message.
2. The MGC translates the media description into descriptions for one or more streams to be supported by the terminations of the context. It may apply local policy (e.g. a limit on bandwidth based on what the subscriber has paid for) during this translation, but there may be a range of choice left after the translation is complete. The choices may involve different codecs or, especially for video, different sets of attribute values to be applied to the operation of a specific codec.
3. The MGC presents these stream descriptions with their choices to the MG.
4. The MG determines what it can support and returns the result to the MGC. This result includes the address(es) and port(s) at which it is willing to receive the incoming media flows.
5. The MGC formats the returned stream descriptions into a session description to be sent as an offer in a SIP INVITE to the peer SIP device at the far end.

6. The SIP peer returns an answering session description. This answer differs from the offer at a minimum by providing the receiving address(es) and port(s) at the far end. It may also indicate support for additional codecs and reject some of the codecs in the offer.
7. The MGC reformats the returned session description into stream descriptions and sends them to the MG.
8. The MG determines what it can support and returns the result to the MGC. The MG now has the information it needs to send and receive media.

As described in the previous section, the H.248 Local descriptor indicates what the MG is prepared to receive from the termination (i.e. from the SIP peer). The Remote descriptor indicates what the MG may send to the termination (i.e. to the SIP peer). In a strict sense, the MGC should negotiate both the local and remote media descriptions with the MG in steps 2 to 4. This way, the offer sent to the SIP peer in step 5 can confidently propose outgoing as well as incoming flows. In practice, however, the MGC negotiates only the Local descriptor at that point. The Remote descriptor can be provided in step 7 and joint support for it and the local description confirmed in step 8.

The negotiation sequence can be viewed as having two phases in another sense. In the first phase, represented by steps 2 to 4, the MG can be invited to return a range of choices for the media description. This maximises the probability that the far end can support at least one of the choices. If the far-end narrows the set of choices in its answer, then in step 7, in addition to forwarding the remote description to the MG, the MGC may also update the local description to release some of the resources reserved earlier.

To support the reservation of resources for multiple choices at once, H.248 offers the following three degrees of freedom:

- The ability to specify multiple codecs on a single media ("m=") line of a session description. This is normal SDP usage.
- The ability to specify multiple session descriptions in the same media description for a given stream. This is unique to H.248. It was originally introduced to provide the possibility of allowing the MG to choose between complex bundles of attributes such as those need to describe video flows. However, it is also a practical necessity to specify alternatives such as voice (media type "audio") versus facsimile (media type "image"), since SDP supports only one media type per media line.
- The ability to wildcard, specify a range for or a list of alternatives for a particular parameter within a session description. This capability is more meaningful for some parts of SDP than for others.

When the MGC presents a session description to the MG with multiple codecs on a single line or with multiple session descriptions for the same stream description, the semantics are potentially ambiguous. The MGC may ask the MG to choose one of the alternatives or ask the MG to reserve resources for as many of the alternatives as it can. The desired semantics really depend on whether the MGC is in the first phase (steps 2–4) or the second phase (steps 7–8) of the negotiation. H.248 provides two control properties in the LocalControl descriptor to resolve the ambiguity. ReserveValue tells the MG to reserve for all codecs on a media line if it is true, or select a single codec if it is false. Similarly, when multiple session descriptions are provided, ReserveGroup = true indicates that the MG should reserve resources for as many of them as it can. ReserveGroup = false asks it to choose a single one of the alternative session descriptions.

Summing up: if a stream supports a bidirectional flow, H.248 requires that the media description for that stream include both a Local and a Remote descriptor. Negotiations between the MGC and its remote peer determine the final content of both descriptors. To allow the MGC to provide as much choice as possible to its peer in an initial offer, the protocol allows it to request reservation of resources for multiple session descriptions and multiple codecs within a session description. The MGC can also tell the MG to choose from an MGC-specified set of alternatives.

4.5.2.7 Signals and Events

Recall the progress of a telephone call on an ordinary telephone line. At certain points in the call, the line generates events which must be processed by the local exchange. Two examples of such events are transitions between on-hook and off-hook and transmission of digits in the form of DTMF tones. Similarly, the local exchange must transmit signals to the caller or the called party. Again, two examples of these are the dial tone returned to the caller and the ringing of the called telephone. Notice that some events and signals are in-band, meaning in H.248 terms that they are associated with specific streams. Others are out-of-band, meaning that they are associated with the termination as a whole. Moreover, some make sense only for certain termination types, while others can apply to any termination, physical or ephemeral.

Almost all H.248 events and signals are defined in H.248 extensions.[27] Events and signals can be considerably more abstract than the simple examples just cited. Some examples of events that have been defined are as follows:

- detection of on-/off-hook;
- detection of digits received using analogue inter-exchange signalling (recall Section 4.2.2);
- detection that packet loss rate in a media stream has exceeded a threshold;
- detection of modem, facsimile or textphone signals in a media stream;
- detection of key-down/key-up state changes on a termination representing the keypad of a business telephone set;
- detection that a period of inactivity (no commands received or events detected) has exceeded a threshold for a given termination.

The H.248 view of signals is even more abstract. In effect, as the protocol evolved, signals became the equivalent of standardised procedure calls on terminations. Some examples of signals that have been defined are to request the following:

- transmission of dial tone;
- transmission of signals within an analogue inter-exchange signalling protocol (again recalling Section 4.2.2);

---

[27] See the discussion of H.248 packages in Section 4.5.2.9.

- transmission of calling number to be displayed at a telephone set;
- indication that all processing associated with a given termination is complete, as a substitute for the V5.x end-of-signalling procedure (recall Section 4.2.1)[28];
- playing of a specified prompt and collection of recorded voice in response (to be used, e.g. in a voice mail application);
- storage of the voice recording in the previous example, using a specified identifier.

The specification of which events to report and which signals to play out on a termination is accomplished through the setting of the Events and Signals descriptors respectively. As for other properties, default values for these descriptors can be determined by configuring the MG, but normally they will be set explicitly by the Add, Move and Modify commands. These descriptors not only identify the events to be monitored and signals to be played out at any given time but also provide additional parameter values that affect how the MG handles them. Some of the added parameters are common to all events or signals, while others are specific to the individual event or signal.

The list of events in the Events descriptor determines which events get reported at any given time. H.248 defines two event-handling modes. The event-handling mode for a particular termination is determined by the value of the EventBufferControl property in the TerminationState descriptor. If EventBufferControl is set to "OFF", then the events in the current Events descriptor are reported as they are detected. There may be multiple events in the same report if they are detected within the same small period of time. If the message rate in this mode is too high, the MGC can set EventBufferControl to "LockStep" and provide a list of events in the EventBuffer descriptor. In this case, each time an event in the EventBuffer descriptor is detected, it is queued for reporting. Queued events are popped off the queue whenever the Events descriptor changes. If the event at the head of the queue is not in the new Events descriptor, it is discarded. This continues until an event is found that is in the new Events descriptor. This event is reported, and buffering resumes. Note that the MGC may modify the contents of the EventBuffer descriptor in the same command that changes the Events descriptor.

H.248, like MGCP, allows for reflex actions. The MGC may specify a new Events descriptor as a parameter to an event, meaning that if the MG detects the first event, the Events descriptor immediately changes. One example where this might be used would be when monitoring the hook state of an idle line. When the line goes off-hook, a new Events descriptor is activated which now starts watching for on-hook. This avoids loss of an on-hook event due to the time lag involved in notifying the MGC and receiving new instructions. H.248 also allows the MGC to specify a new Signals descriptor that should be activated when a specific event is observed. An example of the use of this (but not typical practice) might be the automatic activation of the dial tone when off-hook is detected.

The MG reports events to the MGC using the NTFY request. The NTFY request contains a request identifier which can be used to correlate the report with a particular version

---

[28] The definition of this particular signal in ITU-T Recommendation H.248.34 [60] (a V5.x emulation package) was controversial. It represents the completion of the emulated V5.x protocol state machine for a given termination, but is otherwise unnecessary in H.248 terms. In most cases, the state machine could be naturally terminated with the subtraction of the termination from a context. In certain cases, however, the emulated V5.x signals are applied when the termination is in the null context, so no subtraction occurs in the natural flow of H.248 commands.

of the Events descriptor. It may contain a timestamp and will contain an ObservedEvents descriptor identifying the events being reported as well as any reportable parameters associated with those events. As an example, the on-hook event of the analogue line supervision package (H.248.1 Annex E.8) has a configuration parameter, "strict", that was defined to overcome a potential race between the configuration of the event by an Add, Move or Modify command and the actual occurrence of the transition to on-hook. Depending on how "strict" is set, the NTFY will return an additional parameter with the event report, indicating either that a transition from off-hook to on-hook was observed after the event was configured or that the line was already on-hook when the event was configured.

Like MGCP, H.248 supports the use of digit maps to reduce the amount of messaging required to collect digits. The details differ from those of MGCP. The digit map procedure defined in the first version of H.248 was modified with experience. H.248.1 (2002) changed the base procedure, and H.248.16 [61] provided an alternative DTMF digit map package to the one specified in H.248.1 Annex E.

It is often desirable for events to interact with signals. For example, if a line receiving ringing goes off-hook, the ringing should stop. H.248 provides that when the MG recognises that some event specified in the Events descriptor has occurred, it stops the playout of any signals in progress unless the KeepActive keyword parameter is set for the event.

H.248 requires that if multiple signals are configured on a termination at the same time, they must be played out simultaneously. However, H.248 also allows the MGC to specify signal lists. A signal list contains a sequence of signals to be played out one at a time. An example of the use of such a sequence might be to play out a sequence of digits in inter-exchange signalling.

4.5.2.8 Control Associations

MGCP took the philosophy that, as much as possible, the control association was between the MGC and the individual end points. In contrast, H.248 was designed to incorporate a very explicit relationship between the MG and its controlling MGC. For a start, the MG is modelled as a special termination, the ROOT termination. Like other terminations, the ROOT termination can have properties – for example, a value indicating the maximum number of terminations the MG supports per context. The ROOT termination also supports events. However, it does not support media streams and cannot appear in a context.

A control association is initiated when the MG sends a ServiceChange request to an MGC, indicating that the ROOT termination is coming into service. The ServiceChange request contains further information relating to the state of the MG and its control association. The key distinction is between situations where

- the MG has no active state (has just rebooted);
- it has active state but has not had contact with an MGC for some period of time (loss of connection or MGC failure detected);
- the MG has been directed by one MGC to transfer control to another.

The ServiceChange request also contains information needed to negotiate the session startup: an indication of which version of H.248 the MG wishes to use and optionally the name of a profile the MG supports. A profile is a specification of which particular protocol options the MG understands. Implicitly, the startup ServiceChange request also contains another piece of information: the transport protocol the MG wishes to use for the association. The core specification allows the options of UDP and TCP. Extensions document the use of Stream Control Transport Protocol (SCTP) (described below) and ATM-based transport.

The MGC completes the startup registration by responding to the ServiceChange request. This response may modify the protocol version or the address and port to which the MG should send requests or direct the MG to try a different MGC.

Once the initial registration is completed, it is normal practice for the MGC to audit the state of the MG's terminations. It is the established convention that when a MG first establishes a control association after it starts up or has a cold reboot, the MGC can assume that there are no active contexts and that the MG's physical terminations are all in service unless otherwise advised. However, even in this case the MGC has to learn what default properties have been configured on the physical terminations. In other startup cases, there will be active contexts and ephemeral terminations. The MGC may be aware of this initial state to a greater or lesser extent and must ensure that it is synchronised with the MG before it goes on.

H.248 allows for the possibility of dividing up a physical MG into multiple virtual MGs, each controlled by a different MGC. To prevent contention between these controllers, any one physical termination can belong to only one virtual MG. The virtual MG concept is especially simple to implement with packet-to-packet gateways.

4.5.2.9 Packages and Profiles

H.248 has a formal extension mechanism: the package. Packages provide detailed capabilities needed to support particular applications. A number of basic packages appear in Annex E of the base protocol specification, H.248.1. Other package definitions appear as Recommendations in the H.248.x series or as documents published by other organisations such as ETSI. It is also possible for private organisations to define their own packages. To avoid collisions between package names used in the protocol, these names can be registered with the Internet Assigned Numbers Authority (IANA).[29]

A package may define new properties (in the narrow sense), events, signals and statistics and must specify the protocol procedures associated with them. Some packages define properties and events for the ROOT termination and others, for terminations of a specific type. Many deal with specific applications, such as sending or detecting tones, playing out announcements or signalling on analogue inter-exchange circuits. Sometimes, multiple packages have been defined to achieve the same result. For example, H.248.10 [62], H.248.11 [63] and H.248.32 [64] all deal with overload reporting from the MG.

This leads us into the topic of profiles. The idea of a profile was built into H.248 from the beginning, but not fully formalised until H.248.1 (2005).[30] Historically, the first H.248 profile was specified in RFC 3054 [65], for an IP business phone. The profile

---

[29] IANA's web site is at http://www.iana.org. At the time of writing, the Megaco/H.248-specific registrations were to be found at http://www.iana.org/assignments/megaco-h248.

[30] H.248.1 (2005) Appendix III is a lengthy template for and checklist of items that may be dealt with in a profile.

specifies an architecture from which a set of termination types is derived. It defines a
naming convention for the terminations, lists the packages that must be supported for
them and indicates the encoding and transport protocol that must be supported for the
overall operation of the protocol. The basic purpose of this and any profile is to assure
interoperability for a specific application. The MultiService Forum (MSF), referred to in
earlier chapters, is one of the organisations that has defined H.248 profiles.

Packages and profiles can have versions. Some packages can extend others. This means
that they may add more properties, events, signals or statistics or extend the range of values
that individual parameters within the original package can take on.

4.6 The Sigtran Protocols

Recall from Section 4.3.1 that the IETF created two Working Groups to follow through
on their PSTN interworking architecture. The output of the Megaco Working Group,
developed in combination with ITU-T Study Group 16 with requirements fed in from
other bodies, has just been described. This section considers the work done by the IETF
Sigtran Working Group on the transport of PSTN signalling across the IP network.

The protocol requirements considered by the Sigtran Working Group were as follows:

- reliable, in-order delivery;
- secure as the SS7 network in the PSTN;
- message transfer times should be comparable with those achieved in the PSTN;
- capable of carrying a variety of PSTN-signalling protocols. In fact, Sigtran solutions
  have emerged for the transport of ISUP, Q.931, MTP 3 and TCAP from the ISDN
  suite, but also DPNSS (a PBX protocol) and V5.2 (described in Section 4.2.1).

The Working Group quickly decided on a layered protocol architecture. The lower layer
would be a general-purpose reliable transport protocol, above which a protocol-specific
adaptive layer would provide the specific interface expected by each signalling protocol.
The next section describes the general-purpose transport protocol that emerged from this
effort: SCTP. The following section takes a look at the "User Adaptation Layers", to use
the Sigtran terminology.

4.6.1 The Stream Control Transmission Protocol (SCTP)

STCP (RFCs 2960 [66] and 3309 [67]) began as one of several protocol proposals made
to Sigtran for use as the lower layer. The basic idea came from Motorola engineers Randall
Stewart and Qiaobing Xie: to minimise the time required to recover from network failures
by including multiple signalling paths within the scope of a single transport association.
Figure 4.18 illustrates the concept.

The starting assumption is that the signalling end points concerned are multi-homed,
so that more than one packet transmission path exists between them. The two sides of
the figure show the amount of effort required to re-establish the signalling flow if the
active path fails. The left side of Figure 4.18 shows the situation one would have if one
uses TCP/IP. The failure has to be detected and then a new TCP/IP connection has to be
initiated. This takes time. The right side of the figure shows the situation if one uses a
new protocol which maintains multiple paths at once within a single transport association,

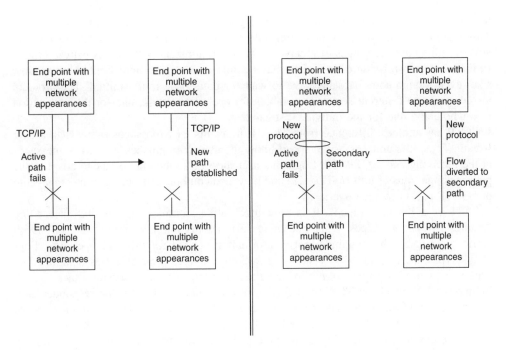

Figure 4.18 Reliability through managing multiple paths in one association

even if just one path is active at any given time. Failure of the active path still requires some time to detect, but after that the signalling flow can be redirected over another path within the association with a minimum of effort. Moreover, the task of keeping track of which messages the other end has already received and which must be retransmitted because of the failure can now be delegated to the protocol stack, simplifying application development.

The basic structure of SCTP emerged from a meeting of the Sigtran design team in early 1999. Like TCP, it is connection oriented. The word "association" was used above and continues to be used to describe the relationship between the two SCTP end points. An SCTP association is defined by a set of one or more IP addresses and a single SCTP port number at each end point.

SCTP has further differences from TCP. One is indicated in the first part of its name[31]: it supports multiple streams rather than just one. This feature was added to get around TCP's head-of-line blocking characteristic; that is, TCP delivers data in the order it was received, but at the cost of delaying everything until the data at the head of the queue is delivered successfully. Typically, signalling entities like telephone exchanges and Call Agents serve many calls at the same time. The messages relating to each call must be delivered in order, but the messages for different calls are independent of each other. In support of this, SCTP allows the end points to indicate the number of streams they wish

---

[31] The rest of SCTP's name was a deliberate and lighthearted transposition of TCP: "Transmission Control Protocol" with a twist!

to use. Messages are delivered in order[32] in a given stream, but no stream blocks another. This feature was achieved by designing into the protocol the ability for the receiver to acknowledge multiple ranges of received data chunk numbers with breaks between them. In this way, transmission of new data may be able to continue (until limited by window size) even though some streams have to wait for data to be retransmitted. The protocol also carries two different sequence numbers for each data chunk, one for the association as a whole and one for the individual stream.

SCTP has another difference from TCP. It transmits user messages rather than a byte stream. Thus, it is unnecessary to specify how the user distinguishes successive messages at the receiving end. SCTP also fragments user messages when necessary to adapt to the path message transfer unit (MTU) size and may also combine multiple user messages into one SCTP packet to save overhead.

SCTP is a binary protocol. It is structured into a common header and succeeding "chunks" of various types. Only one of these types (as initially defined in RFC 2960) carries user data. The remainder carry information needed for protocol operation. Four chunk types are exchanged in successive packets to start up an SCTP association. The startup sequence is designed to minimise vulnerability to a denial of service attack like the infamous SYN attack on TCP, by deferring resource commitment at the responder until the initiator confirms its initial request. Other chunk types are exchanged to close down an association gracefully. One pair, HEARTBEAT and HEARTBEAT ACK, are used to verify the continued availability of inactive paths and to form an estimate of round-trip time along them. The chunk approach to protocol structure provides great flexibility for extensions to the protocol.

As with any transport protocol, the IETF required SCTP to compete fairly with TCP for bandwidth and to reduce its bandwidth consumption in the face of congestion. Like TCP, SCTP calculates transmission window sizes using estimates of path round-trip time.

SCTP was originally designed to operate on top of UDP, RFC 768. However, as the design progressed, it became clear that SCTP had more general application than just signalling transport. IETF management decreed that it should be redesigned to operate directly on top of IP, like UDP and TCP. This set the completion of the design back by nearly a year. RFC 2960 was published in October 2000. Shortly afterward, it was determined that the checksum formula used to maintain packet validity in the face of bit errors in the network was less effective than desired. RFC 3309 was issued to provide a better formula.

4.6.2 User Protocol Adaptive Layers

The Sigtran Working Group defined a number of user protocol adaptive layers to run between SCTP and various signalling protocols. These adaptive layers are substitutes for the upper layers of the protocol stacks used to carry that same signalling in the PSTN. As such, they were designed to provide similar services.

The most complex of these upper layers is MTP 3 (Message Transfer Part layer 3) in the SS7 signalling stack. MTP 3 provides the routing functions within an SS7 signalling network. In addition, it provides routing management capabilities, at a number of

---

[32] Since its original design, SCTP has been extended to support unreliable streams, meaning that messages placed in those streams may be lost, but are never delayed while waiting for others to be delivered.

levels of granularity related to SS7 network architecture. The M3UA adaptive layer (RFC 3332 [68], about to be replaced by an updated version), can be used to transport ISUP between the SG and the MGC.[33] In addition, it provides the MGC with remote access to the management capabilities of MTP 3 at the SG. This is illustrated in Figure 4.19. In this arrangement, the SG is a point on the SS7 network with an SS7 network address. The MGC has remote access to that network; it is reached via the SG's routing tables rather than SS7 routing.

Figure 4.20 provides a complete picture of the user adaptation layers developed as replacements for different parts of the SS7 stack. M2UA (RFC 3331 [69]) and M2PA

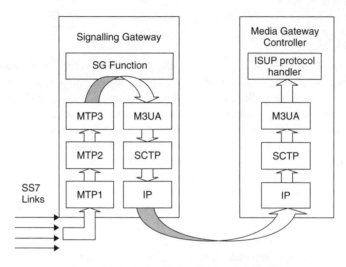

Figure 4.19   Transfer of ISUP using M3UA

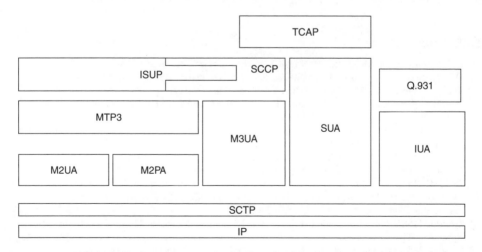

Figure 4.20   Relationship of the ISDN-related user adaptation layers

---

[33] Recall the architecture described in Section 4.3.1.

(RFC 4165 [70]) are substitutes for MTP2 and allow the transport of MTP3 as well as the ISUP on top of it. M2UA is an alternative to M3UA for carrying signalling between an SG and an MGC. M2PA is a substitute for the lower layers of SS7 for carrying signalling between two points in an SS7 signalling network.

SUA (RFC 3868 [71]) is a substitute for the SS7 Signalling Connection Control Part. It supports TCAP, the database query/response protocol mentioned in Section 4.2.2. Finally, on the access side, IUA (RFC 4233 [72]) is a substitute for Ref. 73, the transport on which Q.931 messages are carried.

As mentioned already, Sigtran has also defined user adaptation layers for V5.2 and the PBX protocol DPNSS.

4.7 Summary

This chapter has looked at the architectural and protocol developments needed to permit the NGN to interoperate with the PSTN. The chapter began with a review of transmission and signalling in the PSTN. Signalling and supervision on the access side were distinguished from inter-exchange signalling and supervision. The development of the ISDN and the various parts of Signalling System Number 7 (SS7) was described.

Section 4.3 went on to describe the development of the architecture that the NGN has adopted for interworking between the NGN and the PSTN. It considered the topics of PSTN bridging, interworking between SIP and ISUP, PSTN emulation and simulation and transport of DTMF signalling.

Section 4.4 provided a general description of MGs and described the requirements of a protocol for controlling them. It illustrated these requirements with an extended example of MG application. Section 4.5 went on to describe the development and structure of the two gateway control protocols now in widespread use, MGCP and Megaco/H.248. Finally, Section 4.6 described the protocols developed by the Sigtran Working Group of the IETF to carry various PSTN signalling protocols across IP networks.

References

1. G.711. "Pulse Code Modulation (PCM) of Voice Frequencies", November 1988.
2. H.320. "Narrow-Band Visual Telephone Systems and Terminal Equipment", March 2004.
3. Q.23. "Technical Features of Push-Button Telephone Sets", November 1988.
4. G.964. "V-interfaces at the Digital Local Exchange (LE) – V5.1 Interface (based on 2048 kbit/s) for the Support of Access Network (AN)", March 2001.
5. G.965. "V-interfaces at the Digital Local Exchange (LE) – V5.2 Interface (based on 2048 kbit/s) for the Support of Access Network (AN)", March 2001.
6. Q.931. "ISDN User-Network Interface Layer 3 Specification for Basic Call Control", May 1998.
7. H.324. "Terminal for Low Bit-Rate Multimedia Communication", September 2005.
8. ISO/IEC 11572:2000. "Information Technology – Telecommunications and Information Exchange Between Systems – Private Integrated Services Network – Circuit Mode Bearer Services – Inter-Exchange Signalling Procedures and Protocol", February 2001.
9. ISO/IEC 11582:2002. "Information Technology – Telecommunications and Information Exchange Between Systems – Private Integrated Services Network – Generic Functional Protocol for the Support of Supplementary Services – Inter-Exchange Signalling Procedures and Protocol", January 2004.
10. H. Schulzrinne, T. Taylor. "Definition of Events for Channel-Oriented Telephony Signalling", draft-ietf-avt-rfc2833biscas-03.txt, November 2005.
11. Q.761. "Signalling System No. 7 – ISDN User Part Functional Description", December 1999.

12. Q.762. "Signalling System No. 7 – ISDN User Part General Functions of Messages and Signals", December 1999.
13. Q.763. "Signalling System No. 7 – ISDN User Part Formats and Codes", December 1999.
14. Q.764. "Signalling System No. 7 – ISDN User Part Signalling Procedures", December 1999.
15. Q.700. "Introduction to CCITT Signalling System No. 7", March 1993.
16. H.323. "Packet-Based Multimedia Communications Systems", July 2003.
17. F. Andreasen, B. Foster, RFC 3435. "Media Gateway Control Protocol (MGCP) Version 1.0", January 2003.
18. L. Ong, I. Rytina, M. Garcia, H. Schwarzbauer, L. Coene, H. Lin, I. Juhasz, M. Holdrege, C. Sharp, RFC 2719. "Framework Architecture for Signaling Transport", October 1999.
19. M. Arango, C. Huitema. "Simple Gateway Control Protocol (SGCP), Version 1.1", draft-huitema-sgcp-v1-02.txt, expired, July 1998.
20. H. Schulzrinne, S. Petrack, RFC 2833. "RTP Payload for DTMF Digits, Telephony Tones and Telephony Signals", This will be replaced within the next few months by three documents, currently known as draft-ietf-avt-rfc2833bis-12.txt, draft-ietf-avt-rfc2833bisdata-06.txt, and the next document. May 2000.
21. Q.1902.1. "Bearer Independent Call Control Protocol (Capability Set 2): Functional Description", July 2001.
22. Q.1902.2. "Bearer Independent Call Control Protocol (Capability Set 2) and Signalling System No. 7 ISDN User Part: General Functions of Messages and Parameters", July 2001.
23. Q.1902.3. "Bearer Independent Call Control Protocol (Capability Set 2) and Signalling System No. 7 ISDN User Part: Formats and Codes", July 2001.
24. Q.1902.4. "Bearer Independent Call Control Protocol (Capability Set 2): Basic Call Procedures", July 2001.
25. Q.1990. "BICC Bearer Control Tunnelling Protocol", July 2001.
26. Q.765.5. "Signalling System No. 7 – Application Transport Mechanism: Bearer Independent Call Control (BICC)", April 2004.
27. Q.1950. "Bearer Independent Call Bearer Control Protocol", December 2002.
28. Q.1970. "BICC IP Bearer Control Protocol", July 2001.
29. G.723.1. "Dual Rate Speech Coder for Multimedia Communications Transmitting at 5.3 and 6.3 kbit/s", March 1996.
30. G.729. "Coding of Speech at 8 kbit/s Using Conjugate-Structure Algebraic-Code-Excited Linear Prediction (CS-ACELP)", March 1996.
31. H. Schulzrinne, S. Casner, R. Frederick, V. Jacobson, RFC 3550. "RTP: A Transport Protocol for Real-Time Applications", July 2003.
32. E. Zimmerer, J. Peterson, A. Vemuri, L. Ong, F. Audet, M. Watson, M. Zonoun, RFC 3204. "MIME Media Types for ISUP and QSIG Objects", December 2001.
33. S. Donovan, RFC 2976. "The SIP INFO method", October 2000.
34. A. Vemuri, J. Peterson, RFC 3372. "Session Initiation Protocol for Telephones (SIP-T): Context and Architectures", September 2002.
35. G. Camarillo, A.B. Roach, J. Peterson, L. Ong, RFC 3398. "Integrated Services Digital Network (ISDN) User Part (ISUP) to Session Initiation Protocol (SIP) Mapping", December 2002.
36. A. Johnston, S. Donovan, R. Sparks, C. Cunningham, K. Summers, RFC 3666. "Session Initiation Protocol (SIP) Public Switched Telephone Network (PSTN) Call Flows", December 2003.
37. G. Camarillo, RFC 3959. "The Early Session Disposition Type for the Session Initiation Protocol (SIP)", December 2004.
38. G. Camarillo, H. Schulzrinne, RFC 3960. "Early Media and Ringing Tone Generation in the Session Initiation Protocol (SIP)", December 2004.
39. G. Camarillo, A.B. Roach, J. Peterson, L. Ong, RFC 3578. "Mapping of Integrated Services Digital Network (ISDN) User Part (ISUP) Overlap Signalling to the Session Initiation Protocol (SIP)", August 2003.
40. Q.1912.5. "Interworking Between Session Initiation Protocol (SIP) and Bearer Independent Call Control Protocol or ISDN User Part", March 2004.
41. C. Jennings, J. Peterson, M. Watson, RFC 3325. "Private Extensions to the Session Initiation Protocol (SIP) for Asserted Identity within Trusted Networks", November 2002.
42. A.B. Roach, RFC 3265. "Session Initiation Protocol (SIP)-Specific Event Notification", June 2002.
43. J. Rosenberg. "A Framework for Application Interaction in the Session Initiation Protocol (SIP)", draft-ietf-sipping-app-interaction-framework-05.txt, Will eventually be an RFC when related work on which it depends is completed. July 2005.

44. E. Burger, M. Dolly. "A Session Initiation Protocol (SIP) Event Package for Key Press Stimulus (KPML)", draft-ietf-sipping-kpml-07.txt, Will eventually be an RFC when related work on which it depends is completed. December 2004.
45. H.248.1. "Gateway Control Protocol Version 1", 2000.
46. F. Cuervo, N. Greene, A. Rayhan, C. Huitema, B. Rosen, J. Segers, RFC 3015. "Megaco Protocol Version 1.0", November 2000.
47. Q.732.2. "Call Diversion Services", December 1999.
48. T.38. "Procedures for Real-Time Group 3 Facsimile Communication over IP Networks", September 2005.
49. N. Greene, M. Ramalho, B. Rosen, RFC 2805. "Media Gateway Control Protocol Architecture and Requirements", April 2000.
50. M. Handley, V. Jacobson, RFC 2327. "SDP: Session Description Protocol", This is about to be replaced by draft-ietf-mmusic-sdp-new-026.txt, which will be published as an RFC within the next month or two. April 1998.
51. P. Faltstrom, M. Mealling, RFC 3761. "The E.164 to Uniform Resource Identifiers (URI) Dynamic Delegation Discovery System (DDDS) Application (ENUM)", April 2004.
52. H.248.1. "Gateway Control Protocol Version 2", 2002.
53. H.248.1. "Gateway Control Protocol Version 3", 2005.
54. H.245. "Control Protocol for Multimedia Communication", October 2005.
55. X.680. "Information Technology – Abstract Syntax Notation One (ASN.1): Specification of Basic Notation", July 2002.
56. D. Crocker, Ed., P. Overell, RFC 4234. "Augmented BNF for Syntax Specifications: ABNF", October 2005.
57. X.690. "Information Technology – ASN.1 encoding rules: Specification of Basic Encoding Rules (BER), Canonical Encoding Rules (CER) and Distinguished Encoding Rules (DER)", July 2002.
58. H.248.19. "Gateway Control Protocol: Decomposed Multipoint Control Unit, Audio, Video and Data Conferencing Packages", March 2004.
59. R. Kumar, M. Mostafa, RFC 3108. "Conventions for the Use of the Session Description Protocol (SDP) for ATM Bearer Connections", May 2001.
60. H.248.34. "Gateway Control Protocol: Stimulus Analogue Lines Package", January 2005.
61. H.248.16. "Gateway Control Protocol: Enhanced Digit Collection Packages and Procedures", November 2002.
62. H.248.10. "Gateway Control Protocol: Media Gateway Resource Congestion handling package", July 2001.
63. H.248.11. "Gateway Control Protocol: Media Gateway Overload Control Package", November 2002.
64. H.248.32. "Gateway Control Protocol: Detailed Congestion Reporting Package", January 2005.
65. P. Blatherwick, R. Bell, P. Holland, RFC 3054. "Megaco IP Phone Media Gateway Application Profile", January 2001.
66. R. Stewart, Q. Xie, K. Morneault, C. Sharp, H. Schwarzbauer, T. Taylor, I. Rytina, M. Kalla, L. Zhang, V. Paxson, RFC 2960. "Stream Control Transmission Protocol", October 2000.
67. J. Stone, R. Stewart, D. Otis, RFC 3309. "Stream Control Transmission Protocol (SCTP) Checksum Change", September 2002.
68. G. Sidebottom, K. Morneault, J. Pastor-Balbas, (Editors). RFC 3332. "Signaling System 7 (SS7) Message Transfer Part 3 (MTP3) – User Adaptation Layer (M3UA)", This is about to be replaced by the document currently known as draft-ietf-sigtran-rfc3332bis-06.txt. September 2002.
69. K. Morneault, R. Dantu, G. Sidebottom, B. Bidulock, J. Heitz, RFC 3331. "Signaling System 7 (SS7) Message Transfer Part 2 (MTP2) – User Adaptation Layer", September 2002.
70. T. George, B. Bidulock, R. Dantu, H. Schwarzbauer, K. Morneault, RFC 4165. "Signaling System 7 (SS7) Message Transfer Part 2 (MTP2) – User Peer-to-Peer Adaptation Layer (M2PA)", September 2005.
71. J. Loughney, Ed., G. Sidebottom, L. Coene, G. Verwimp, J. Keller, B. Bidulock, RFC 3868. "Signalling Connection Control Part User Adaptation Layer (SUA)", October 2004.
72. K. Morneault, S. Rengasami, M. Kalla, G. Sidebottom, RFC 4233. "Integrated Services Digital Network (ISDN) Q.921-User Adaptation Layer", January 2006.
73. Q.921. "ISDN User-Network Interface – Data Link Layer Specification", September 1997.

5

Evolution of Mobile Networks and Wireless LANs

5.1 Introduction

Service Providers are currently facing a very competitive market with ever-changing traffic demands and frequent variations in the requirements for capacity, which must be met rapidly.

Most multimedia services are presently operated over separate voice and data networks, employing network overlay architectures. BT's current access network is a typical example in that it is not a single network, but a range of different networks overlaid on top of each other. The networks have different but sometimes overlapping service capabilities, which leads to greater complexity and increased operating costs [1].

There is a need to decrease operational expenditures by evolving legacy services towards a common architecture, which is also applicable to new services and can be deployed on a single network. For this reason, although there are many different new, as well as legacy, network access technologies, network cores increasingly appear similar; that is they are converging, and most industries are looking at the IP Multimedia core network Subsystem (IMS) model as their preferred future architecture.

The Service Provider's ability to survive and grow in the market rests on being able to quickly and easily implement new IP-based services onto the network – this is currently the Service Provider's most important objective, and the IMS is designed with exactly this in mind. The essential prerequisites for the IMS architecture are the ability to do the following:

- Support the launch of new IP multimedia services easily without compromising the quality of existing services
- Provide a common applications environment
- Offer unified call control
- Separate control from delivery
- Impose Quality of Service (QoS)
- Provide a framework for improved traffic management techniques

Converged Multimedia Networks   Juliet Bates, Chris Gallon, Matthew Bocci, Stuart Walker and Tom Taylor
© 2006 John Wiley & Sons, Ltd

- Allow users to access services from different locations over different types of access network.

The IMS model was first introduced by the Third Generation Partnership Project[1] (3GPP) in Release 5 [2], and further enhanced in 3GPP Release 6 [3]. Traditionally, 3GPP develops standards for mobile voice and data networks[2], and the term "3G" generally means a 3GPP system using a Radio Access Network (RAN), but 3GPP's IMS architecture is unique because it caters for any type of access network, not just a RAN. The IMS is therefore seen as the key enabler for interoperability between future fixed and mobile networks and the Internet, and the future promise of seamless converged services.

The IMS architecture allows services to be made up from multiple media components, which can be added, or dropped, during the same session, and all under the control of a single signalling session. Separate IP flows are used to carry the data from each media component and each of these flows can have different QoS characteristics. When setting up a connection, the IMS makes it possible for a network to disassociate between the negotiation of sessions, and the flows within a single session, and also to disassociate between the bearers, which are needed to support the flows. The architectural rules relating to the separation of the bearer level, the session control level and the service level, in an end-to-end solution, are expressed in 3GPP TS 23.221 [4]. At the same time, the IMS also provides a framework, which will allow policies to be applied to enable a network to handle resources more efficiently.

3GPP adopted a layered approach in designing the IMS, which ensures that the IMS is access independent, so that IMS services can be provided over any IP-based network, and minimum dependence between the services permits the easy addition of new types of access. This is central in the move towards convergence, and now each access type is being "enabled" to interwork with an IMS core, be it Digital Subscriber Line (DSL), Wireless LAN (WLAN), the General Packet Radio Service (GPRS) (used in mobile access) or an emerging technology, such as WiMAX.

3GPP's solution for the support of IP multimedia applications consists of mobile terminals, GERAN (GSM or EDGE features – see Section 5.6), or Universal Terrestrial Radio Access Network (UTRAN) radio access networks, the GPRS-evolved core network, as well as the IMS. The next sections of this chapter are intended to show the basic components of these networks and illustrate how they will be combined with the IMS.

This chapter begins with an explanation of the development of mobile network architectures. The mobile subscriber's handset joins the mobile core network via an air interface, which has traditionally offered a comparatively low bandwidth of 9.6 kbps (in GSM the Global System for Mobile Communications). In recent years, new methods of modulation and coding have been developed which have helped mobile operators to provide more bandwidth in the RAN to support multimedia services.

An essential aspect of a mobile RAN is the ability to "hand over" a subscriber from one area of the network (cell) to another. New broadband data wireless technologies do not currently provide "handover", but do allow for "hot spot" access, that is, radio coverage of a specific area. These technologies can be considered to be "enablers" for convergence, because they can also be used to bridge the gap between fixed and mobile

---

[1] Information on 3GPP Releases 99 (3), 4, 5, 6 and beyond can be found at http://www.3gpp.org/specs.
[2] 3GPP specifications include GSM, GPRS and EDGE and W-CDMA, UTRAN and UMTS.

networks, but they are not truly in the same category as mobile networks, because they can only provide services to the subscriber in a specific limited area. The range of these technologies, such as Worldwide Interoperability for Microwave Access (WiMAX), is being extended to wider areas and can offer more bandwidth than even the 3G RAN of the mobile networks, while providing a roaming user with an alternative way to access network services in public areas. Broadband wireless is therefore a hugely significant player in the race towards fixed and mobile convergence, because as well as enhancing broadband access to voice and data, over both global mobile networks (through greater bandwidth) and fixed networks (extending range), it can also, uniquely, provide a point of "handover" between the existing fixed (e.g. Public Switched Telephone Network (PSTN)) and mobile (e.g. 3GPP) networks, and this will enable network operators to start the process of merging the core of these giant infrastructures to provide converged multimedia networks.

5.2 1G and 2G Mobile Networks

Mobile voice and data networks have evolved gradually to cater for the basic transport of data, as well as voice, over the limited bandwidth of the air interface between a mobile handset and a radio station. The first-generation analogue mobile networks (1G) began in the 1980s. The second-generation (2G) digital Global System for Mobile Communications (GSM) has existed since 1991, when the world's first public GSM call was made (in Finland), and GSM is presently the most widely used standard for mobile radio networks.

There are four different cell sizes in a GSM network – macro, micro, pico and umbrella cells – and the coverage area of each cell is different (with pico cells being the smallest). The cell radius varies depending on antenna height, antenna gain and propagation conditions, from a couple of hundred meters to several tens of kilometres. The longest distance that the GSM specification supports in practical use is 35 km, or 22 mi. (There is also a concept of an extended cell, where the cell radius could be more than doubled.) One of the key features of GSM is the Subscriber Identity Module (SIM) card which contains the user's subscription information and phone book.

5.3 Development of 3G

Development of a Third Generation mobile network (3G) began in 1998, when the European Telecommunications Standards Institute (ETSI) put forward a proposal that a new 3GPP should be formed to focus on a GSM technology. As this was a European Initiative, a parallel Partnership Project 3GPP2 was also established by the American National Standards community (ANSI-41).[3] In Europe and Japan, 3G is known as the Universal Mobile Telecommunications System (UMTS).

3GPP initially decided to prepare specifications every year, with Release 99 being the first set. Release 99 laid the foundations for future high-speed traffic transfer in both Circuit Switched (CS) and Packet Switched (PS) modes. The next release, 3GPP R00 was subsequently divided into 3GPP R4 and 3GPP R5.

---

[3] Although discussions did take place between ETSI and the ANSI-41 community with a view to consolidating collaboration efforts for all ITU "family members," it was deemed appropriate that a parallel Partnership Project be established – "3GPP2" which was based on the International Telecommunications Union (ITU's) International Mobile Telecommunications initiative (IMT-2000).

5.4 Release 99 UMTS Architecture

Currently, most cellular mobile networks are still 2G or 2.5G networks, that is, they are not truly 3G because they have a limited multimedia capability and are broadly equivalent to Release 99 (or earlier) of the 3GPP UMTS architecture, as illustrated in Figure 5.1. 2G GSM provides voice access in the Circuit Switched (CS) Domain and GPRS (2.5G) provides data transport over the Packet Switched (PS) Domain.

As illustrated in Figure 5.1, the architecture can be simply described as being divided into three major parts:

- The air interface (from the mobile handset)
- The RAN
- The core network.

A mobile handset communicates over the air interface with a radio network, which is provided by a Base Station Subsystem (BSS). In the BSS, a Base Station Controller (BSC) controls a Base Transceiver Station (BTS). The core network from a functional point of view is divided into a PS domain and a CS domain. Legacy 2G GSM voice is carried over the "A" interface to a Mobile Switching Centre (MSC) in the CS domain. IP packets are transported over the Gb PS interface by GPRS. A Home Location Register (HLR) stores information about mobile subscribers and the services they are entitled to, as well as details about whether they are post-payers or use prepaid services.

GSM networks have traditionally used FDD/TDMA (Frequency Division Duplex/Time Division Multiple Access) systems, that is, frequency is used to separate uplink and downlink traffic, and time division is used to separate mobiles using the same frequency. In a standard GSM network, a user is assigned one timeslot for the downlink and one for the uplink and this symmetric pipe originally provided 9.6 kbps. Further optimisation of the channel coding – a method called High Speed Circuit Switched Data (HSCSD) can produce rates of 14.4 kbps (under the best conditions) through a single timeslot which would normally only carry 9.6 kbps.

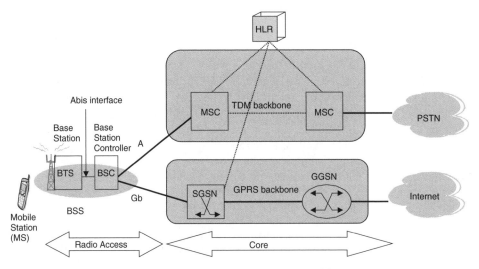

Figure 5.1   3GPP UMTS Release 99

A further innovation of HSCSD is the ability to use up to four timeslots at the same time, for a single user, to provide a maximum throughput of up to 57.6 kbps.

However, the symmetric CS GSM interface is not the best medium for data transfer which is typically asymmetric, that is, in the transfer of packetised data, the data rates used from the user to the network (the uplink) are usually much lower than the rates in the reverse direction on the downlink (from the network to the user).

5.5 General Packet Radio Service (GPRS)

In 2001, the General Packet Radio Service (GPRS) was added to an extension to GSM to provide an increase in data rates. GPRS uses unused channels in the GSM network allowing for faster browsing, web surfing and e-mail through an increased throughput and allowing connections to be asymmetric when required. Timeslots are shared between the CS domain and the PS domain of the network. The proportion and division of timeslots is controlled by the operator and can be static or dynamic. The speed of GPRS is dependent on the number of TDM timeslots available, and generally the connection speed reduces logarithmically as the distance from the base station increases.

The theoretical maximum throughput of GPRS is 160 kbps (using eight timeslots and the highest throughput per slot of 20 kbps) but realistically the bit rate on the downlink is a maximum of 80 kbps because at most four timeslots can be used.

In the CS Domain, a connection reserves a circuit for the lifetime of that connection. GPRS is packet switched which means that multiple users share the same transmission channel, transmitting only when they have data to send. Whereas circuit switched users are billed per second, GPRS data users are typically billed per kilobyte (kB).

Circuit switched data is usually given priority over packet switched data and there may be relatively few seconds when no data are being transferred. HSCSD may be preferred over GPRS for reliability for downloading, and operators have generally priced GPRS relatively cheaper compared to HSCSD.

GPRS is supported by two enhanced routers in the core network: a Serving GPRS Support Node (SGSN) and a Gateway GPRS Support Node (GGSN). The SGSN communicates with the GGSN, which is an IP border router and Gateway (GW) operating on a uni-directional flow of packets, that is, in either an upstream or downstream direction. The gate consists of a packet classifier and a gate status (open/closed). When a gate is closed, all the packets in the flow are dropped. When a gate is open, the packets in a flow are accepted and can be classified. The GGSN checks the source or the destination address in the basic IP header of incoming or outgoing packets, against a set of packet-filtering information established during the set-up of the GPRS/UMTS session.

The mobile network must keep track of where a mobile subscriber is located. When the subscriber is in the home network, the location of the mobile handset is stored in a location register, which is handled by the HLR. A mobile handset in the home domain identifies itself in a GPRS attach using an IMSI (International Mobile Subscriber Identity) which is stored in the SIM. The IMSI consists of a three-digit Mobile Country Code (MCC), a two- or three-digit Mobile Network Code (MNC) and a Mobile Subscriber Identity Number (MSIN) so that the total IMSI does not exceed 15 digits. The HLR authenticates services to both the CS and PS domains, through the implementation of the Authentication Centre (AuC) function, which carries out the IMSI authentication. Additionally, the HLR manages some of the mobility aspects of a subscriber, by providing routing information for calls to the subscriber, for example, when a call can be successfully connected or when the call should be directed to voicemail.

A roaming mobile handset in a visited domain identifies itself using a TMSI (Temporary Mobile Subscriber Identity) if in the CS domain, or a P-TMSI (Packet-TMSI) in the PS domain. A (P-)TMSI is a 32-bit local identifier stored either by the Visitor Location Register (VLR), or by the SGSN to which the mobile station is temporarily attached. A location register function in the SGSN also stores subscription information and location information for each registered subscriber.

The GGSN stores information about every subscriber (whether home based or roaming) for which it has at least one active Packet Data Protocol (PDP) context. A location register function in the GGSN stores subscription information and routing information, which are needed to tunnel packet data traffic to the SGSN. The GGSN also has methods of allocating IP addresses and protects the integrity of the PS core network through an externally facing firewall. Optionally, the GGSN can apply Network Address Translation (NAT) and also implement some billing functions.

5.6 Enhanced Data Rates for GSM Evolution (EDGE)

EDGE (Enhanced Data Rates for GSM Evolution) further increases the transmission rate, in both the CS and PS domains of GSM networks, allowing operators to deliver high-speed mobile Internet and multimedia service. A mobile BTS can be upgraded to apply a modulation technique, called *Octagonal Phase Shift Keying* (*OPSK*), which can treble the data throughput of GSM and GRPS. Using standard GSM timeslots, one air interface symbol carries a combination of three information bits. In EDGE Phase 1, the rate in the PS domain can be increased to up to 384 kbps.[4] (EDGE Phase 2 provides guidelines on how this service can also be achieved in the GSM CS domain.) In the 3GPP organisation, the term GSM now implies a 3GPP network using a GERAN RAN, that is, including both GPRS and EDGE features.

5.7 Release 4 UMTS Architecture

Figure 5.2 illustrates Release 4, which followed Release 99, showing how a split was introduced, separating the function of the Media Gateway (MGW) from the function of an MSC Call Agent. In the same release, the concept of an "All IP" network was initially introduced. This concept was extended in 3GPP Release 5 with the standardisation of the IP Multimedia core network Subsystem (IMS). Release 5 specifies the basics of IMS but the subsequent Release 6 provided a more complete solution with the enabling of service interworking with CS and other IP networks. At the time of writing, 3GPP Release 7 is being defined.

In 3GPP's Release 4, as illustrated in Figure 5.2, the Release 99 Mobile Radio Base Station has become a Node B, a transmitter, which controls a single cell, that is, one area of radio coverage. Typically, a short link (possibly microwave) will link several Node B's into an exchange where leased lines are used to connect the Node B's to Radio Network Controllers (RNCs) in regional centres. Within the UTRAN, all transmission

---

[4] Each channel (or timeslot) delivers 48 kbps (instead of 9.6 kbps as in GSM) and each GPRS terminal gets eight channels, which collectively provide up to 384 kbps.

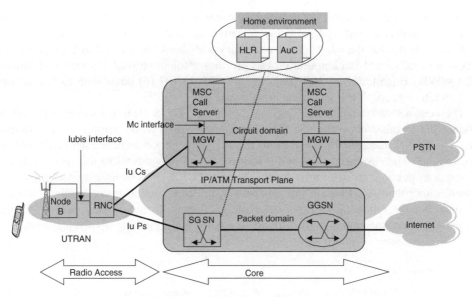

Figure 5.2   3GPP UMTS Release 4

takes place over an ATM-switched network and an RNC is an ATM switch, controlling one or a number of Node B's. The RNC supports services for CS connections, over the Iu CS interface, and PS connections, over the Iu PS interface and multiplexes/de-multiplexes packets and circuits together. Traditional GSM voice is transported from the RNC to the Mobile Switching Server/MGW in the CS domain, using ATM Adaptation Layer 2 (AAL2) over the Iu CS interface. IP Packets are carried to the SGSN over the Iu PS interface using ATM Adaptation Layer 5 (AAL5). AAL5 is also used to carry signalling. This book does not delve into further detail on the protocol stacks involved but further information on these and recommended optimum settings for the traffic descriptors, to meet a targeted Cell Loss Ratio (CLR), in the presence of compression and silence detection, can be found in Bates [5].

The Radio Link Control (RLC) protocol runs over the UTRAN, which stretches from the Node B to the RNC. The UTRAN's main job is to set up Radio Access Bearers (RAB) for communication between the User Equipment (UE) and the Core Network (CN). The UMTS protocol stack of a mobile handset consists of a PHY (Physical) Layer, a MAC (Medium Access Control) layer and a RLC layer, which implements an ARQ (Automatic Repeat Request) mechanism for ensuring reliable data transmission.

There are four different QoS classes for UMTS:

- Conversational class
- Streaming class
- Interactive class
- Background class.

The main distinguishing factor between these QoS classes is how delay sensitive the traffic is: Conversational class is meant for traffic which is very delay sensitive while Background class is the most delay-insensitive traffic class. Conversational and Streaming classes are mainly intended to be used to carry real-time traffic flows and the main difference between them is how delay sensitive the traffic is. Real-time services, like video telephony, are the most sensitive to delay and should be carried in Conversational class. Interactive and Background classes are meant for traditional Internet applications like WWW, E-mail, Telnet, FTP and News. 3GPP 23.107 [6] deals with QoS concepts and architecture for 3GPP Releases 99, 4, 5 and 6.

The core network implements QoS on top of the UMTS Bearer Service (BS). When connecting to other networks, the QoS requirements need to be mapped to an external BS. The GGSN manages the connections towards other packet switched networks such as the Internet. The current goal of 3GPP Release 5, and onwards, is to use IP networking as much as possible and all traffic coming from the UTRAN is eventually going to be IP based, rather than ATM based. For this reason, the next section looks briefly at the CS Domain but the following sections cover the PS domain in more detail, setting out the essential components and also showing the protocols and signalling involved in transport.

5.7.1 Circuit Switched Domain

In the Circuit Switched (CS) domain of a Release 4 core network, Mobile Switching Centre Call Agents (MSC-S) and Media Gateways (MGWs), perform two roles.

When traditional voice call and data session set-up is required in the CS domain, the MGW receives RANAP (Radio Access Network Application Part) signalling from the UTRAN and tunnels this directly to the MSC Call Agent over the Mc interface.

The MSC Call Agent also uses the Mc interface, connecting it to its pairing MGW, to control the bearer in the core network using the H.248 protocol. H.248 uses contexts to describe the topology of the connections within a MGW, that is, who hears/sees whom, and the media mixing and/or switching parameters, if more than two media streams are involved in the association. (BICC [7] protocols are also specified for the Mc interface, in addition to further packages defined in 3GPP TS 29.232 [8].)

The architectural principles here are exactly the same as the relationship between a Call Agent and a MGW in a fixed network, as explained in more detail in Chapter 4, on the Next Generation Network (NGN) and the PSTN. In a fixed network, the call control element, the Media Gateway Controller (MGC) which controls MGWs can be a separate component, or can be provided with a Call Agent/Softswitch. H.248 operates between the Call Agent and the MGW and allows for the possibility of dividing up a single physical MGW into multiple virtual MGWs, each controlled by a different MGC.

In the mobile network, the MSC-S and MGW can also provide a point of access to, and from, the PSTN. In this case, an MSC-S (and its pairing MGW) must interwork with ISDN User Part (ISUP) signalling which is used in the PSTN. The MGW may also need to translate in the bearer plane, between PSTN codecs (e.g. G.711) and mobile codecs (e.g. Adaptive Multi-Rate (AMR)). When in this role, the names of the MSC and MGW are often appended with a "G" to indicate that they act as a Gateway to the PSTN, that is, GMSC-S and GMGW.

When a call to a fixed-line (black phone) ends, the MSC and the GMSC, involved at the point of access to the PSTN, produce Charging Data Records (CDRs) with a record of the call between the called and calling party, and the CDRs are forwarded to a billing server for processing.

5.7.2 Packet Switched Domain

In the Packet Switched (PS) core, the SGSN performs encryption and compression of the media transmitted, on the uplink, and the routing of IP Packets, on the downlink, to the appropriate RNC in the access network. The SGSN also performs mobility and data session management and keeps track of the location of an individual mobile handset and performs some security functions including access control.

When the user wishes to send IP packets, the user activates a PDP context by signalling to the SGSN (Serving Gateway Support Node) and the GGSN to set up the context for a packet transfer session. The PDP context defines a logical association that begins at the mobile handset and details many network aspects such as Routing, QoS, Security, Billing, and so on. When the PDP context is active, the mobile handset acquires an IP address and then IP data packets are carried over the air interface, between the mobile handset and the RNC, by the Packet Data Convergence Protocol (PDCP).

From the RNC to the SGSN, IP packets are tunnelled over the Iu PS interface, within GTP-U (GPRS Tunnelling Protocol-User Plane) tunnels. GPRS uses the PDP encapsulated within GTP-C (GPRS Tunnel Protocol-Control Plane) to establish contexts with different QoS requirements. The GTP tunnels[5] between the RNC and the SGSN are set up by the RANAP protocol while those between the SGSN and GGSN are established by GTP-C signalling (see Figure 5.3).

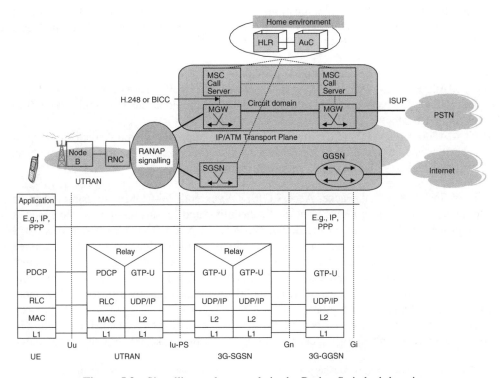

Figure 5.3   Signalling and protocols in the Packet Switched domain

---

[5] Note: GTP tunnelling incurs large overheads (IP header 20 bytes + UDP header 8 bytes + GTP header 20 bytes).

A tunnel identifier in a GPRS Tunnelling Protocol (GTP) header field identifies the packets as belonging to a particular PDP context, and packets associated with this tunnel achieve a certain level of QoS. QoS is negotiated at context initiation with the subscriber application requesting the QoS required, and this being either being accepted or adjusted by the GGSN, based on the network-held subscriber attributes and available network resources.

3GPP has also defined the use of DiffServ[6] to differentiate IP packets in the Iu PS transport network, where an RNC can map the DiffServ TOS (Type of Service) bit to the selected transport bearer requirements.

5.8 Wideband Code Division Multiple Access (W-CDMA)

In 1998, the ETSI Special Mobile Group selected Wideband Code Division Multiple Access (W-CDMA) and Time Division CDMA (TD-CDMA) as the best technologies for a UMTS network, partly to align with Japanese standardisation.

Time Division Synchronous Code Division Multiple Access (TD-SCDMA) is a new 3G standard used in China, employing time division, in contrast to the frequency division used by W-CDMA. The number of timeslots used for the downlink and uplink can be dynamically adjusted to carry asymmetric traffic. The standard has been adopted in 3GPP Release 4 and is known as the ULTRA TDD 1.28 Mbps option.

The International Telecommunication Union (ITU) has selected another type of CDMA system called *CDMA2000*. CDMA2000 is standardised by 3GPP2[7] and includes: CDMA2000 1x, CDMA2000 1xEV-DO and CDMA2000 1xEV-DV.

- CDMA2000 1x is the core standard providing a peak rate up to 144 kbps.
- CDMA2000 1xEV-DO (Evolution-data optimised) is a 3G standard supporting down-link (Handset to Network) data rates of up to 3.1 Mbps and uplink (Network to Handset) of up to 1.8 Mbps.
- CDMA2000 1xEV-DV (1xEvolution-Data/Voice) supports downlink data rates up to 3.1 Mbps and uplink of up to 1.8 Mbps as well as concurrent operation of legacy voice, data and high-speed data users, within the same channel.

High Speed Downlink Packet Access (HSDPA) extends W-CDMA by using Adaptive Modulation and Coding (AMC) and fast scheduling and retransmissions at the Node B to improve the throughput on the downlink. When a Node B decides which users will receive data in the next frame, it also decides which channelisation code(s) will be applied for each user, and data can be sent to a number of users simultaneously using different channelisation codes. (This differs from CDMA2000 1xEV-DO where data is sent to only one user at a time.) HSDPA can increase the available data rate by a factor of 5 or more to typically 2 Mbps or even more.

High Speed Uplink Packet Access (HSUPA) is a data access protocol for mobile networks, with very high upload speeds of up to 5.76 Mbps. Like HSDPA, it is considered

---

[6] The Differentiated Services Architecture is based on a simple model where traffic entering the network is classified and possibly conditioned at the boundaries of the network and assigned to different behaviour aggregates. Each behaviour aggregate is identified by a single Diffserv codepoint.

[7] 3GPP2 have created a Multimedia Domain (MMD) solution, which uses IMS as a base, to allow CDMA2000-based access networks to provide 3G mobile services.

to be 3.75G. The specifications for HSUPA are included in UMTS Release 6. A packet scheduler deals with requests to send data from mobile handsets and decides when and how many handsets will be permitted to do so.

5.9 Introduction to IMS

The Internet Multimedia core network Subsystem (IMS) comprises of all the essential core networks elements that are needed for the provision of multimedia services. This includes both the signalling and the bearer (media transport) related network elements. The elements of the IMS, are functions and more than one function can be combined in the same node (hardware box). Also, within a network, there may be several instances of the same function, for redundancy, load balancing, and so on.

IMS users connect to an IMS network, using Session Initiation Protocol (SIP) over the Internet Protocol (IP). Legacy systems, such as POTS (Plain Old Telephone Service) are connected to the IMS through gateways.

Within the IMS model, there is a Call Session Control Function (CSCF), which controls a single session between itself and the User Equipment (UE) in the access network, over the PS domain. Routing of SIP signalling uses SIP URIs like *sip:amy.doe@example.com* within the IMS (or other non-SIP Absolute URIs when the IMS user is communicating with external networks). A few of these PS services now provide an equivalent to some of the services, which have previously only been offered in the CS domain.

To ensure the convergence of, and access to, voice, video, messaging, data and web-based technologies for wireless or mobile users and combine the growth of the Internet as well as maintaining continuing interoperability with fixed networks, the IMS conforms to IETF (Internet Engineering Task Force) "Internet Standards" and the Session Initiation Protocol (SIP), defined in RFC 3261, is central to the operation of the IMS model.

A separate Media Gateway Control Function (MGCF) controls the GW to the PSTN. A Media Resource Function (MRF) applies the bearer resource when it is invoked. A

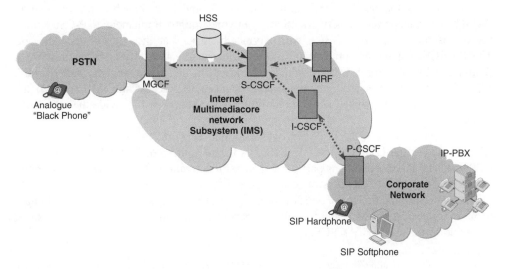

Figure 5.4 The IMS essential core network functions

Signalling Gateway (SGW) translates between SIP signalling, which is being used in the IMS, and external signalling, for example, ISUP signalling[8] which is used in the PSTN.

It is important to note that the IMS does not provide complete service equivalence to the mobile network's CS domain. The IMS is independent of the CS domain, although some network elements may be common with the CS domain. This means that it is not necessary to deploy a CS domain in order to support an IMS based network [9]. The "All IP" network concept does not consider the evolution of the CS domain but recognises that it will still be necessary to interwork with the CS domain for some time to come [10]. The migration towards an IMS-only network, completely replacing a mobile circuit switched core network, can currently only be seen as a long-term vision. As the "All IP" network becomes more and more widespread, the conditions for phasing out the CS domain may eventually be achieved, and this would lead to a further simplification of the core network architecture as well as a reduction in OPEX costs. The evolution of an All IP Network also envisages the PS domain, including the IMS, working together with many other access areas, for example, WLAN.

3GPP specifications have designed the user interface to exclusively support IPv6 for the connection to the IMS, although some early IMS implementations are actually based on IPv4. For the future, 3GPP TS 23.221 [4] compromises by specifying that the IMS architecture should "make optimum use of IPv6", but also gives advice on how users attached to IPv4 SIP-based networks, can communicate with an IPv6 IMS, by configuring tunnels. 3GPP provides guidelines on interworking with an IPv4-based IMS in TR 23.981 [11].

5.9.1 The Proxy Call Session Control Function (P-CSCF)

The CSCF, in the initial simple IMS model, has developed into three separate roles: The Proxy Call Session Control Function (P-CSCF), the Interrogating CSCF (I-CSCF) and the Serving Call Session Control Function (S-CSCF).

A P-CSCF is the user's entry point to the IMS, and there can be one, or several, P-CSCFs in an operator's network. Some networks might implement a P-CSCF in a Session Border Controller (SBC), which typically resides at the edge of the network. The P-CSCF is a server application, and its primary role is to send SIP requests to a destination P-CSCF (or S-CSCF), which is "closer" to the destination or "called user". When the subscriber is roaming, the mobile handset will access the IMS through the nearest P-CSCF in the visited network, and the home network will provide the other IMS functions.

If IMS services are to be provided over a mobile network, using GPRS access, over GERAN and/or the UTRAN, the UE discovers the IP address of its P-CSCF following a GPRS attach, either after, or as part of, a successful activation of a PDP context, through the SGSN to the GGSN as illustrated in Figure 5.5.

The DHCP (Dynamic Host Control Protocol) can also provide the user with the domain name of its nearest P-CSCF and the address of a Domain Name Server (DNS) to resolve the name of the selected P-CSCF.

---

[8] ISUP signalling has different variants for example, ETSI ISUP is used in Europe. UK-ISUP is used in the United Kingdom (as mandated by Ofcom).

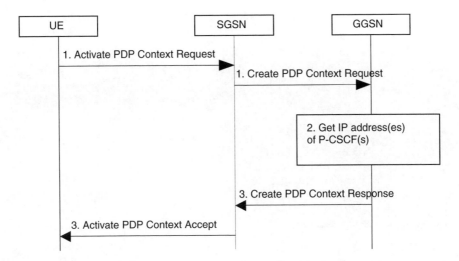

Figure 5.5 Acquiring the IP address of the P-CSCF in GPRS access

The P-CSCF registers with an I-CSCF by forwarding a SIP register request received from the UE to the I-CSCF determined using the home domain name, which has been provided by the UE.

As part of the registration process the I-CSCF feeds back the name of the Serving CSCF (S-CSCF) to the P-CSCF, which then forwards further SIP messages received from the UE to the S-CSCF.

The P-CSCF is also responsible for the generation of Call Data Records (CDR), which can be used for billing purposes. To save bandwidth, the P-CSCF should perform some SIP message compression and decompression and should also implement security between itself and the user.

Figure 5.6 shows how the user communicates with the nearest P-CSCF over the Gm interface, and the P-CSCF communicates with the S-CSCF and the I-CSCF over the Mw interface.

The P-CSCF is central to the working of the IMS and so before moving on, we will summarise the functions of the P-CSCF, which can be either in the home network or in a visited network:

The functions performed by the P-CSCF [3] are as follows:

- Forwarding the SIP register request received from the UE to an I-CSCF determined using the home domain name, as provided by the user.
- Forwarding the SIP messages received from the UE to the S-CSCF in the home network, whose name the P-CSCF has received as a result of the registration process with the I-CSCF.
- Forwarding a SIP request or response to the user.
- Detecting and handling an emergency session establishment request as per error-handling procedures.
- Generating CDRs.
- Maintaining a Security Association between itself and each UE (3GPP TS 33.203).

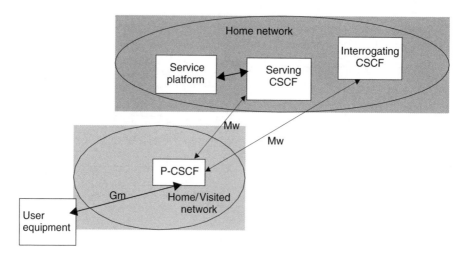

Figure 5.6   The P-CSCF, S-CSCF and I-CSCF

- Performing SIP message compression/decompression.
- Authorising bearer resources and QoS management. (3GPP TS 23.207 [9]).[9]

5.9.2 The Interrogating Call Session Control Function (I-CSCF)

When a user first registers with the P-CSCF to use IMS services, the P-CSCF (in the home or visited network) forwards the initial SIP register requests received from the user to an I-CSCF in the home domain, using the home domain name provided by the user and other information about where the user is located.

The main role of an I-CSCF is to assign an appropriate S-CSCF to the user. The I-CSCF needs some information on the location, capabilities and availabilities of each S-CSCF to help it to nominate an appropriate S-CSCF for the IMS subscriber. The I-CSCF determines which S-CSCF the user should register with by querying the Home Subscriber Server (HSS), over the Cx interface[10] using the Diameter protocol (See Table 5.1).

The HSS (which is similar to the HLR) contains information about the user's profile and entitlement to services. When there are multiple HSSs, the I-CSCF and the S-CSCF query a Subscription Locator Function (SLF) to find the appropriate HSS, which will provide information about the capabilities the user will need to have for services, as well as other information about the preferred operator for that user. Each subscribed IMS user can have multiple service profiles defined in the HSS. The HSS checks that the user is allowed to register in the originating network and returns an S-CSCF name and capability, if this is the case. The HSS then provides the I-SCSF and the S-CSCF with specific information about the user's subscription data in terms of the user's location and authentication data.

---

[9] 3GPP[TM] TSs and TRs are the property of ARIB, ATIS, ETSI, CCSA, TTA and TTC who jointly own the copyright in them. They are subject to further modifications and are therefore provided to you "as is" for information purposes only. Further use is strictly prohibited.

[10] 3GPP has defined the Cx interface application of Diameter in 3GPP 29.228.

5.9.3 The Serving Call Session Control Function (S-CSCF)

The previous sections described how the S-CSCF provides initial registration services to the user. Once the I-CSCF has selected an S-CSCF and provided its name to the P-CSCF, the P-CSCF and the allocated S-CSCF store each other's names for future use. The I-CSCF function can then normally be removed from the signalling path.[11]

The principal job of the S-CSCF is to provide the IMS user with access to the full service platform. The IMS architecture is based on the principle that service control for Home-subscribed services (even for a roaming subscriber) is always based in the Home network, and this where a S-CSCF is always located [3] even though the IMS Service Platform could be located in a visited network. For example, in Figure 5.6, the service platform was shown as being based in the Home Network but it could also be located in an External Network.

Figure 5.7 illustrates the topology of mobile access to IMS, and Figure 5.8 shows the process of registration of an IMS subscriber. An IMS call or media session between two users may involve two S-CSCFs, that is, one S-CSCF in the home network of the user at each end of the call or the session. When a user is using a phone number as the dialled address, the phone number in the E.164 format can be converted into a SIP URI by the S-CSCF which translates the E.164 number using an ENUM DNS translation mechanism (as specified in IETF RFC 376). If this translation fails, then depending on the configuration selected by the network operator, either the session will be routed to the PSTN or an appropriate notification will be sent to the caller.

An S-CSCF also provides resources to Application Servers (AS). For example, a typical Application Server might organise the management of information for multimedia conference booking, such as the start time, the duration, the list of the participants, and so on.

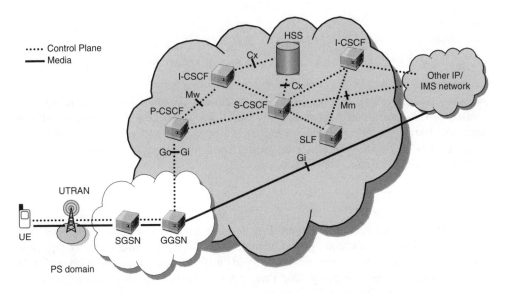

Figure 5.7   Example UTRAN access using IMS CSCFs

---

[11] The exception to this is if the THIG (Topology Hiding Inter-network Gateway) function of the I-CSCF is being used.

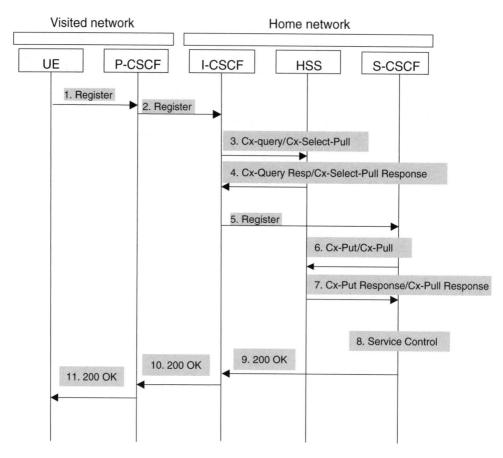

Figure 5.8 IMS Registration of a Subscriber

Hosting and executing services on an AS, which can be located in the home network or in an external third-party network, allows the easy addition and deployment of new value-added services. The Application Server can operate in SIP Proxy mode, SIP User Agent (UA) mode or SIP B2B (Back-to-Back) User Agent mode. (See Chapter 2, on Call Control, for further explanation of these roles.)

5.9.4 IMS Subscriber Identities

Every IMS subscriber has one or more Private User Identities stored by the HSS. The Private User Identity is a unique global identity defined by the Home Network Operator, which is used within the home network, to identify the user's subscription and IMS service entitlement and capability. The Private User Identity identifies the subscription, not the user, and it is used for registration, authorisation and accounting. The S-CSCF needs to obtain and store the Private User Identity and upon registration, the relevant service profile for the IMS subscriber is downloaded from the HSS to the S-CSCF [3]. An IMS subscriber can also have one, or several, Public User Identities, which must all

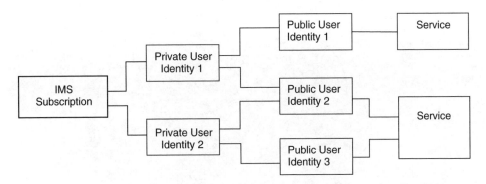

Figure 5.9 IMS Private and Public Subscriber Identities and Service Profiles [3][13]

be registered at a single S-CSCF, at one time. Each Public User Identity in the S-CSCF can also only be associated with a single service profile at any one time.[12]

The Private User Identity might represent a telephone and the Public User Identities, the people who share its use. Amy (Public User Identity 2) and Eric (Public User Identity 3) have a different service profile to that of their teenage daughter (Public User Identity 1), because their service profile is set up with the additional permission to make international calls. A Public User Identity may be simultaneously registered from multiple user locations that use different Private User Identities and different contact addresses as illustrated in Figure 5.9.

It is time to place the three CSCFs in the context of the IMS architecture. The IMS is a separate domain with its own services. Inter-domain roaming, to another domain or to the PSTN CS Network, allows for user services, which are not available in this IMS domain, to be provided from elsewhere. To enable this, the three CSCFs (P-CSCF, I-CSCF and S-CSCF) are supported by several other new components as illustrated in Figure 5.10. Note that Figure 5.10 does not show the Subscription Locator Function (SLF).

5.9.5 The Breakout Gateway Control Function (BGCF)

The Breakout Gateway Control Function (BGCF) is used when a call or a media session needs to connect to the PSTN. The BGCF selects the network in which the PSTN domain breakout is to occur. If breakout is to occur in the same network as the one in which the BGCF is located, the BGCF will forward the SIP signalling to the MGCF within the network. If the breakout takes place in another network, the BGCF forwards the SIP signalling, via the I-CSCF, towards the BGCF of the other network.

A MGW and a SGW (or a combination of the two) provide a point of connect to the PSTN. The MGW is controlled (opened and closed) by the MGCF over an H.248 interface. The MGCF receives ISUP signalling from the PSTN and converts it into SIP signalling for transport to the UE at the edge of the network.

---

[12] The Public User Identity/Identities take the form of a SIP URI, as defined in RFC 3261 and RFC 2396 or for the "tel:"-URI format see RFC 3966.

[13] 3GPP™ TSs and TRs are the property of ARIB, ATIS, ETSI, CCSA, TTA and TTC who jointly own the copyright in them. They are subject to further modifications and are therefore provided to you "as is" for information purposes only. Further use is strictly prohibited.

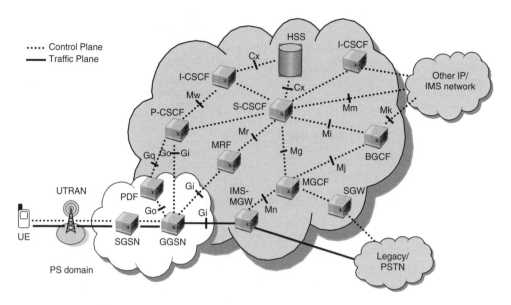

Figure 5.10 UTRAN access to the Internet and PSTN via IMS

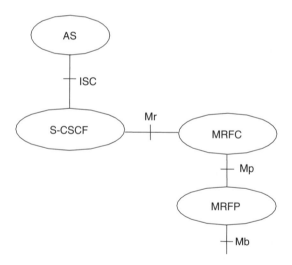

Figure 5.11 Architecture of the MRF [3][14]

5.9.6 The Media Resource Function (MRF)

In Stage 2 of the IMS architecture, the MRF, see Figure 5.11, is divided into a Multimedia Resource Function Processor (MRFP), which is controlled by a Multimedia Resource Function Controller (MRFC), as illustrated in Figure 5.12. The MRFP is controlled by

---

[14] 3GPP™ TSs and TRs are the property of ARIB, ATIS, ETSI, CCSA, TTA and TTC who jointly own the copyright in them. They are subject to further modifications and are therefore provided to you "as is" for information purposes only. Further use is strictly prohibited.

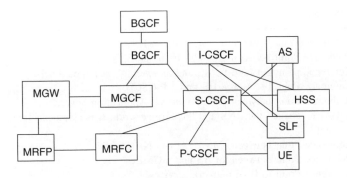

Figure 5.12   IMS Stage 2

the MRFC using H.248. The main job of the MRFP is to control a bearer. The S-CSCF has access to the MRFC and MRFP functions using SIP (as defined by RFC 3261).

The tasks of the MRFC are the following [3]:

- Controlling the media stream resources in the MRFP.
- Interpreting information coming from an Application Server and the S-CSCF (e.g. the Session Identifier) and control of the MRFP accordingly.
- Generation of CDR for example for billing.

The tasks of the MRFP include the following [3]:

- Control of the bearer on the Mb reference point.
- Providing resources to be controlled by the MRFC.
- Mixing of incoming media streams (e.g. for multiple parties).
- Media stream source (e.g. for multimedia announcements).
- Media stream processing (e.g. audio transcoding, media analysis).
- Floor Control (i.e. management of access rights to shared resources in a conferencing environment).

Figure 5.12 illustrates the components of the IMS and their interrelationships in Stage 2. The CSCF, BGCF and MGCF nodes are identified by a SIP URI (Host Domain Name or Network Address) in the header field of SIP messages on the interfaces supporting the SIP protocol, (e.g. Gm, Mw and Mg). Table 5.1 lists the notation of the main IMS interfaces, some of which were shown in Figure 5.10.

5.9.6.1 IMS Forking [3]

The IMS has the capability to fork requests to multiple destinations; that is, to choose to forward a SIP INVITE to multiple locations in parallel. This capability is subject to rules for forking proxies defined in RFC 3261 [12] and 3GPP TS 23.228 [3] describes how this works.[15]

---

[15] 3GPP[TM] TSs and TRs are the property of ARIB, ATIS, ETSI, CCSA, TTA and TTC who jointly own the copyright in them. They are subject to further modifications and are therefore provided to you "as is" for information purposes only. Further use is strictly prohibited.

Table 5.1 IMS Interfaces

Cx	Reference Point between a CSCF and an HSS.
Dx	Reference Point between an I-CSCF and an SLF.
Gi	Reference point between GPRS and an external packet data network.
Gm	Reference Point between a UE and a P-CSCF.
ISC	Reference Point between a CSCF and an Application Server.
Iu	Interface between the RNS and the core network. It is also considered as a reference point.
Ix	Reference Point between IMS ALG and NA(P)T-PT.
Le	Reference Point between an AS and a GMLC.
Mb	Reference Point to IPv6 network services.
Mg	Reference Point between an MGCF and a CSCF.
Mi	Reference Point between a CSCF and a BGCF.
Mj	Reference Point between a BGCF and an MGCF.
Mk	Reference Point between a BGCF and another BGCF.
Mm	Reference Point between a CSCF and an IP multimedia network.
Mr	Reference Point between a CSCF and an MRFC.
Mw	Reference Point between a CSCF and another CSCF.
Mx	Reference Point between a CSCF and IMS ALG.
Sh	Reference Point between an AS (SIP-AS or OSA-CSCF) and an HSS.
Si	Reference Point between an IM-SSF and an HSS.
Ut	Reference Point between UE and an Application Server.

A Public User Identity can be registered from multiple contact addresses, (as defined in RFC 3261) and this, together with the S-CSCF's support for forking, enables an incoming SIP request, for example a phone call, to be addressed to a Public User Identity which can be proxied to multiple registered contact addresses. An example of this is where a household phone has a Private User Identity. The people who live in the house have a separate Public User Identity. Each person can be contacted on several devices, depending on their location, that is, at home or at work, and so on. SIP Forking is also explained in detail in Chapter 2.

5.10 GPRS Access to IMS

The P-CSCF is the entry point to the IMS and in order to make use of IMS services the mobile UE must obtain the IP address of its local P-CSCF from the GGSN as shown in Figure 5.13. Before the mobile handset can send any SIP messages to the P-CSCF, the UE first applies the normal GPRS procedure to set up an initial bidirectional PDP context for SIP signalling (as defined in 3GPP TS 23.228 [3]) between its local SGSN and the GGSN, and then uses this PDP context to acquire its own IP address from the GGSN, which allocates the address dynamically.

Normally, the user also uses this initial PDP context to find out the IP address, of the P-CSCF, and then activates a further PDP context to the SGSN and GGSN, in preparation for sending SIP-signalling messages to enable the setting up of a call or a media session to another party. The mobile UE can have several active separate PDP contexts concurrently, or SIP signalling and media streams can share the same PDP context.

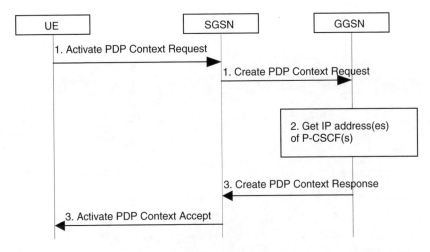

Figure 5.13 P-CSCF discovery using PDP Context Activation signalling [3]

In order to provide IP end-to-end QoS to subscribers, it is necessary to manage the QoS across each domain in the network path. A Policy Decision Function (PDF) mediates between requests for bandwidth, from the P-CSCF, at each end of the connection. A P-CSCF sends a request for resources (bandwidth) to its PDF. At the same time, the P-CSCF provides the PDF with details of the IP bearer level policy. When a PDF makes a decision to allocate resources, it maps the policy set-up information received from the P-CSCF into IP QoS parameters. The PDF communicates with the GGSN in the underlying core network, over the Go interface, as illustrated in Figure 5.14 and an IP BS Manager within the GGSN controls the external IP BS. A GGSN has a single master, that is, it reports to only one PDF. The GGSN can store the policy decisions made by the PDF, which allows the GGSN to make further admission control decisions without requiring additional interactions with the PDF.

The PDF is in the same domain as the GGSN but the P-CSCF may be either in the same domain or in a different domain. Also the PDF may be a logical entity of the P-CSCF, or it may be located within the IP BS Manager in the GGSN or a completely separate physical node. If the PDF is implemented in a separate physical node, the interface between the PDF and the P-CSCF is the Gq interface standardised in 3GPP TS 23.207 [9].

The Diameter protocol is used over the Gq interface between the P-CSCF and the PDF. Diameter is an Authentication, Authorisation and Accounting (AAA) protocol [13].[16] Terminating and originating P-CSCFs both authorise resources through their respective PDFs, the terminating one first.

The protocol framework chosen for the Go interface, between a PDF and a GGSN, is the IETF COPS (Common Object Policy Service) protocol [14]. The COPS protocol supports a client/server interface between a Policy Decision Point (PDP), which in this case is in the PDF and Policy Enforcement Point (PEP), which is in the GGSN.

---

[16] Diameter is loosely based on the Remote Authentication Dial In User Service (RADIUS) developed in RFC 2865 from the IETF.

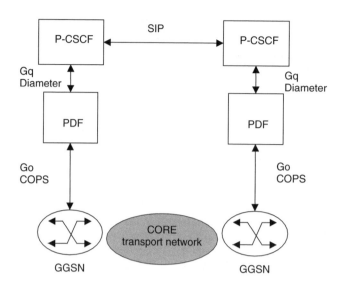

Figure 5.14 The PDF and the GGSN

The "Go" interface conforms to IETF COPS (as a requirement) and allows service-based local policy and QoS interworking information to be "pushed to" the GGSN by the PDF, and/or "pulled" or requested, from the PDF by the GGSN. 3GPP TS 23.207.6.6.0 [9] further describes how the PDF is instrumental in providing QoS in IMS sessions and 3GPP TS 23.228 [3] defines how the PDF can instruct the GGSN to allocate or release resources.

The following steps summarise the authorisation of a mobile subscriber applying for IMS services through a GPRS attach, and these steps are also illustrated in Figure 5.15.

(1) The User sends a second Activate PDP Context message to the SGSN with the GERAN or UMTS QoS parameters.
(2) The SGSN forwards a corresponding Create PDP Context message to the GGSN.
(3) The GGSN sends a COPS REQuest message with the Binding Information over the Go interface to the PDF to request the policy information.
(4) The PDF sends an authorisation request to the P-CSCF over the Gq interface.
(5) At this stage, the PDF will generate an authorisation token to keep track of the authorisation status information of the requesting user.
(6) If the session description is consistent with the operator policy rules defined in the PDF, the PDF preliminarily authorises the required QoS resources for the session and waits for information on the IP bearer level policy from the P-CSCF.
(7) The PDF installs the IP bearer level policy, received from the P-CSCF, in its internal database. The PDF's token is passed back to the user to indicate that the request for resources has been accepted and the PDF sends a COPS DECision message back to the GGSN.
(8) The GGSN sends a COPS RePorT message to the PDF, which may also trigger a report message to be sent from the PDF to the P-CSCF.

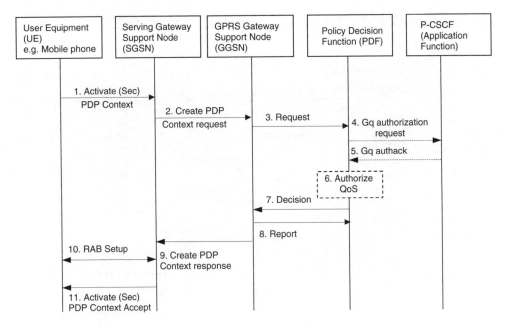

Figure 5.15 Setting up a PDP Context

(9) The GGSN maps IP flow–based policy information into the PDP context–based policy information and uses the PDP context–based policy information to accept the PDP activation request and sends a Create PDP Context Response message back to the SGSN.

(10) The set-up of the Radio Access Bearer (RAB) is carried out by the RAB Assignment procedure.

(11) The SGSN sends a message back to the User accepting its Activate (Secondary) PDP Context message (which was sent in the first step of this process).

5.10.1 Creating a Session

Figure 5.16 shows the detailed sequence of events involved in setting up a media session and the steps are numbered for clarity. The authorisation of resources by a P-CSCF (through its PDF and GGSN), as described in the previous example, is now abbreviated to a single step, as in Step 9, shown as "Authorise QoS Resources" in Figure 5.16.

Step 1 A SIP Invite, containing an initial Session Description Protocol (SDP) unit, is sent by the user to the local P-CSCF.

Steps 2 and 3 The P-CSCF forwards the invite to the S-CSCF, which validates the user's service profile and also invokes other services if appropriate (which may involve sending the INVITE to an Application Server).

Steps 4–6 The S-CSCF then forwards the INVITE request to the I-CSCF, which performs a location query procedure with the HSS to acquire the address of the destination (called) user's S-CSCF and forwards the INVITE to this address.

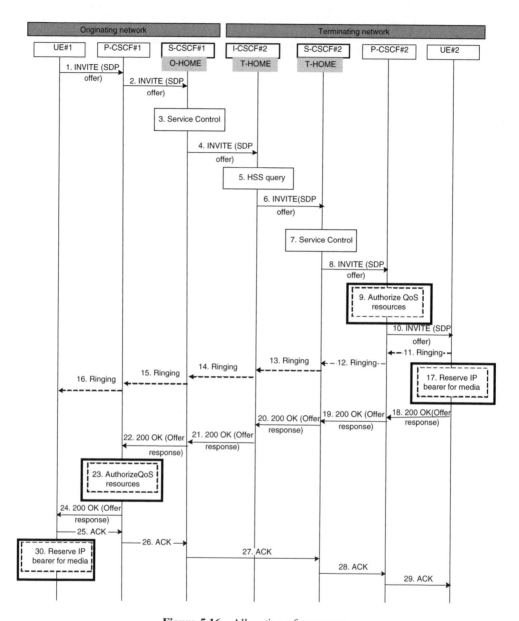

Figure 5.16 Allocation of resources

Steps 7–10 The destination S-CSCF, receiving the INVITE, validates the user's service profile and forwards the INVITE again to the destination P-CSCF, which uses a PDF, to authorise the resources for this session and then forwards the INVITE request to the destination user. The INVITE can now include the authorisation token issued by the PDF.

Steps 11–17 The user at the called destination can generate a ringing message (optionally) back towards the originator of the call. The destination user can reserve a dedicated bearer in the access network on the basis of the parameters it has received in the SDP offer, or it may just decide to use an existing bearer if a new bearer is not needed.

Steps 18–22 The destination user also sends a 200 OK response back to its P-CSCF and this response travels all the way back to the originator of the call.

Steps 23–30 The P-CSCF at the start of the call also has its own PDF, which authorises the resources for the initial segment of the call and generates the authorisation token, which can be included in the 200 OK with an acknowledgement, which is sent to the destination user.

5.10.2 Authorisation and Reservation of an IP bearer

Figure 5.16 illustrates how the P-CSCF, the I-CSCF and the S-CSCF interact and work with each other, using SIP signalling, to ensure that resources are authorised and reserved. There are a few key points which should be highlighted:

- The I-CSCF and the S-CSCF are always found in a user's home domain, within the respective originating and terminating networks.
- The authorisation and reservation of the IP bearer takes place at the call destination first, and the originator of the call only seeks authorisation after the called party has completed the authorisation and reservation process and has accepted the call.
- Allocation of resources at the calling end is the very last step of the process.

5.10.3 Storage of Session Paths

During registration and session initiation, there are SIP mechanisms to determine the path the session will take through the network. The session path can be stored by the P-CSCF and the S-CSCF, that is, the nodes that are controlling the service so that subsequent session requests can be routed through the same path.

As all initial requests to or from the user go via an assigned S-CSCF, this S-CSCF can use a "Record-route" mechanism defined in RFC 3261 [12] to remain in the signalling path for subsequent requests too and in this way the S-CSCF is able to record routes. This is considered to be the default behaviour for all IMS communication. However, if Application Servers under operator control can guarantee the home control of the session, then it may not be necessary for all subsequent requests to traverse the S-CSCF, and in such cases the operator may choose that the S-CSCF does not "record route".

Once the IMS session has started, the P-CSCF has control of the flows, in terms of the source and destination IP addresses of the communicating endpoints, and decides when the user plane traffic between the end-points of the SIP session may start, and will stop, and this is then synchronised with the session charging. In a mobile 3GPP network, the actual enforcement point of this control is within the GGSN. If the user becomes unreachable, perhaps because the access network has either modified or suspended the bearer associated with a session, the access network informs the P-CSCF.

5.11 Broadband Data Wireless Access

In the future, mobile operators will deliver Triple Play and Quad Play services to their subscribers using a variety of diversified radio access methods, such as W-CDMA, CDMA2000, HSDPA and HDUPA, which have already been described in this chapter, and provide a way to increase the bandwidth on the air interface. The problem is that GPRS does not offer sufficient bandwidth for 3G services. The essential aim of these other technologies is to increase bandwidth in the 2G and 3G RAN to enable roaming subscribers to download large volumes of data, such as video clips or e-mail files, over their mobile handsets or laptops.

The idea of also combining Broadband Data Wireless access with a mobile cellular service has captured the attention of both mobile and fixed service providers looking to grow and protect their existing business. A single handset adapter for multiple access, with one number and one voicemail, can provide an "all inclusive" service, seamlessly encompassing wireless, mobile and fixed access.

Many new types of service are being deployed in wireless "hotspot" zones. In large public areas, such as railway stations and airports, a number of wireless access points can enable a user to hold a voice or video call, while moving around, in a limited area, and essentially by-passing the existing cellular network.

Other short-range wireless protocols are used to complement the wireless solution for "hotspots". Within a Personal Area Network (PAN), protocols such as Bluetooth and WiFi enable "short-range" connection between devices such as mobile handsets, laptops, PCs and printers, and the Internet. Bluetooth allows for devices to exchange information over a distance of up to 100 m and implementations with Versions 1.1 and 1.2 can reach speeds of 723 kbps. Version 2.0 implements Bluetooth Enhanced Rate and can reach a bandwidth of circa 2 Mbps.

The lowest rate of WiFi provides a throughput of 1 Mbps. A typical WiFi home router using 802.11b or 802.11g can have a range of 45 m indoors and (90 m) or 300 ft outside.

UMA (Unlicensed Mobile Access) is an industry collaboration to extend GSM and GPRS services into the residential home and customer sites by utilising unlicensed radio technologies such as WiFi (Wireless Fidelity) and Bluetooth.

Wireless coverage for voice in the home is not new. Many residential users have had DECT (Digital Enhanced Cordless Telecommunications) phones for some time.[17]

The UMA specifications have been approved by 3GPP for Generic Access to the A/Gb interface (See 3GPP Release 6 TS 43.318) and UMA has been renamed as the Generic Access Network (GAN).

GAN (UMA) specifications enable users to roam and handover between cellular networks and public and private unlicensed wireless networks. A mobile subscriber, with a dual-mode handset, connects by default, using Bluetooth or WiFi, to an unlicensed Wireless Access Point or WLAN, when within range.

Upon connecting, the dual-mode handset contacts the UMA Network Controller (UNC) over the broadband IP access network to allow the user to be authenticated and authorised

---

[17] DECT is an ETSI standard for digital portable phones, commonly used for domestic or corporate purposes which, like GSM, is a cellular system. A major difference between the systems is the cell radius, DECT cells have a radius of 25 to 100 m, while GSM cells have a radius of 2 to 10 km.

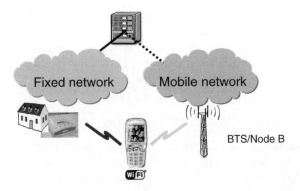

Figure 5.17   Wireless – bridging the gap between fixed and mobile

to access GSM voice and GPRS data services. If approved, the subscriber's current location information, which is stored in the broadband core network, is updated and from that point on all mobile voice and data traffic is routed to the handset via the Unlicensed Mobile Access Network (UMAN) rather than a cellular RAN.

When out of range of the wireless "Hot Spot", the handset connects automatically to a wide area GSM (or GPRS) mobile network in the normal way. GAN (UMA) supports seamless in-call handover to an access point from GSM and from GSM to an access point. (Note: CDMA is not supported.)

The use of GAN (UMA) demonstrates how wireless technologies can successfully bridge the gap between mobile and fixed networks as shown in Figure 5.17. A call on a mobile handset can be transferred seamlessly from a mobile network to a wireless network, to a fixed network. The next step is to create one number and one bill for all types of calls. A voice Virtual Private Network (VPN) can help in this implementation. Chapter 7 provides further information on the role and implementation of VPNs.

5.12 Wireless LAN Interworking

Small businesses can run voice and data services over a shared wireless network, and access to voice and data applications by Wireless LAN behaves functionally in the same way as attachment by GSM (or GPRS), that is, IP address assignment, hosted VPN services, and direct network connections, and so on, all work equally well across Wireless Access.

The IEEE 802.11b standard describes sharing a transmission media at the link level. However, any QoS that has been built in at the layer 3 IP level can easily be lost in the LAN CSMA/CD[18] environment through collisions and lack of transmission opportunity. For example, as in GPRS, there is no way to give a voice call priority over the web browsing. To do this would require QoS co-ordination at the link layer and this is a problem that the IEEE 802.11e standard is currently trying to solve.

---

[18] Collision Sense Multiple Access/Collision Detection.

Other IP-based wireless technologies such as WiMAX 802.16x, and 802.20x, are evolving towards higher broadband rates[19] over increasing ranges. WiMAX can support point-to-multipoint wireless networking over areas of between 10 to 30 miles. Better support for continuous mobility over a wider area is being fostered by industry alliances as well as standards development organisations.

5.12.1 3GPP Release 6 Integration of Wireless LANs

3GPP Release 6 defines the standardisation of 3GPP-WLAN Inter-working [15]. A Wireless LAN Access Gateway (WAG), as illustrated in Figure 5.18, defines a GW between the WLAN Access Network and 3GPP PS services, which are accessed via a Packet Data Gateway (PDG). The PDG can support connections from WLAN users using both IPv4 and IPv6 local addresses, through the implementation of a dual IP stack.

In the non-roaming case, the WAG and the PDG both reside in the home mobile network. In the roaming case, they both reside in the visited network. The WAG is used to generate charging information. The WAG collects per tunnel accounting information for inter-operator settlements and forwards the charging information to a 3GPP AAA Proxy, which resides in the Visited 3GPP network.

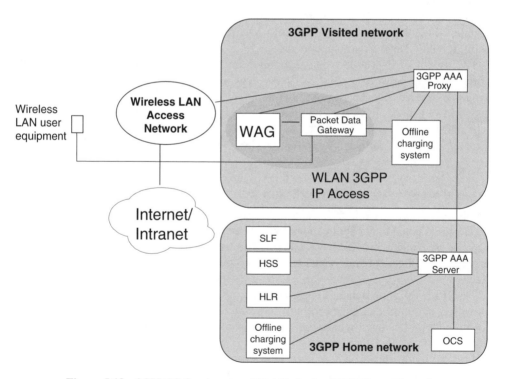

Figure 5.18   3GPP PS Services to a WLAN via the 3GPP Home Network

---

[19] 802.11a – 5 GHz – ratified in 1999 802.11b – 11 Mb 2.4 GHz – ratified in 1999 (11 Mbps 100–150 feet, 5.5 Mbps 150–250 ft, 2 Mbps 250–350 ft) 802.11 g – Higher Data rate (>20 mBps) 2.4 GHz.

It is the WAG's responsibility to make sure that all packets from the user are routed to the appropriate PDG and that the user only receives packets from other authorised WLAN users. The WAG can also implement policy enforcement before a tunnel is established to enhance a FW against packets from unauthorised WLAN users and to prevent the roaming WLAN user from trying to establish a tunnel to any other network other than its own home network. A policy bound to a user's traffic allows the WAG to drop unauthorised packets sent to and from the user.

The 3GPP AAA Proxy functions include relaying the AAA information between the WLAN and the 3GPP AAA Server [15]. A SLF is located within the 3GPP subscriber's home network and enables a 3GPP AAA Server to find the address of the Home Subscriber Service, which holds the subscriber data for a given user identity. The SLF is used in the same way for a wireless LAN, as it is for IMS, as specified in TS 23.228 [3] An Offline Charging System and an Online Charging System are both located within the 3GPP network. The mapping of the Offline Charging System in a Release 6 charging architecture is specified in TS 32.252 [16] and the Online Charging System is described in TS 32.296 [17].

5.13 Mobile TV and Video

Broadcast video and television on a handheld mobile device will be a reality in the next two or three years, but it is not yet clear how they will be delivered. High capacity mobile broadcast services can deliver content to groups of users, with some degree of personalisation and once a mobile handset supports broadcasting capabilities, a large range of customer services can be made available:

- Free mobile "live" TV programs
- Pay mobile "live" TV programs
- Interactive mobile TV: gambling, subscription and feedback during a "live" mobile TV program
- Snapshot TV: short mobile videos specifically created for mobile consumption
- Carrousel audio, video and data
- Push and store: content is broadcast to all subscribers (pushed), stored on the terminals, and then played/run on demand, providing an illusion of instant transmission; examples of this kind of service are downloading software updates and video and music content download for playing after payment (pay per view)
- Local broadcast: museum information and shopping mall news
- Group messages: fleet management.

Today, mobile video streaming and downloading are implemented as part of 2.5G and 3G mobile services but the delivery cost, based on unicast communication (communication with a single receiver), is delaying mass-market deployment.

Mobile operators have decided to deliver mobile TV and video services by leveraging the complementary nature of both cellular and broadcast technologies and benefiting from their respective strengths. This means optimisation of current digital TV broadcasting networks based on technologies such as Terrestrial (Digital Video Broadcasting Handheld – DVB-H) and Satellite (Satellite Digital Multimedia Broadcast – S-DMB).

Adaptation of their capabilities to the mobile environment should provide the most cost-effective way to offer sufficient bandwidth on the downlink channels to mobile handsets to enable the efficient delivery of bandwidth-consuming video applications.

DVB systems distribute data by the following:

- Satellite
- Cable
- Terrestrial Television (DVB-T) and
- Terrestrial Television for handhelds (DVB-H).

DVB-H (Handheld) is a technical specification for delivering broadcast services to handheld receivers and was formally adopted as an ETSI standard, EN 302 304, in November 2004. In contrast to other DVB transmission systems, which are based on the DVB Transport Stream, adopted from the MPEG-2 standard, the DVB-H base-band interface is an IP interface and therefore can be combined with other IP-based services.

The DVB-H system provides another way to broadcast services to mobile handsets. DVB-H can offer a maximum capacity of 11 Mbps on an 8 MHz channel. Depending on the configuration, an average number of 60 video streams at 128 kbps should be possible. The term handheld terminal includes multimedia mobile phones with colour displays, as well as Personal Digital Assistants (PDAs) and pocket PC type of equipment. All these kinds of devices have a number of features in common: small dimensions, lightweight and battery operation. These properties are a precondition for mobile usage but can also imply several severe restrictions on the transmission system. The terminal devices lack an external power supply in most cases and have to be operated with a limited power budget, and low power consumption is necessary to obtain reasonable usage and standby cycles.

The typical user environment of a DVB-H handheld terminal is very similar to a mobile radio environment but DVB-H needs to provide the same geographic coverage and mobility, so that access to services is possible from almost all indoor and outdoor locations and also while on the move, for example in a vehicle at high speed.

The handover between adjacent DVB-H radio cells must happen imperceptibly when moving over larger distances, but fast varying DVB-H channels are very error prone. Also, an additional antenna (second aerial) has to be added to a mobile handset, as DVB does not work on existing 2G, 2.5G or 3G phones. Handheld devices have limited dimensions and cannot be pointed at a transmitter if the terminal is in motion. Interference with GSM mobile radio signals can also occur. As a result of these problems, the provision of downstream access of several megabits per second to handheld terminals using DVB-H is currently a demanding task. Nevertheless at the time of writing, several mobile operators are conducting trials with live TV being delivered to mobile handsets equipped with tiny digital TV receivers, receiving terrestrial and satellite broadcasts via an implementation of the DVB-H standard.

Mobile operators who are willing to deliver their own mobile broadcast content could use the S-DMB network as a complementary overlay to DVB-H networks. The S-DMB functionality can be implemented at a marginal cost on a 3G handset, and combined with 3G DVB-H handsets, it could represent a very cost-effective solution which will also ensure seamless nationwide service continuity of the mobile operator's broadcast services.

T-DMB, which stands for "Terrestrial-Digital Multimedia Broadcasting", is another system for broadcasting different types of digital content to mobile devices, like cell phones. The system was first proposed in 2002 and is based on the Eureka 147 standard used by Digital Audio Broadcasting (DAB) which it uses as the broadcast channel to deliver TV, audio, video and data to mobile devices, enabling consumers to choose from multiple channels of DTV and then view it on their mobile phones, PDAs or cars. DAB digital radio has already had some success and is being rolled out in the United Kingdom, Germany, Belgium, Italy, Spain, Switzerland, Holland and Scandinavia. The technology for mobile TV using T-DMB is based on the same basic infrastructure as that used for DAB. The technology for DAB is considered to be ideally suited to mobile TV because it was originally designed to enable robust reception in a mobile environment while keeping power consumption to the minimum. South Korea is at the forefront of the development of T-DMB, with commercial services having been launched in the third quarter of 2005.

The MBMS (Multimedia Broadcast Multicast Service) is also believed to be currently under active evaluation by Mobile Network Operators as a means of delivering Mobile Television to the mass market. MBMS is well suited as a complement to dedicated broadcast networks, by delivering local content over limited coverage areas to a limited audience.

Although MBMS offers only limited capacity that is, three channels at 128 kbps in the 5-MHz band, the impact on the design of the network is minimal since no new transmitters need to be deployed and the ratio between unicast and multicast can be dynamically adjusted on a cell-by-cell basis. With the advent of the HSDPA for unicast transmission and the opening of the 3G extension bands, the freed up capacity could be used for more video content delivery, and it is expected that pre-MBMS equipment will be available as of 2006, while fully compliant infrastructure could be available by 2007.

The viability and success of all these technologies depends on spectrum availability[20] and the cost of building new broadcast infrastructure.

5.14 Related Work in other Standards Bodies

The Internet is being recognised as a critical infrastructure and many standards development organisations are developing standards to improve the Internet's availability and ease of use. 3GPP has already completed standardisation for access to IMS from UMTS (and GERAN) networks, and specifications for Wireless LAN access are evolving, as described in the last section. Other organisations are looking at the implementation of other types of access.

5.14.1 ETSI

ETSI (http://www.etsi.org) is an independent, non-profit organisation, which produces telecommunications standards and maintains a membership from 55 countries. ETSI is also defining a NGN and a substantial part of this is based on IMS.

---

[20] The phased switch-off of the United Kingdom's analogue TV signal begins in 2008, but is not due to be completed until 2012. The analogue switch-off will release up to 122 Mhz of spectrum in the UHF (Ultra High Frequency) band. The spectrum is particularly sought after because of its combination of high capacity and range. At the time of writing, Ofcom is seeking views on how the spectrum could be used. Mobile operators want to use it for DVB-H, while TV broadcasters want to use if for standard and high definition television. Wireless ISP's want the spectrum to deliver to remote areas.

TISPAN (Telecoms & Internet converged Services and Protocols for Advanced Networks) within ETSI, develops standards for converged networks. TISPAN aims to migrate existing PSTN functionality to an IP core where IMS can provide an equivalent to current PSTN services through PSTN simulation [18]. A number of requirements have also been developed to identify further extensions to SIP [19] for this purpose.

TISPAN has designed its own Resource and Admission Control Subsystem (RACS) to provide applications with a mechanism to request and reserve resources from an access network, as shown in Figure 5.19. RACS is responsible for elements of policy control, resource reservation and admission control as well as for including support for NAT and FW traversal.

TISPAN's RACS architecture provides a clear separation between layers so that an application service can run over different access networks (without impacting an application's capabilities), as long as the required resources are available.

Figure 5.19 illustrates the RACS architecture. An Application Function (AF) offers services that use IP bearer resources. The P-CSCF in the IMS model is an example of a type of AF. The AF maps the application layer QoS information, for example, defined in SDP, into QoS request information to be sent via the Gq' interface to the Service Policy Decision Function (SPDF).

The SPDF provides a single point of contact for an AF and authorises resources to ensure QoS by making policy decisions, using policy rules for Service-based Policy Control (SBP) and communicating these decisions to the Access-Resource and Admission Control Function (A-RACF) over the Rq interface.

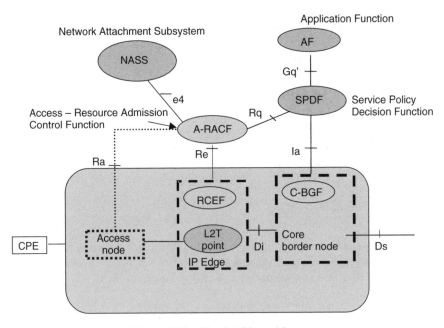

Figure 5.19 The RACS Architecture

The function of the A-RACF is to implement resource reservation and admission control. In the access network, the A-RACF interacts with a Network Attachment Subsystem (NASS), via the e4 reference point. The A-RACF has different instances for different types of access and the Ra reference point is indicated as a dotted line in Figure 5.19 to illustrate that there are different methods by which the A-RACF can interact with access and aggregation networks.

The A-RACF may also receive resource requests from the IP Edge via the Resource Control Enforcement Function (RCEF) over the Re interface. Resource requests over the Re interface can be triggered at the IP Edge by the receipt of a resource reservation request from the CPE. The RCEF is a logical element in the transport layer, enforcing the policies, which enable RACS to guarantee the availability of resources such as gate control, packet marking and policing.

The Layer 2 Termination Function (L2TF) is the point where the Layer 2 communication with the CPE is terminated. The RCEF and the L2TF are usually physically implemented in an IP Edge Node.

For gate control and NAT control, the SPDF communicates with the Core BGF (Border Gateway Function). The C-BGF is located anywhere in the transport network. It is generally found between an access network and a core network or between two core networks. The role of the BGF is to implement NAT, Gate Control, Packet marking, Usage metering and Traffic policing.

Figure 5.20 illustrates how RACS can interwork with an IMS. In the RACS model, the AF becomes the P-CSCF, identifying the need to request RACS services and sending the service request information to the SPDF within RACS.

The SPDF authorises the request and communicates with an appropriate BGF. The BGF allocates the necessary resources to create an RTP (Real-time Protocol) relay function and confirms this action to the SPDF. The SPDF then forwards the results received from the BGF to the AF.

An Interconnect Border Control Function (I-BCF) (not shown in Figure 5.19), which may also be thought of as a type of AF, deals with interconnect between networks and controls the media exchanged across an operator boundary. The functions of the I-BCF include the provision of Network Address and Port Translation (NAPT) and FW policies for signalling, policing of signalling, topology hiding and conversion between IPv4 and IPv6 addressing.

Other standards bodies, as well as ETSI are also tracking the progress of IMS. The following sections briefly cover the responses of the ITU, ATIS and the IETF to the IMS architecture.

5.14.2 ITU-T

As described in previous sections, one of the key aspects of the IMS is the ability to support various access technologies, and the International Telecommunications Union-Telecommunications (ITU-T) Standardization sector have endorsed the IMS (defined by 3GPP and 3GPP2) as a candidate for the core network for session management for NGN. Study Group 13 within the ITU-T leads research on NGN where its focus is mainly

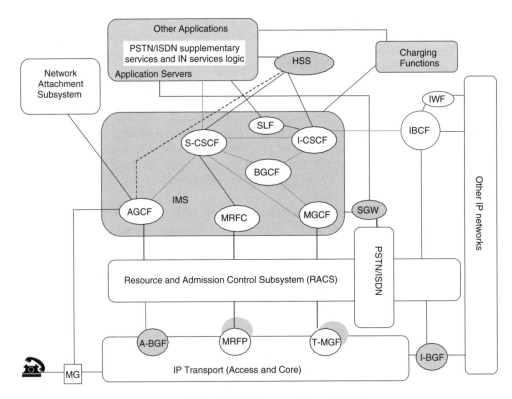

Figure 5.20 RACS interworking with IMS

on broadband communication with mobility as an inherent feature. IMS does not provide support for all types of NGN services, and, in particular, services that are not session based, such as broadcast services, need further research if they are to be transferred to IMS.

5.14.3 ATIS

The Alliance for Telecommunications Industry Solutions (ATIS)[21] standardisation activities for wireless and wireline networks include interconnection standards, number portability, improved data transmission, Internet telephony, toll-free access, telecom fraud and order and billing issues, among others. ATIS sees IMS as "a good foundation for moving towards a robust IP-based network" but "considers that it needs more standards development before it will be robust enough to deliver a full slate of data, voice and video across a converged wireline and wireless network."

5.14.4 IETF

Today, the IETF is a large, international community open to any interested individual, concerned with the evolution of the Internet architecture and the smooth operation of the Internet.

---

[21] ATIS is accredited by the American National Standards Institute (ANSI).

Table 5.2   IETF workgroups

IP Telephony (VoIP)	
AVT	The Audio/Video Transport Working Group was formed to specify a protocol for real-time transmission of audio and video over unicast and multicast UDP/IP. This is the Real-time Transport Protocol, RTP, together with its associated profiles and payload formats.
IPTEL	The focus of the IP Telephony (IPTEL) group is on the problems related to naming and routing for Voice over IP (VoIP) protocols. Naming is accomplished through the use of the tel URI, which specifies a URI for telephone numbers. The tel URI was originally defined in RFC 2806, which was developed outside of any IETF working group.
MMUSIC	The Multiparty Multimedia Session Control (MMUSIC) Working Group was chartered to develop protocols to support Internet teleconferencing and multimedia communications
SIP	The Session Initiation Protocol (SIP) working group is chartered to maintain and continue the development of SIP currently specified as proposed standard RFC 3261 and its family of extensions.
SIPPING	The Session Initiation Protocol Project INvestiGation (SIPPING) working group is chartered to document the use of SIP for several applications related to telephony and multimedia and to develop requirements for extensions to SIP needed for those applications.
Interoperation with the Circuit Domain	
ENUM	The ENUM working group has defined a DNS-based architecture and protocol [RFC 3761] by which an E.164 number, as defined in ITU Recommendation E.164, can be expressed as a Fully Qualified Domain Name in a specific Internet Infrastructure domain defined for this purpose
MEGACO	The Megaco media gateway control protocol, RFC 3015 [20] (also published as ITU-T Recommendation H.248 [21]), was developed by the Megaco Working Group in close cooperation with ITU-T Study Group 16. The protocol responds to the requirements documented in RFC 2805 [22].
PINT	The PSTN/Internet Interfaces (PINT) WG addresses connection arrangements through which Internet applications can request and enrich PSTN (Public Switched Telephone Network) telephony services. An example of such services is a Web-based Yellow Pages service with the ability to initiate PSTN calls between customers and suppliers.
SIGTRAN	The primary purpose of this working group is to address the transport of packet-based PSTN signalling over IP Networks, taking into account functional and performance requirements of the PSTN signalling. For interworking with PSTN, IP networks will need to transport signalling such as Q.931 or SS7 ISUP messages [23] between IP nodes such as a Signalling Gateway and Media Gateway Controller or Media Gateway.

(continued overleaf)

Table 5.2 (*continued*)

SPIRITS	The Services in the PSTN/IN Requesting InTernet Services (SPIRITS) Working Group addresses how services supported by IP network entities can be started from IN (Intelligent Network) requests, as well as the protocol arrangements through which PSTN (Public Switched Telephone Network) can request actions to be carried out in the IP network in response to events (IN Triggers) occurring within the PSTN/IN.

Instant Messaging and Presence

SIMPLE	This working group focuses on the application of the Session Initiation Protocol (SIP, RFC 3261) to the suite of services collectively known as instant messaging and presence (IMP). The IETF has committed to producing an interoperable standard for these services compliant to the requirements for IM outlined in RFC 2779 (including the security and privacy requirements there) and in the Common Presence and Instant Messaging (CPIM) specification, developed within the IMPP working group.

Traffic Engineering

MPLS	The MPLS working group is responsible for standardising a base technology for using label switching and for the implementation of label-switched paths over various packet-based link-level technologies, such as Packet-over-Sonet, Frame Relay, ATM and LAN technologies (e.g. all forms of Ethernet, Token Ring, etc.). This includes procedures and protocols for the distribution of labels between routers and encapsulation

The IETF is working on many extensions to SIP, which will ultimately be incorporated in new versions of existing 3GPP specifications. Work is ongoing in many areas including IP Telephony, Interoperating with the Circuit Domain, Instant Messaging and Presence, QoS [24] and so on. The SIP working group within the IETF has defined numerous RFCs (Requests For Comments) dealing with proposed extensions to SIP. For example, RFC 2976 proposes an extension to SIP to add the INFO method to the SIP protocol where the intent of the INFO method is to allow for the carrying of session-related control information that is generated during a session. One example of such session control information is ISUP and ISDN signalling messages used to control telephony call services. RFC 3325 describes private extensions to SIP to enable a network of trusted SIP servers to assert the identity of authenticated users. There are many workgroups within the IETF where research will help further define the interfaces, which are needed for integration of the IMS.

Table 5.2 lists some of the IETF workgroups where the reader can track developing standardisation, which will help underpin the evolution of IMS.

Summary

Chapter 5 introduced the IMS architecture, which most industry bodies have now recognised as their preferred future architecture, and showed how the IMS is designed to be access independent, so that IMS services can be provided over any IP-based network. The IMS model was first introduced by 3GPP and standards for UMTS and GERAN access to IMS are now well established. The chapter shows how 3GPP developed its mobile network architecture beginning with Release 99, to UMTS Release 4, where a split was introduced, separating the function of the MGW from the function of an MSC Call Agent established, and finally to the first introduction of the IMS model in UMTS Release 5. Methods of mobile user identification, tracking of location and signalling and bearer protocols are explained. Examples of GPRS attachment to IMS are provided. The P-CSCF is central to the workings of the IMS and its role and those of the other supporting components within the IMS are covered in detail. Diversified radio access methods, such as W-CDMA, CDMA2000, HSDPA and HDUPA, have been described in this chapter and provide a way to increase the bandwidth over the air interface in a cellular network. 3GPP Release 6 defines the standardisation of 3GPP-WLAN Inter-working and Wireless LAN access to IMS is explained. The Internet is being recognised as a critical infrastructure and many standards development organisations are developing standards to improve the Internet's availability and ease of use as well as looking at the implementation of other types of access to IMS. The work within other standards bodies such as ETSI TISPAN, the ITU-T, IETF and ATIS is summarised and important standards are referenced to help the reader explore other areas of research.

Appendix

3GPP Specifications

The following specs can be downloaded from http://www.3gpp.org/specs/numbering.htm.
   The list below is a small selection from 3GPP specifications

TS 21.905 Vocabulary for 3GPP Specifications
TS 22.066 Support of Mobile Number Portability (MNP); Stage 1
TS 22.101 Service Aspects; Service Principles
TS 22.141 Presence Service; Stage 1
TS 22.228 Service requirements for the IP multimedia core network subsystem; Stage 1
TS 22.250 IMS Group Management; Stage 1
TS 22.340 IMS Messaging; Stage 1
TS 22.800 IMS Subscription and access scenarios
TS 23.002 Network Architecture
TS 23.003 Numbering, Addressing and Identification
TS 23.008 Organisation of Subscriber Data
TS 23.107 Quality of Service (QoS) principles
TS 23.125 Overall high level functionality and architecture impacts of flow based charging; Stage 2
TS 23.141 Presence Service; Architecture and functional description; Stage 2
TS 23.167 IMS emergency sessions
TS 23.207 End-to-end QoS concept and architecture
TS 23.218 IMS session handling; IM call model; Stage 2
TS 23.221 Architectural Requirements
TS 23.228 IMS stage 2
TS 23.234 WLAN interworking
TS 23.271 Location Services (LCS); Functional description; Stage 2
TS 23.278 Customized Applications for Mobile network Enhanced Logic (CAMEL) – IMS interworking; Stage 2

TS 23.864 Commonality and interoperability between IMS core networks

TR 23.867 IMS emergency sessions

TS 23.917 Dynamic policy control enhancements for end-to-end QoS, Feasibility study

TS 23.979 3GPP enablers for Push-to-Talk over Cellular (PoC) services; Stage 2

TR 23.981 Interworking aspects and migration scenarios for IPv4-based IMS implementations (early IMS)

TS 24.141 Presence Service using the IMS Core Network subsystem; Stage 3

TS 24.147 Conferencing using the IMS Core Network subsystem

TS 24.228 Signalling flows for the IMS call control based on SIP and SDP; Stage 3

TS 24.229 IMS call control protocol based on SIP and SDP; Stage 3

TS 24.247 Messaging using the IMS Core Network subsystem; Stage 3

TS 26.235 Packet switched conversational multimedia applications; Default codecs

TS 26.236 Packet switched conversational multimedia applications; Transport protocols

TS 29.162 Interworking between the IMS and IP networks

TS 29.163 Interworking between the IMS and Circuit Switched (CS) networks

TS 29.198 Open Service Architecture (OSA)

TS 29.207 Policy control over Go interface

TS 29.208 End-to-end QoS signalling flows

TS 29.209 Policy control over Gq interface

TS 29.228 IMS Cx and Dx interfaces: signalling flows and message contents

TS 29.229 IMS Cx and Dx interfaces based on the Diameter protocol; Protocol details

TS 29.278 CAMEL Application Part (CAP) specification for IMS

TS 29.328 IMS Sh interface: signalling flows and message content

TS 29.329 IMS Sh interface based on the Diameter protocol; Protocol details

TS 29.962 Signalling interworking between the 3GPP SIP profile and non-3GPP SIP usage

TS 31.103 Characteristics of the IMS Identity Module (ISIM) application

TS 32.240 Telecommunication management; Charging management; Charging architecture and Principles

TS 32.260 Telecommunication management; Charging management; IMS charging

TS 32.299 Telecommunication management; Charging management; Diameter charging applications

TS 32.421 Telecommunication management; Subscriber and equipment trace: Trace concepts and requirements

TS 33.102 3G security; Security architecture

TS 33.108 3G security; Handover interface for Lawful Interception (LI)

TS 33.141 Presence service; security

TS 33.203 3G security; Access security for IP-based services

TS 33.210 3G security; Network Domain Security (NDS); IP network layer security

TS 33.978 Security aspects of early IP Multimedia Subsystem (IMS)

3GPP Technical Specifications for MBMS

MBMS Bearer Service (Distribution Layer)

3GPP TS.23.146 Multimedia Broadcast/Multicast Service (MBMS); Stage 1

3GPP TS 23.246 Multimedia Broadcast/Multicast Service (MBMS); Architecture and functional description

3GPP TS 25.346 Introduction of the Multimedia Broadcast/Multicast Service (MBMS) in the Radio Access Network (RAN); Stage 2

3GPP TS 43.246 Multimedia Broadcast/Multicast Service (MBMS) in the GERAN; Stage 2

3GPP TR 25.803 S-CCPCH performance for Multimedia Broadcast/Multicast Service (MBMS)

MBMS User Service (Service Layer):

3GPP TS 22.246 Multimedia Broadcast/Multicast Service (MBMS) user services; Stage 1

3GPP TS 26.346 Multimedia Broadcast/Multicast Service (MBMS); Protocols and codecs

3GPP TR 26.946 Multimedia Broadcast/Multicast Service (MBMS) user service guidelines

3GPP TS 33.246 3G Security; Security of Multimedia Broadcast/Multicast Service (MBMS)

3GPP TS 32.273 Telecommunication management; Charging management; Multimedia Broadcast and Multicast Service (MBMS) charging

IETF Specifications

RFC 2327 Session Description Protocol (SDP)

RFC 2748 Common Open Policy Server protocol (COPS)

RFC 2782 a DNS RR for specifying the location of services (SRV)

RFC 2806 URLs for telephone calls (TEL)

RFC 2915 the naming authority pointer DNS resource record (NAPTR)

RFC 2916 E.164 number and DNS

RFC 3087 Control of Service Context using SIP Request-URI

RFC 3261 Session Initiation Protocol (SIP)

RFC 3262 reliability of provisional responses (PRACK)

RFC 3263 locating SIP servers

RFC 3264 an offer/answer model with the Session Description Protocol

RFC 3265 SIP-Specific Event Notification

RFC 3310 HTTP Digest Authentication using Authentication and Key Agreement (AKA)

RFC 3311 update method

RFC 3312 integration of resource management and SIP

RFC 3319 DHCPv6 options for SIP servers

RFC 3320 signalling compression (SIGCOMP)

RFC 3323 a privacy mechanism for SIP

RFC 3324 short term requirements for network asserted identity

RFC 3325 private extensions to SIP for asserted identity within trusted networks

RFC 3326 the reason header field

RFC 3327 extension header field for registering non-adjacent contacts (path header)

RFC 3329 security mechanism agreement

RFC 3420 Internet Media Type message/sipfrag

RFC 3428 SIP Extension for Instant Messaging

RFC 3455 private header extensions for SIP

RFC 3485 SIP and SDP static dictionary for signaling compression

RFC 3515 the SIP REFER method
RFC 3550 Real-time Transport Protocol (RTP)
RFC 3574 Transition Scenarios for 3GPP Networks
RFC 3588 DIAMETER base protocol
RFC 3589 DIAMETER command codes for 3GPP release 5 (informational)
RFC 3608 extension header field for service route discovery during registration
RFC 3665 SIP Basic Call Flow Examples
RFC 3680 SIP event package for registrations
RFC 3725 best current practices for Third Party Call Control (3pcc) in SIP
RFC 3824 using E164 numbers with SIP
RFC 3840 indicating user Agent Capabilities in SIP
RFC 3841 caller preferences for SIP
RFC 3842 SIP event package for message waiting indication and summary
RFC 3856 SIP event package for presence
RFC 3857 SIP event template-package for watcher info
RFC 3858 XML based format for watcher information
RFC 3891 the SIP Replaces Header
RFC 3911 the SIP Join Header
RFC 4028 session timers in SIP
RFC 4235 an INVITE-Initiated dialog event package for SIP

References

1. A.F. Cameron, D.J. Thorne, K.T. Foster, S.I. Fisher. Fixed Access Network Technologies. *BT Technology Journal*, 22(2), 48–59, 2004.
2. 3GPP TS 21.101 V5.10.0. "3rd Generation Partnership Project; Technical Specification Group Services and System Aspects; Technical Specifications and Technical Reports for a UTRAN-based 3GPP System". 2005–06.
3. 3GPP TS 23.228 V6.9.0. "3rd Generation Partnership Project; Technical Specification Group Services and System Aspects; IP Multimedia Subsystem (IMS) Stage 2 (Release 6)". 2005–03.
4. 3GPP TS 23.221 V6.3.0. "3rd Generation Partnership Project; Technical Specification Group Services and System Aspects; Architectural Requirements". 2004–06.
5. J. Bates "Optimizing Voice in ATM/IP Mobile Networks". McGraw Hill Telecom Engineering, ISBN 0-07-139594-6 2002.
6. 3GPP TS 23.107. "3rd Generation Partnership Project; Technical Specification Group Services and System Aspects; Quality of Service (QoS) Concept and Architecture" 2006–03.
7. ITU-T Recommendation Q.1950. "Specifications of Signaling Related to Bearer Independent Call Control (BICC)" 07/2001.
8. 3GPP TS 29.232 V6.1.0. "3rd Generation Partnership Project; Technical Specification Group Core Network; Media Gateway Controller (MGC) – Media Gateway (MGW) Interface Stage 3 (Release 6)". 2005–03.
9. 3GPP TS 23.207 V6.6.0. "3rd Generation Partnership Project; Technical Specification Group Services and System Aspects; End-to-end Quality of Service (QoS) Concept and Architecture (Release 6)". 2005–09.
10. 3GPP TR22.978-7103GPP TR 22.978 V7.1.0. "3rd Generation Partnership Project; Technical Specification Group Services and Systems Aspects; All-IP Network (AIPN) feasibility study (Release 7)". 2005–06.
11. 3GPP TR 23.981 V6.4.0. "3rd Generation Partnership Project; Technical Specification Group Services and System Aspects; Inter-Working Aspects and Migration Scenarios for IPv4 based IMS Implementations (Release 6)". 2005–09.
12. J Rosenberg, H Schulzrinne, G. Camarillo, A. Johnston, J. Peterson, R. Sparks, M. Handley, E. Schooler. RFC 3261 SIP. "Session Initiation Protocol". June 2002.
13. RFC 3588. "Diameter Base Protocol". IETF September 2003.
14. RFC 2748. "The COPS (Common Open Policy Service Protocol)". IETF January 2000.

15. 3GPP TS 23.234 V6.6.0. "3GPP 3rd Generation Partnership Project; Technical Specification Group Services and System Aspects; 3GPP system to Wireless Local Area Network (WLAN) Inter-Working; System Description (Release 6)". 2005–09.
16. 3GPP TS 32.252 V6.0.0. "3rd Generation Partnership Project; Technical Specification Group Service and System Aspects; Telecommunication Management; Charging Management; Wireless Local Area Network (WLAN) Charging (Release 6)". 2005–09.
17. 3GPP TS 32.296 V6.2.0. "3rd Generation Partnership Project; Technical Specification Group Service and System Aspects; Telecommunication Management; Charging Management; Online Charging System (OCS): Applications and Interfaces (Release 6)". 2005–09.
18. "Input Requirements for the Session Initiation Protocol (SIP) in support for the European Telecommunications Standards Institute (ETSI) Next Generation Network (NGN) simulation services", draft-jesske-sipping-tispan-requirements-01.txt (work in progress), June 2005.
19. R. Jesske, D. Alexeitsev, M. Garcia-Martin. 3G Offered Traffic Characteristics Report Report No. 33 Report from the UMTS Forum. November 2003.
20. F. Cuervo, N. Greene, A. Rayhan, C. Huitema, B. Rosen, J. Segers. RFC 3015. "Megaco Protocol Version 1.0". November 2000.
21. ITU-T Recommendation H.248.1. "Gateway Control Protocol Version 2". 05/2002.
22. N. Greene, M. Ramalho, B. Rosen. RFC 2805. "Media Gateway Control Protocol Architecture and Requirements", April 2000.
23. R.P. Ejzak, C.K. Florkey, R.W. Hemmeter. Network Overload and Congestion: A Comparison of ISUP and SIP. *Bell Labs Technical Journal*, 9(3), 173–182, 2004.
24. J. Robert Ensor, J.O. Esteban. On SIP Performance Mauricio Cortes. *Bell Labs Technical Journal*, 9(3), 155–172, 2004.
25. 3GPP TS 23.002 version 6.7.0 Release 6/ETSI TS 123 002 V6.7.0. "Digital Cellular Telecommunications System (Phase 2+); Universal Mobile Telecommunications System (UMTS); Network Architecture". 2005–03.

6

Value-added Services

6.1 Introduction

This chapter describes how higher-level "value-added" services are delivered through a converged multimedia network. In the current competitive environment, value-added services are of increasing importance for Service Providers as they strive to maintain and increase the ARPU (Average Revenue Per User) against the erosion of revenues from voice and connectivity services alone.

No industry-wide definition of what constitutes a value-added service exists but for the purposes of this chapter, the following definition is used. A value-added service is where application (business) logic exerts influence over the real-time progression of activity within the converged multimedia network. The focus is on "externalised" application logic, in addition to the underlying network infrastructure. Thus, the presentation or restriction of the Calling Line Identifier (CLI) would not be considered a value-added service since it is embedded within the network infrastructure itself, whereas simple number translation (such as 1–800 services) would be considered value-added services since the logic for such services (the translation function) resides in an external application platform.

6.2 Service Creation and Delivery Technologies

This section focuses on the different techniques and technologies available for creating and delivering value-added services across a converged multimedia network.

6.2.1 Service Delivery in the PSTN

6.2.1.1 Intelligent Networks

Intelligent Networks (INs) appeared in the early 90s as a model for Service Providers to deliver value-added services apart from traditional TDM networks. The main driving forces behind the evolution of the IN model was the realization that it would not be possible to continue to build more and more services into the digital switches themselves, to speed up the time taken to create and launch new services by decoupling the service

---

Converged Multimedia Networks Juliet Bates, Chris Gallon, Matthew Bocci, Stuart Walker and Tom Taylor
© 2006 John Wiley & Sons, Ltd

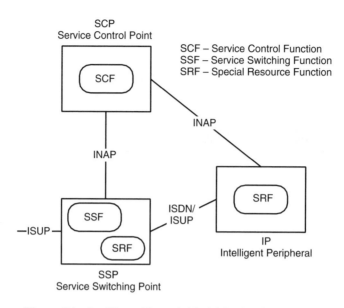

Figure 6.1   Intelligent Network Model for Service Delivery

development from switch development (which typically had long lead times) and the introduction of newer, off-the-shelf IT technology into the service development process.

The IN model physically separated the session/call control functions and the value-added service functions as shown in Figure 6.1.

The Service Switching Point is a digital switch with additional functionality (SSF and Special Resource Function (SRF)) that allows it to engage the services of an external application platform at certain points in the call processing. Both Bellcore and ETSI developed call state models and defined "trigger points" within the call model at which the application platform (SCP) could be engaged. Trigger points could be "armed" on a variety of criteria such as per subscriber, per trunk group or number prefix. Call set-up processing is paused when an armed trigger point is reached and a request is sent to the application platform (SCP); the SSP awaits instructions from the application platform (SCP) to determine how to proceed with call set-up. An additional protocol, INAP – Intelligent Network Application Part, was defined by Bellcore and ETSI (each organisation defined its own version) for the SSP to SCP interface, although in practice many vendor and network proprietary extensions to the standard INAP protocols emerged. INAP defined a set of operations that the SCP could perform on the call, allowing it to modify call parameters (such as changing the destination address), to arm further "dynamic" triggers, allowing the application to be aware of subsequent events that occur on the session and to perform set-up and tear down functions on the session itself. The IN model also defined a SRF to provide interaction functions such as the playing of tones and announcements and the collection of keypad (DTMF) input. The SRF was either co-located with the SSF or located in a separate network element, the Intelligent Peripheral. When the SRF was located in a separate network element, then in order to make use of the function, the SCP would instruct the SSP to establish a temporary bearer connection to the IP and then control the IP using the INAP protocol.

Typically, a Service Creation Environment (SCE) would be provided with the SCP in order to allow the rapid creation of services. A SCE would generally comprise a graphical tool that allows the construction of services by combining a number of pre-defined building blocks through a decision tree. Some standardisation occurred in defining the building blocks termed *SIBBs (Service Independent Building Blocks)* in ETSI and DGN (Decision Graph Nodes) in Bellcore. Many vendors also included additional proprietary building blocks with their SCEs.

The IN model delivered a number of extremely successful services and is the most common delivery mechanism for prepaid mobile services, which account for over half of all mobile phones worldwide. For further information on INs, a number of textbooks on the subject are available such as Magedanz and Popescu-Zeletin [1]

6.2.1.2 Service Nodes

Service Nodes were another model for service delivery that appeared in a similar time-frame as the IN model (described above) as shown in Figure 6.2. The Service Node model is the co-location of the Service Switching Function, Service Control Function and SRF into a single network element.

The combining of the three functions into a single network element often meant that a greater degree of flexibility was possible since the vendor of the Service Node was in full control of all the functional elements and not constrained by a "standards-based" call model or INAP protocol. Similar to the Service Control Points of the IN model, Service Nodes often also included graphical SCE. In the case of Service Nodes, the building blocks used in the SCE were almost always proprietary in nature. The downside of the Service Node model was its centralised nature; in order to access the services it provided, calls had to be routed to it, resulting in the "tromboning" of calls through the Service Node. This "tromboning" increases the cost of service delivery through increased transport costs and increases the number of switching "ports" used per call. Service nodes were

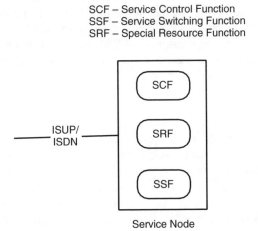

Figure 6.2   Service Node Model for Service Delivery

widely deployed for call-terminating services such as dial in conferencing, voicemail and televoting.

6.2.2 SIP Application Servers

In converged multimedia networks, where SIP is the ubiquitous core network signalling protocol (a detailed description of the SIP protocol can be found in Chapter 2, value-added services are predominantly delivered through SIP Application Servers, sometimes termed *feature servers*.

From a signalling standpoint, SIP Application Servers are somewhat analogous to the service nodes of traditional networks. SIP Application Servers form part of the SIP-signalling path and act upon that signalling in order to provide services to the session. Since SIP has a natural separation of the signalling path and media stream, this model does not incur the cost of tromboning the media through the application network element and does not require a separate protocol to be defined for applications to interact with the session.

The following sections look at the different technologies used to create SIP Application Servers, the three recommended by the IETF (viz Call Processing Language (CPL), SIP Common Gateway Interface and SIP Servlets) along with another technology from the JAIN (Java for Advanced Intelligent Networks) family. Parlay and Parlay X also feature as important service delivery technologies in converged multimedia networks; these are covered in some detail later in the chapter.

6.2.2.1 CPL

CPL is described and defined in RFC 2824 [2] and RFC 3880 [3]. CPL is designed to work with both SIP and H.323 protocols and is based on a model that allows users to create services written in a simple, static and non-expressionally complete language. CPL is an XML-based scripting language designed to be interpreted by a lightweight parser and can be considered safe in that it does not allow the definition of variables and loops, and the CPL scripts do not have the ability to execute an external program.

The CPL scripts could be created by the Service Provider, the end subscribers themselves or by trusted third parties. Typically, a CPL script will be created either by hand or by a GUI tool (similar to the creation of HTML pages). Once created, the CPL scripts are downloaded to a SIP-signalling server, effectively making it an Application Server. The CPL scripts control the actions of the Application Server in the SIP-signalling path, which can proxy, redirect or reject the SIP INVITE. Generally, a CPL script is associated with a particular end point or subscriber identified by a SIP address.

The CPL language itself is an XML-based scripting language that allows the definition of a directed service graph or service tree. The elements within the service graph are termed *nodes*; each node may have a number of outputs, which in turn determine the next node to process. Nodes take two basic forms, decisions (termed *switches*) and actions.

There are four switches defined in CPL, Address, String, Time and Priority. The Address switch allows a decision to be made against the calling and/or called address or its sub-element. The String switch allows free-format text-based checks against other identified data elements in the call set-up signalling (INVITE in SIP), for example, Organisation, Subject, User Agent and Display. The Time switch allows for data- and time-dependent decisions and Priority allows decisions to be made on the priority associated with the call set-up signalling.

There are four main actions defined in CPL, these being proxy, redirect, reject and log. The proxy action allows the propagation of call set-up signalling to the destination address (which can be modified by the CPL script), the node allows for further continuation nodes determined by the outcome of the call set-up attempt, namely, busy, no answer, redirection, failure, timeout, default (any other condition). The redirect node will cause the Application Server to send a SIP Redirect in response to the received INVITE; the node has no continuations since the redirect node effectively removes the Application Server from the SIP transaction. The reject node will send an appropriate error response (the error value is specified by the CPL script) and, like the redirect node, has no continuations. The log node allows the storage of session information into non-volatile storage on the Application Server.

CPL does allow for extensions to the language to be defined, any extensions must have an appropriate XML namespace assigned to it.

The example CPL script below (taken from RFC 3880 [3]) acts such that it rejects anonymous calls. Since the Address switch does not define an "otherwise" clause then if an INVITE is received that does not contain an anonymous FROM header, then the Application Server would apply its default behaviour, that is to proxy the INVITE towards the destination.

```
<?xml version="1.0" encoding="UTF-8"?>
<cpl xmlns="urn:ietf:params:xml:ns:cpl"
   xmlns:xsi="http://www.w3.org/2001/XMLSchema-instance"
   xsi:schemaLocation="urn:ietf:params:xml:ns:cpl cpl.xsd ">
   <incoming>
     <address-switch field="origin" subfield="user">
       <address is="anonymous">
         <reject status="reject" reason="No anonymous calls"/>
       </address>
     </address-switch>
   </incoming>
</cpl>
```

6.2.2.2 SIP-CGI

The Common Gateway Interface for SIP (SIP-CGI) is defined in RFC 3050 [4] and is an adaptation of the CGI interface used by web servers to create dynamic content for web pages. SIP-CGI provides considerable control of the SIP interface by the applications and services. All the SIP headers received in the SIP message are available for the application; presented as environment variables, the application can make use of or ignore any of the headers as it sees fit. The application also has full control of the SIP requests and responses that it issues (the headers, status codes, reason phrases and message bodies). SIP-CGI is an interface definition and does not mandate any particular programming language for the applications themselves. Applications can be written in a variety of programming languages, typically Perl, C or C++. Figure 6.3 shows the main components of a SIP-CGI-based Application Server. The SIP-CGI Server itself handles the SIP interface and will invoke an application program upon receipt of a SIP message. The determination

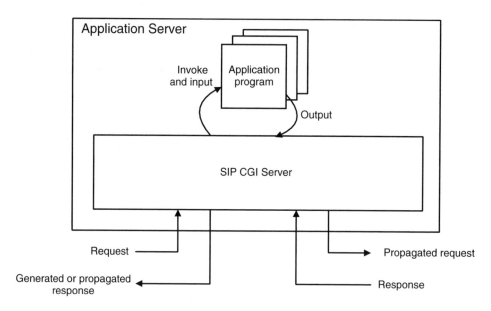

Figure 6.3   SIP-CGI Application Server

of which application program to invoke is made from information contained in the SIP message received but there is no standard definition of which information is used or how the determination is made. RFC 3050 [4] only defines the interface between the SIP-CGI Server and the application program. Vendor implementations of SIP-CGI servers will typically engage applications for either originating or terminating sessions from/to the subscriber. The subscriber is typically identified from the content of the FROM header in the case of originating sessions and the TO header or Request-URI in the case of terminating sessions.

Once invoked, the application program can perform four basic functions towards the SIP-CGI Server, these being the following:

Proxy Request – The request is received by the application program, which can add, modify or remove any headers, and their values, before forwarding the request towards the destination.

Return Response – After receiving the response from the destination client, the application program propagates the response back to the sending client; the application program can control the addition, deletion and modification of headers to include in the response.

Generate Request – A new request is created, originating at the Application Server; the application program creates the header values and body to include in the request.

Generate Response – After receiving a request, the application program generates a response back to the sending client. The application program creates the header values to include in the response.

Depending on the function of the application program, several of these functions may be performed at once, for example, generating a provisional response to an INVITE back towards the sending client while the request itself is proxied on towards the destination client.

The SIP-CGI model supports stateful application programs. An application program executes when a message is received and terminates once the output has been returned to the SIP-CGI server. The application program can instruct the SIP-CGI server upon its termination that it should be invoked for any subsequent messages received for this session. If re-invocation is required, the application program passes a unique opaque tag to the SIP-CGI server. This tag is returned unmodified in an environment variable when the application program is re-invoked upon receipt of a subsequent message for the session. The tag can be used by the application program as an identifier to access appropriate session context information. The use of the opaque tag to enable stateful applications is very similar to the concept of cookies in HTTP.

6.2.2.3 SIP Servlets

The SIP Servlet Application Programming Interface (API) was defined through the Java Specification Request 116, started in April 2001 with its final release in February 2003. The specification can be accessed from the Java Community Process site [5]. In a manner similar to SIP-CGI (described in the previous section), SIP Servlets extend the Servlet model developed for HTTP and apply it to the SIP protocol, one goal of which is to enable mainstream IT developers to create SIP-based network applications.

A SIP Servlet is a Java-based application component, which is managed by a SIP Servlet container. Being Java based, the application components are compiled into platform-neutral byte code that is loaded dynamically into the Java-enabled SIP Application Server (itself running inside the Java Virtual Machine, JVM). SIP Servlet applications should be highly portable across different Java-enabled platforms.

The SIP Servlet model is similar in concept to the SIP-CGI model, but rather than delivering a SIP request to a separate stand-alone application program, the message is passed to a SIP Servlet class running on the SIP Application Server. SIP Servlet applications register an interest in the events that they wish to be triggered by. When a SIP message arrives, the Servlet container checks if any Servlet-triggering rule has been registered that matches the characteristics of the arriving message. If so, one of the Servlet's methods is invoked (e.g. doInvite() upon receipt of an INVITE request).

Access to the SIP Headers and their contents are provided through the getHeader() and setHeader() methods, giving the application full flexibility on the handling of headers in propagated and created SIP requests and responses.

The SIP Servlet container also provides a timer service that allows an application to schedule timers and receive notification of their expiry (similar to SIP requests and responses, an application registers interest in timer events). The application invoked by the timer can take any appropriate action such as sending SIP requests and responses. An example of a timer-invoked application might be a prepaid application tearing down a session upon expiration of a timer, the timer having been calculated from the remaining balance in the account.

The SIP Servlet model also has a more sophisticated context model than SIP-CGI. Similar to SIP-CGI application programs, SIP Servlets are stateless but access to session state information is provided through the SIP Servlet container. The SIP Servlet model defines two kinds of sessions for context purposes. The protocol session essentially relates all SIP messages and SIP Servlet invocations related to a SIP dialogue (sharing the same SIP Call Id). The application session contains one or more protocol sessions that are related at an application level; context information is stored against the application session and is available to the SIP Servlet whichever protocol session within the application session it is invoked against.

6.2.2.4 JAIN SIP/SIP-Lite

Two other Java-based technologies for creating SIP application have also emerged, JAIN SIP and SIP-Lite.

JAIN SIP is part of the JAIN (Java for Advanced Intelligent Networks) family of API definitions. The goals of JAIN are as follows:

Service Portability – standardised APIs permit the portability of applications between one vendor's equipment and another.
Network Convergence – the abstraction of the API definitions allow an application to behave in the same manner irrespective of the underlying network technology.
Service Provider Access – permit applications that run outside the network operator's domain to access network capabilities in a controlled and secure manner.

Only the first of these goals, Service Portability, applies to the protocol-level JAIN SIP API. The API provides a low-level interface (little abstraction) to the SIP protocol. As part of the JAIN family JAIN SIP includes a downloadable reference implementation and test suite (termed *Technical Compatibility Kit – TCK*). JAIN SIP is defined through JSR-000032.

SIP-Lite is a high-level API that provides a high degree of abstraction from the SIP protocol. SIP-Lite is aimed at applications that have SIP as their underlying protocol but obviating the need for the application developers to understand the intimate details of the protocol. The goal is to allow the wider community of Java developers to create SIP-based network applications. SIP-Lite is defined through JSR 125.

6.2.2.5 Extending IN – IM-SSF

The delivery of existing (within the PSTN) IN services into SIP-based networks can be achieved through the IM-SSF function (IP Multimedia Service Switching Function – the acronym was defined by Third Generation Partnership Program (3GPP) as part of their IP Multimedia Subsystem). The IM-SSF acts as a SIP Application Server towards the other SIP network elements (typically as a proxy or Back-to-Back User Agent) but acts as a traditional IN Service Switching Point towards the IN application; from the IN application perspective, the IM-SSF appears as an SSP.

Logically, the IM-SSF function comprises three main functions (shown in Figure 6.4). The SIP stack interfaces to the SIP network elements under the control of the SSP State Machine. The SSP State Machine replicates the call state model of the SS#7 switches (or its elements that map to SIP) and at the relevant triggering conditions, engages the IN applications through appropriate TCAP/INAP messages. The IM-SSF function is uni-directional in that it is possible to engage an IN application from the SIP network but not

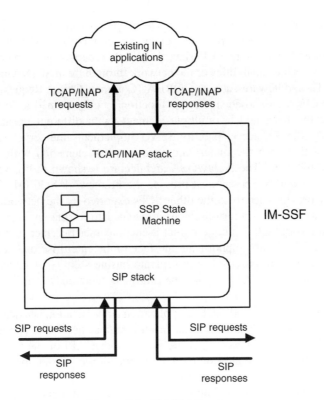

Figure 6.4   IM-SSF function

the converse (engage a SIP application from the TDM network). This is largely because SIP is effectively a peer protocol of ISUP and the IM-SSF works by "peering" SIP to ISUP in the SSP State Machine.

The IM-SSF function is often not realised as a stand-alone network entity but is embedded into the Call Agent or Softswitch itself; this can lead to some limitation on how services are engaged onto a subscriber session.

6.2.3 Parlay

The Parlay group was formed in 1988 by a community of operators, IT vendors, network equipment providers and application developers to define a set of APIs that would combine the best of both the telecom and IT worlds. The aim was to enable telecom services to be developed using common off-the-shelf IT technology rather than specialised network technology, opening up enhanced service creation to a wider community of developers. Through the Parlay Joint Working Group (JWG), other standards organisations have input into the API definitions, including the 3GPP, The Third Generation Partnership Program 2 (3GPP2) and the European Telecommunications Standards Institute (ETSI). 3GPP has adopted the Parlay APIs under the name Open Service Access (OSA) but somewhat confusingly use a different release numbering for the API versions; OSA Rel 4 corresponds to ETSI ES 201 915/Parlay 3.2, whereas OSA Rel 5 corresponds to ETSI 202 915/Parlay 4.0. The latest Parlay specification is 5.0, which is backwards compatible with earlier versions.

The APIs are grouped into sets termed *Service Capability Features* (*SCF*) which export control of some aspect of network or associated systems behaviour such as Call Control or Charging (in earlier versions of Parlay, these API groups were termed *Services*). An application accesses the capabilities of the network through these SCFs which are exported over standard IT middleware such as CORBA (Common Object Request Broker Architecture). The SCFs are exposed through a functional entity termed a *Service Capability Server* (physically this could be multiple elements with different instances supporting different SCFs). The Service Capability Server implements the server side of the API (often by interacting with the underlying core network elements) with the application itself acting as the client. The Parlay/OSA architecture is shown in Figure 6.5.

One Service Capability Server must provide the Framework SCF [6]. The Framework SCF essentially provides access to the other SCFs exported by the network; SCFs register with the Framework which in turn provides a mechanism for the applications to discover them. The Framework SCF is unique in that it encompasses interface to both the applications and other SCFs (it also supports an interface to the enterprise operator for scenarios in which applications are provided by a separate business entity other than the network operator). The Framework SCF supports the following functions towards the applications.

Authentication. The Framework SCF to application authentication is a peer-to-peer model although it does not need to be mutual in all circumstances. The application must be authenticated by the Framework SCF before any other API of the framework or SCF can be used. The interface also supports authentication of the Framework SCF by the application in scenarios that require this.

Discovery. Once authenticated the application can use the discovery functions to obtain information on the SCFs it is authorised to utilise.

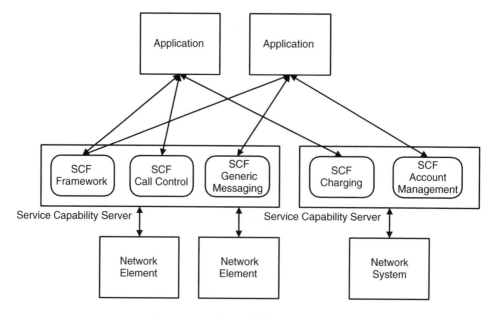

Figure 6.5 Parlay/OSA Architecture

Service Agreement. Before an application can utilise the other SCFs a service agreement must be established. This interface represents the online aspect of any service agreement (there may be offline aspects also) and the agreement must be "signed" by the application before it can access any other SCFs.

SCF Access. The Framework SCF provides access-control functions to police the access to SCFs or Service Data for any API method invoked by an application.

The Framework SCF supports the following function towards the other SCFs.

Registration. The SCFs offered by the Service Capability Servers are registered with the Framework SCF; this enables the Framework SCF to support the discovery functions towards the applications.

The Framework SCF supports the following function towards an enterprise operator (a provider of applications which is a separate business entity to the network provider).

Service Subscription Function. This represents a contractual agreement between the enterprise operator (providing the applications) and the Framework SCF. In the subscription business model, the enterprise operator takes the role of a customer and the client application, that of a user with the framework acting as a retailer. Details on the Framework SCF can be found in ETSI ES 203 915-3 [6].

In addition to the Framework, there are 12 other SCFs currently defined in Parlay; these are summarised in Table 6.1 below.

6.2.4 Parlay X

Parlay X is a set of Web Services that expose a highly abstracted set of network functions for the development of applications using the Web Services model. The higher degree of abstraction and the use of prevailing IT development practices (Web Services) are intended to open the development of applications to IT developers that are not experts in networks and telecommunications. The higher degree of abstraction lends itself to applications which require some underlying network function but are not concerned with the details of the underlying protocols. This approach will not suit all applications as the more esoteric details of the underlying protocols are provided for a reason and the delivery of some services may require their manipulation.

Figure 6.6 shows the relationship between Parlay X Web Services and other Parlay functions and the supporting network elements. It most cases, Parlay X Web Services are implemented by the Parlay X Server invoking functions on the Parlay GW making use of the Parlay APIs. It is also perfectly legitimate for the Parlay X Server to implement the Parlay X Web Services in a different manner such as interfacing with the underlying network elements or service platforms directly.

Since the Parlay X APIs are Web Services based (the interface is defined in XML and transmitted via the Simple Object Access Protocol – SOAP), the Parlay X Applications can be developed in virtually any modern programming language.

Fourteen Parlay X Web Service APIs are defined and these are summarised in Table 6.2.

Table 6.1 Parlay SCF Descriptions

SCF	Summary
Call Control	The Call Control SCF is composed of four sets of APIs, namely, Generic Call Control, Multiparty Call Control, Multimedia Call Control and Conference Call Control. These APIs support a range of capabilities from basic call control of two party calls to the manipulation of multiparty multimedia conference calls. Details of the APIs can be found in ETSI ES 203 915-4-1 [7], ETSI ES 203 915-4-2 [8], ETSI ES 203 915-4-3 [9], ETSI ES 203 915-4-4 [10], ETSI ES 203 915-4-5 [11].
User Interaction	The User Interaction SCF allows applications to exchange information with the end-users, for example, playing announcements and collecting keypad (DTMF) input or sending short messages (SMS). Details of the User Interaction SCF can be found in ETSI ES 203 915-5 [12].
Mobility	The Mobility SCF allows applications to obtain information about an end-users location and current status. Details of the Mobility SCF can be found in ETSI ES 203 915-6 [13].
Terminal Capabilities	The Terminal Capabilities SCF allows an application to obtain information about the end-user's terminal device and the functions and capabilities it supports. This permits smart applications to adapt themselves according to the device from which the user accesses the application. Details of the Terminal Capabilities SCF can be found in ETSI ES 203 915-7 [14].
Data Session Control	The Data Session Control SCF allows an application to exert a degree of control over data sessions. This API set is predominantly aimed at data sessions initiated from mobile terminal devices. Details of the Data Session Control SCF can be found in ETSI ES 203 915-8 [15].
Generic Messaging	The Generic Messaging SCF provides applications with an interface to a "mailbox" function to send, store and retrieve messages. Details of the Generic Messaging SCF can be found in ETSI ES 203 915-9 [16].
Connectivity Management	The Connectivity Management SCF provides applications with an interface to control and influence the bandwidth and Quality of Service (QoS) between end points in the network. Details of the Connectivity Management SCF can be found in ETSI ES 203 915-10 [17].
Account Management	The Account Management SCF provides applications with an interface to query and update account balances (e.g. to support prepaid applications) and to access account transaction histories. Details of the Account Management SCF can be found in ETSI ES 203 915-11 [18].
Charging	The Charging SCF provides applications with an interface to apply charges to the parties using the application. Details of the Charging SCF can be found in ETSI ES 203 915-12 [19].

Table 6.1   (*continued*)

SCF	Summary
Policy Management	The Policy Management SCF provides application with an interface to network policy functions, allowing the applications to register for policy-related events. Details of the Policy Management SCF can be found in ETSI ES 203 915-13 [20].
Presence and Availability Management	The Presence and Availability SCF allows an application to obtain and set information concerning a subscriber's presence and availability. Details of the Policy Management SCF can be found in ETSI ES 203 915-14 [21].
Multimedia Messaging	The Multimedia Messaging SCF allows an application to perform multimedia (voice, text, binary) message management and delivery. Details of the Multimedia Messaging SCF can be found in ETSI ES 203 915-15 [22].

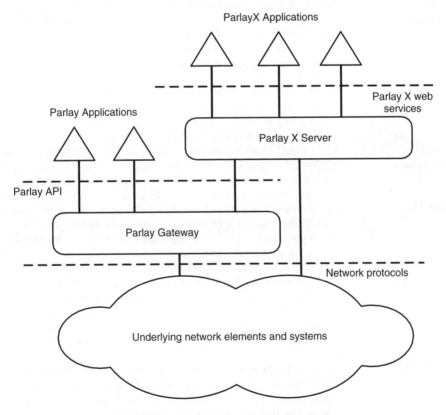

Figure 6.6   Relationship of ParlayX to Parlay

Table 6.2 Parlay X Web Service APIs

Web Service API	Summary
Common	Defines data and service objects that are common across multiple APIs; this includes terminal addresses, charging information and exception handling (both service and policy). Details of the Common Parlay X definitions can be found in ETSI ES 202 391-1 [23].
Third Party Call Control	Supports the creation and management of a call by an application, providing the application with a simple interface to create a call. Details of the Third Party Call Control Web Service can be found in ETSI ES 202 391-3 [25].
Call Notification	Allows application handling of subscriber-initiated calls; the application determines how the call should be treated. Details of the Call Notification Web Service can be found in ETSI ES 202 391-2 [24].
Short Message	A simple interface to allow the sending and receiving of SMS messages. The API supports both a polling and notification model for receiving messages. Details of the Short Message Web Service can be found in ETSI ES 391-4 [26].
Multimedia Message	The Multimedia Messaging API is similar to the Short Message API but more generalised to support a variety of message types including SMS, EMS, MMS, IM and e-mail. This API also supports the notification and polling models for receiving messages. Details of the Multimedia Message Web Service can be found in ETSI ES 202 931-5 [27].
Payment	The Payment API supports payment reservation and both prepaid and post-paid payments. It allows for both volume and currency amounts along with reservation time defaults. Details of the Payment Web Service can be found in ETSI ES 202 931-6 [28].
Account Management	The Account Management APIs provide support for prepaid accounts, allowing the querying of account balances and the recharging of an account either directly or via a voucher. Details of the Account Management Web Service can be found in ETSI ES 202 931-7 [29].
Terminal Status	The Terminal Status API allows an application to retrieve the state (reachable, unreachable and busy) from a terminal or group of terminals. It also allows for application notification in the change of status of a terminal. Details of the Terminal Status Web Service can be found in ETSI ES 202 931-8 [30].
Terminal Location	This API allows an application to request location information for a terminal or group of terminals or to request notification of location change of a terminal or group of terminals. Location is expressed in terms of longitude, latitude, altitude and accuracy. Details of the Terminal Location Web Service can be found in ETSI ES 202 931-9 [31].
Call Handling	The Call Handling API allows an application to set up rules on how calls for a specific number are to be handled. The rules are set up ahead of time and the application is not invoked during the call processing (at least with this API). The rules include Call Accept (White List Screening), Call Reject (Black List Screening), Conditional and Unconditional Call Forwarding and Play Audio. Details of the Call Handling Web Service can be found in ETSI ES 202 931-10 [32].

Table 6.2   *(continued)*

Web Service API	Summary
Audio Call	The Audio Call API provides a flexible way for applications to play prompts to a caller. The API supports three formats of input from the application. Raw text that will be handled by a Text-to-speech engine and played to the subscriber, a Media file (such as a. WAV file) that will be played to the subscriber or VoiceXML which will be interpreted by a VoiceXML browser. Details of the Audio Call Web Service can be found in ETSI ES 202 931-11 [33].
Multimedia Conference	The Multimedia Conference API allows an application to control and manage a multiparty multimedia conference and controlling the conference participants and the individual voice and video media streams. Details of the Multimedia Conference Web Service can be found in ETSI ES 202 931-12 [34].
Address List Management	The Address List Management API actually supports two sets of functions, the management of the lists themselves (creation, deletion, query) and the management of individual addresses within the lists (this API does not allow the creation of addresses themselves). Details of the Address List Management Web Service can be found in ETSI ES 202 931-13 [35].
Presence	The Presence API allows an application to obtain the presence information for one or more subscribers and to register presence for a subscriber or set of subscribers. Details of the Presence We Service can be found in ETSI ES 931-14 [36].

6.3 Service Orchestration

Service Orchestration, as described in this chapter, refers to the integration of multiple applications acting on a single communication session (call instance) without the applications themselves being aware of each other. By insulating the applications from one another, it is possible for a Service Provider to mix and match application offerings from different vendors without requiring the vendors to make custom changes in order to get their particular application to interwork with the other applications that the Service Provider may have chosen to deploy.

The notion of creating larger services or "Metaservices" from smaller functional building blocks is not entirely new. The IN architectures of traditional voice networks used a similar concept of SIBBs being combined in order to create end-user services. There is, however, a critical distinction between the IN architecture and Service Orchestration. The component building blocks of IN, the SIBBs, represented small units of functionality and needed to be combined together in order to provide any form of end-user service. The issue with such an approach is that the SIBBs are tightly coupled with their "combining" environment, which is typically provided by the same vendor. The end-user services are then created within this proprietary environment either by the vendor or by the Service Providers themselves; since it is not credible that a single vendor or Service Provider will create best-in-class applications across the whole application space, the Service Provider faces the choice of deploying inferior services or multiple vendor platforms. This really is no choice at all and inevitably leads to the proliferation of stove pipe applications that we see in most mature Service Provider networks.

The Service Orchestration approach avoids this issue by using complete functional applications as its building blocks. Since the applications can operate as independent stand-alone components, the application vendors are not dependent upon any other vendor's platform or infrastructure. The net result will be a richer "ecosystem" of applications from which the Service Provider is free to pick and choose the most suitable one for its purposes.

The Service Orchestration model has been adopted by both 3GPP, in their IMS architecture, and the MultiService Forum. The similar Service Orchestration models of these two architectures are described below.

6.3.1 IMS Model

The IP Multimedia Subsystem (IMS) is a reference architecture defined by the 3GPP. 3GGP was formed in 1988 by a global consortium of standards bodies to define the next-generation mobile architecture, building upon the success of GSM. IMS is rapidly being adopted as the de facto standard architecture for converged multimedia networks for both wireless and wireline Service Providers. IMS is described in some detail in Chapter 5 and further information can be obtained in books such as Poikselka, Mayer, Khartabil and Niemi [37]. Figure 6.7 shows the components of IMS that are involved in the delivery of value-added services.

The functional elements of IMS involved in service delivery are summarised in Table 6.3 below.

Further details on the IMS functional elements can be found in 3GPP TS 23.288 [38].

The following interfaces are defined between the service delivery functional elements of IMS (as depicted in both Figure 6.7 and Table 6.4).

6.3.1.1 Functional Overview of the IMS Service Delivery Model

When a subscriber registers with the network and is assigned an S-CSCF, his service profile is downloaded from the HSS (subsequent updates to the subscriber's profile in the HSS are pushed down to the S-CSCF). The service profile includes service-triggering information presented in the form of a set of prioritised Initial Filter Criteria (iFC).

Each iFC contains details of the target service that is to be invoked if the set of triggering conditions (Service Points Triggers) are met; the service is identified as a SIP URL. The set of Service Points Triggers can take either the Disjunctive Normal Form (DNF – meaning they conditions are OR'ed together) or the Conjunction Normal Form (CNF – meaning they are AND'ed together). There are four classes of trigger points that can be included in the Service Points Triggers; these are described below.

SIP Method – this indicates the type of request such as an INVITE or NOTIFY.
SIP Header – contains information related to the SIP headers present in the request. The trigger points can be based on the presence or absence of a specific SIP header or the contents of a header.

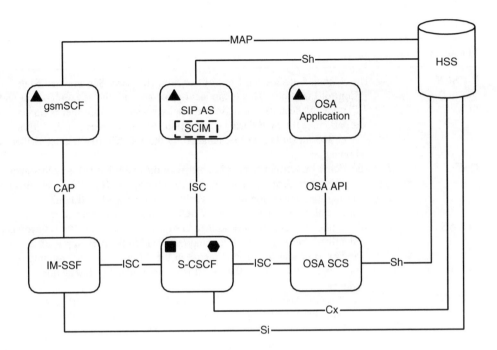

Figure 6.7 IMS Service Delivery Components

Session Case – has three possible values, Originating, Terminating or Terminating Unregistered. The session case of Originating indicates that the Service Point Trigger applies to sessions established by the subscriber (subscriber is calling). The Terminating session case indicates that the Service Point Trigger applies to sessions established to the subscriber (subscriber is being called). The Terminating Unregistered session case indicates that the Service Point Trigger is applied to session attempts to the subscriber when they are not registered (e.g. to forward the call or divert the call to voicemail).

Session Description – defines a Service Point Trigger for the line and content of any line within the Service Delivery Platform (SDP) body.

Additional service information contained in the iFC includes a default behaviour in the event of the service being unavailable; this can take the value of either SESSION_TERMINATE (end the session) or SESSION_CONTINUE (ignore the unavailable service and continue). The iFC can also contain service information which will be transparently transferred from the S-CSCF to the service as it is invoked; further details of this can be found in 3GPP TS23.218 v6.3.0(2005–03) [45].

When a subscriber has registered and the assigned S-CSCF has retrieved the subscriber's service profile, all SIP messages to or from the subscriber are routed through the assigned S-CSCF. When the S-CSCF receives the SIP message, it will compare it against the

Table 6.3 IMS Service Delivery Functional Elements

Functional element	Description
S-CSCF	The Serving Call Session Control Function (formerly Serving Call State Control Function). This component provides both the session control and Service Orchestration functions for subscribers. A subscriber is homed on a particular S-CSCF during registration. All sessions established to or from the subscriber will be handled by the S-CSCF that the subscriber is homed on.
HSS	The Home Subscriber Server is a network database that holds both static and dynamic data elements related to subscribers. The HSS is involved during the registration of a subscriber to the network and holds information regarding which S-CSCF a subscriber is homed on.
SIP AS	SIP Application Server as described earlier in this chapter. 3GPP have also identified, but not defined, a component within the SIP Application Server to handle interactions between the different services running on the Application Server, the SCIM (Service Capability Interaction Manager).
OSA SCS	Open Service Access Service Capability Server, acts as a SIP server towards the S-CSCF and exposes the OSA Service Capability Features (APIs) to the OSA Application. As was described earlier in this chapter, OSA and Parlay are the same.
OSA Application	An application developed using the OSA (Parlay) APIs.
IM-SSF	IP Multimedia Service Switching Function. This acts as a gateway between the SIP signalling used by the IMS application environment and the legacy CAMEL (a GSM standardised version of INAP) services running on the gsmSCF. The IM-SSF will act as a SIP server towards the S-CSCF but as an SSF towards the gsmSCF.
gsmSCF	The GSM Service Control Function (the GSM version of an Intelligent Network Service Control Function described earlier in this chapter). This effectively provides the IN services that were present in the pre-IMS mobile network.

subscriber's iFC which will result in a list of zero or more services to INVOKE for the subscriber. Where the analysis of the SIP message and the iFC result in more than one service, they are ordered on the basis of the priority of the iFC in the subscriber's service profile.

The S-CSCF will then INVOKE the set of services in sequence on the basis of their priority order. The SIP message received by the S-CSCF will be passed to the first service, the output from the first service forming the input to the second service, thus forming a "SIP chain" of services. Figure 6.8 shows a SIP chain of three services being invoked from the SIP INVITE method.

Since the services can be realised by Application Servers acting as Back-to-Back User Agents, the Call Id used by the SIP INVITE messages as they traverse the chain may change and, therefore, a different correlation mechanism is needed. Correlation is achieved through the inclusion of an entry in the Route header. The entry included in the Route header will serve to both ensure that the propagated INVITE from the Application Server

Table 6.4 Interfaces between IMS Service Delivery Functional Elements

Interface	Description
ISC	IMS Service Control interface. ISC is a SIP interface used between the S-CSCF and the different application platforms (SIP AS, OSA SCS and IM-SSF). Further details on the ISC interface can be found in 3GPP TS 24.229 [39].
Cx	The Cx is an interface between the S-CSCF and the HSS (also used by the Interrogating Call Session Control Function which is not shown) in order to access the subscriber's service data. The Cx interface is based on the Diameter protocol defined in RFC 3588 "Diameter base protocol" [52] and RFC 3589 "Diameter Command Codes for Third Generation Partnership Project (3GPP) Release 5" [51]. Further details on the Cx interface can be found in 3GPP TS 29.228 [40] and 3GPP TS 29.229 [41].
Sh	The Sh interface is between the HSS and the SIP AS and OSA SCS applications. It is used by the application to retrieve and update user data stored in the HSS. The Sh interface is also a Diameter-based protocol. Further details on the Sh interface can be found in 3GPP TS 29.328 [42] and 3GPP TS 29.329 [43].
Si	The Si reference point is an interface between the gsmSCF and the HSS based on the MAP (Mobile Application Part) protocol; this replicates the MAP interface to the pre-IMS HLR (Home Location Register). Further details on the Si interface can be found in 3GPP TS 29.278 [44].
OSA API	The OSA API (Parlay API), the interface exposed by the OSA SCS for applications to utilise. An overview of these APIs is presented earlier in this chapter.
CAP	The CAMEL (Customised Application of Mobile Enhanced Logic) Application Part, a GSM standardised equivalent of INAP (Intelligent Network Application Part) carried over an SS#7 signalling interface.

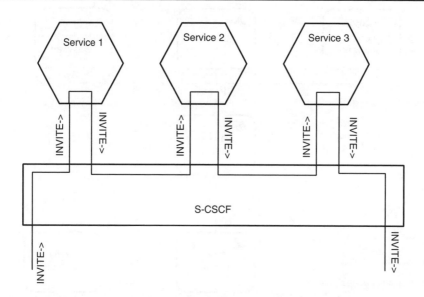

Figure 6.8 S-CSCF sequentially engaging services as a SIP chain

will be back to the S-CSCF instance that sent the INVITE to the Application Server and will contain an embedded token that can be used for correlation purposes by the S-CSCF. Since the S-CSCF both creates and consumes the value of this token, the details of its encoding can be implementation specific; the recommendation, however, is that the token is carried in a character string as the user part of the S-CSCF's URL. For example,

```
Route: <sip:correlationtokenvalue@s-cscf.example.com;lr>
```

Another header used for billing correlation purposes is the P-Charging-Vector (defined in RFC 3455 [46]). The P-Charging-Vector is passed along the SIP "chain", the contents of the P-Charging-Vector is expected to be included in any billing records and events created by the S-CSCF and the Application Servers, allowing the billing events related to a single session to be correlated.

6.3.2 MSF Model

Figure 6.9 shows the components of the MSF (MultiService Forum) Release 2 architecture [53] that are involved in the delivery of services. The functional elements and the interfaces between them are described in Table 6.5 and Table 6.6 respectively.

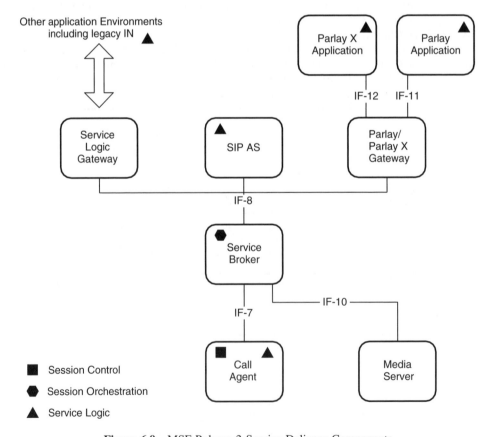

Figure 6.9   MSF Release 2 Service Delivery Components

Table 6.5 MSF Release 2 Architecture Service Delivery Elements

Functional element	Description
Call Agent	The Call Agent provides session set-up, session tear down, session control and routing functions. The Call Agents invoke applications or services via the Service Broker and may themselves provide embedded supplementary services such as Call Waiting. Subscriber terminals are "homed" on a specific Call Agent such that all session set-up requests to and from the subscriber's terminal pass through their "home" Call Agent.
Service Broker	The Service Broker provides Service Orchestration functions for the engagement of one or more applications on a session. The Service Broker enables the interworking of multiple application technologies in a single session, managing feature interaction between different applications and services.
SIP AS	The SIP Application Server provides the service logic and service execution for one or more services; Application Servers are engaged under the orchestration of a Service Broker.
Service Logic Gateway	The Service Logic Gateway function acts as a SIP Application Server towards the Service Broker and exports an application interface towards a non-SIP application function (e.g. an INAP interface to SCP-based applications).
Parlay/Parlay X Gateway	The Parlay/Parlay X Gateway is a type of Service Logic Gateway that exports the Parlay and Parlay X APIs (described in 6.2.3) towards applications.
Parlay Application	The Parlay Application provides the service logic and service execution for one or more services; Parlay Applications are engaged via the Parlay/Parlay X Gateway under the orchestration of a Service Broker.
ParlayX Application	The ParlayX Application provides the service logic and service execution for one or more services; ParlayX Applications are engaged via the Parlay/ParlayX Gateway under the orchestration of a Service Broker.
Media Server	The Media Server provides media functions for the Application Servers; the media functions include the playing of announcements and prompts, tone detection and generation, fax processing, mixing (conference calling), speech recognition and text-to-speech/speech-to-text functions.

The functional elements of the MSF Release 2 architecture involved in service delivery are summarised in the Tables below.

The following interfaces are defined between the service delivery elements of the MSF Release 2 Architecture.

6.3.2.1 MSF Release 2 Service Delivery Overview

The SIP signalling for all calls to and from a subscriber is routed through their "home" Call Agent. The Call Agent can invoke the Service Broker in order to engage and orchestrate

Table 6.6 MSF Release 2 Architecture Service Delivery Interfaces

Interface	Description
IF-7	The Call Agent to Service Broker interface, this SIP interface is defined by MSF-IA-SIP.005-FINAL [47].
IF-8	The Service Broker to Application Server interface, this SIP interface is defined by MSF-IA-SIP.006-FINAL [48].
IF-10	The Service Broker to Media Server interface, this SIP interface is defined by MSF-IA-SIP.009-FINAL [49].
IF-11	This interface represents the Parlay APIs described in 6.2.3.
IF-12	This interface represents the ParlayX APIs described in 6.2.4.

one or more services for the subscriber; the Service Broker is invoked by sending a SIP INVITE to it. The Service Broker is potentially engaged once per subscriber on the session; if both the originating and terminating parties require services, then the Originating Call Agent would engage the originating party's Service Broker and the Terminating Call Agent would engage the terminating party's Service Broker (which could be the same instance). The Call Agent informs the Service Broker if it is being engaged for the originating or terminating party through the appending of a "role" parameter to the Request-URI, "role=orig" indicates that the Service Broker is being invoked for the originating party and "role=term" indicates that the Service Broker is being invoked for the terminating party.

Generally, the Service Broker will propagate the SIP transaction (INVITE) back to the Call Agent for it to perform onward signalling. Since the Service Broker acts as a Back-to-Back User Agent, the Call Agent will need a reliable mechanism to correlate the INVITE received from the Service Broker with the INVITE it sent to it; the Route header mechanism (described in the previous IMS section) is used for this purpose.

Although perhaps not architecturally ideal, the MSF model acknowledges the fact that Call Agents may contain embedded service/application logic. This embedded service logic must be coordinated with the service logic provided by the Application Servers through the Service Broker. As a compromise between flexibility and (signalling path) efficiency, the embedded service logic of a Call Agent can be applied before and/or after the brokered service logic provided by the Application Servers through the Service Broker. A session can therefore be thought of as containing six blocks of service logic shown in Figure 6.10.

Block 1 – Call Agent–embedded service logic applied before any brokered service logic for the originating subscriber.
Block 2 – Service Brokered service logic (potentially provided by multiple disparate Application Servers) for the originating subscriber.
Block 3 – Call Agent–embedded service logic applied after any brokered service logic for the originating subscriber.
Block 4 – Call Agent–embedded service logic applied before any brokered service logic for the terminating subscriber.
Block 5 – Service Brokered service logic (potentially provided by multiple disparate Application Servers) for the terminating subscriber.

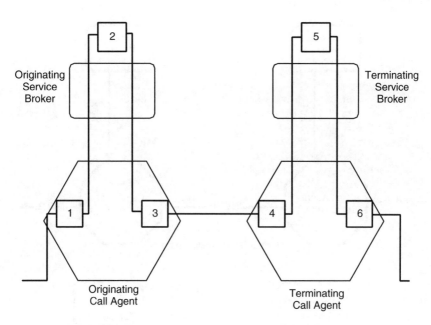

Figure 6.10   MSF Service Logic Distribution

Block 6 – Call Agent–embedded service logic applied after any brokered service logic for the terminating subscriber.

For signalling efficiency in terms of reducing the number of signalling hops, the Originating Call Agent can indicate that the propagated signalling from the Service Broker should be sent directly to the Terminating Call Agent, in case there is no need to apply Block 3 service logic (Call Agent–embedded post Service Broker for the originating subscriber), as depicted in Figure 6.11.

The Originating Call Agent instructs the Service Broker how to route the onward SIP signalling (back to the Originating Call Agent or to the Terminating Call Agent) through the SIP Route header. The Call Agent effectively applies the routing function for the terminating party prior to engaging the Service Broker.

The Service Broker provides the Service Orchestration functions in the MSF architecture, allowing multiple Application Servers to act upon the session in such a manner that the Application Servers themselves do not need to be aware of each other. As previously described, the Service Broker is engaged by the Call Agent for either the Originating or the Terminating subscriber (indicated in the role suffix appended to the Request-URI by the Call Agent sending the SIP INVITE to it). When it is engaged, the Service Broker becomes part of the signalling path for the session.

The Service Broker supports the following four triggering conditions as shown in Table in 6.7 can be applied to either the originating or the terminating sessions.

Different Application Servers will perform different functions on the session depending upon the nature of the services they provide; the MSF model identifies nine capability classes (see Table 6.8). A given Application Server will fulfil one or more of these capability classes.

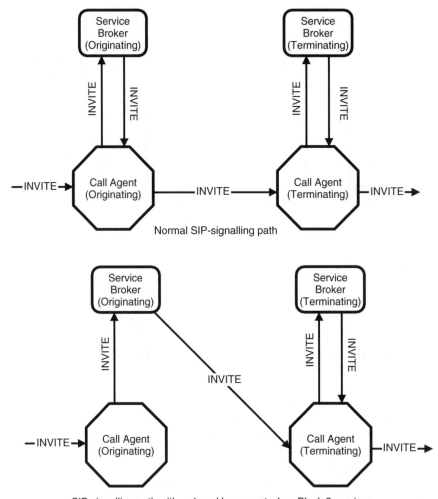

Figure 6.11   MSF SIP-Signalling Paths

Details of how the Service Broker behaves with respect to each of the capability classes are described in the MSF-SIP-IA-SIP.006-FINAL [48].

The Service Broker engages multiple applications onto the session through same SIP-chaining technique employed by the S-CSCF in the IMS model, including using the Route header for correlation purposes and the P-Charging-Vector for billing event correlation. The ordering of applications into the chain is stored in the subscriber service profile; in the MSF model, this is provisioned into the Service Broker from the management layer (whereas in the IMS model the subscriber profile was retrieved from the HSS by the S-CSCF during registration). Another difference between the MSF model and the IMS model of service chaining is that the MSF model allows for application grouping.

Table 6.7 Service Broker Trigger Conditions

Trigger condition	Description
Call Attempt	Service is triggered during the initial call set-up signalling (the SIP INVITE message).
Busy	Service is triggered on a busy condition, identified through the appropriate SIP response (e.g. 486 Busy Here).
No Answer	Service is triggered on a ring-no-answer (RNA) condition, identified either through the appropriate SIP response (480 Temporarily Unavailable) or through a configured timer within the Service Broker.
Hang-up	Service is triggered on a disconnect (SIP BYE message).

Table 6.8 MSF Application Capability Classes

Capability class	Description	Example services
0	Redirection	800 number translation.
1	Basic Call Control	Centrex
2	Multiparty Call Control	Conferencing
3	Referral	Call Transfer
4	User Interaction	Interaction via Media Server
5	Event Notification	Presence, Location
6	Media Terminating	Service Nodes
7	Call Originating	Click to Dial
8	Unified Messaging	Unified Messaging

6.3.2.2 Application Grouping

In order to provide a more efficient signalling path (and hence reduce post-dial delay), the MSF model allows multiple Application Servers to be grouped together and applied into the service chain atomically, reducing the number of signalling hops in the session.

Figure 6.12 shows the SIP-signalling path for three brokered services where Service 1 and Service 2 have been grouped.

6.3.2.3 Application Server Failure

When a service indicates a failure of some kind, either by a SIP error response or through no response at all (which could be the case if the Application Server had failed completely), the Service Broker allows for three treatments to be configured – Continue, Terminate and Net Announcement.

Continue – the failed service is simply ignored and the next service in the chain (if any) is invoked; the session set-up continues without the service.
Terminate – the session is terminated. The error response from the failed service is passed back along the SIP chain.

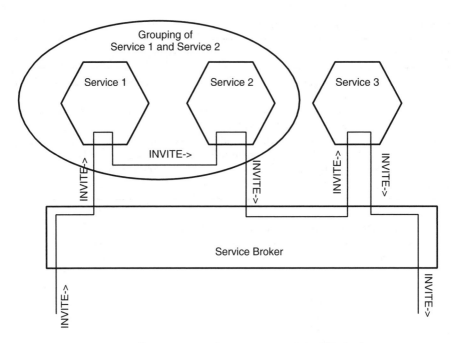

Figure 6.12 Grouping of Applications in the SIP Chain

Net Announcement – the Service Broker invokes a network announcement via a Media Server. The announcement to be played is configurable on a per-service/per-failure-type basis.

These treatments can be configured with varying degrees of granularity such as the following:

A single treatment for all failure types of all services
A specific treatment for all failures type of each service
A specific treatment for each failure type of all services
A specific treatment for each failure type of each service.

6.3.2.4 Multi-service Application Servers

In many cases, a single Application Server may host multiple services. The MSF model allows an individual service within an Application Server to be invoked. The Service Broker may specify an individual service to invoke through the user part of the Request-URI sent to the Application Server. This model follows RFC 3087 [50]. For example, the Service Broker might engage a white list service on an Application Server by using the following Request-URI in the SIP INVITE sent to the Application Server.

```
sip:WhiteList@ScreeningApplicationServer.example.com
```

6.3.2.5 Application Conflict Management

Different services may not be able to function concurrently on the same session because of resource or other conflicts. The Service Broker is responsible for ensuring that conflicting services are not engaged simultaneously on a session. Application conflict resolution is not defined in detail in the MSF model but a simple form of Application conflict resolution (used in the following examples) is the basic service-to-service binary exclusion. For example, service X cannot be engaged on a session that service Y is already engaged on (the rules are directional so the reverse may or may not also be true).

6.3.2.6 Media Server Model

In addition to Service Orchestration, the MSF Service Broker performs a Media Server Brokering function. Media Server Brokering is the dynamic association of a Media Server instance to an Application Server. The Application Server will request a media "capability" from the Service Broker (via a SIP INVITE), which the Service Broker will resolve to a physical Media Server instance and proceed to set up the session to the Media Server instance. Subsequent control of the Media Server by the Application Server will be passively relayed by the Service Broker (if SIP is used as the control protocol). Media Server Brokering allows for a number of efficiencies in Media Server deployment since a pool of Media Servers can be shared across multiple Application Servers; allocation policies can take account of the geographic location of Media Servers relative to the using parties in order to minimise network traffic. In the future versions of the MSF Architecture, the Media Server Brokering function will be extracted from the Service Broker into a separate architectural entity, the Media Resource Broker.

6.4 Service Orchestration Examples

6.4.1 Service Combination Example – IMS

This example shows the Serving Call Session Control Function (S-CSF) combining two discreet applications in order to create an end-user service. The example makes use of two SIP applications, Call Waiting and Call Screening as outlined below.

The Call Waiting application notifies subscribers of an inbound call when they are on a call and allows them to toggle between the two calls. The Application Server operates by being aware of all calls to and from the subscriber (all SIP sessions to and from the subscriber pass through the application server). The notification and toggling functions are performed through SUBSCRIBE/NOTIFY SIP methods; this choice of mechanism was somewhat arbitrary and alternate mechanisms could equally be employed for this purpose.

The Call Screening application acts on inbound calls to a subscriber. Callers are asked to record their name. Call Screening then calls the subscriber and plays the recorded name; the subscriber can then decide either to accept or to reject the call. For the purposes of this example, the Call Accept/Reject function will be performed through in-band DTMF input. The Call Screening application makes use of an independent Media Server for the recording and playback of the name, in addition, to the DTMF collection; this example assumes that the control interface between the Call Screening application and the Media Server is VXML (Voice XML) transported over HTTP, the details of which are not shown.

A combined "Whisper Connect" service can be formed by co-ordinating these two applications such that Call Screening is invoked with the highest priority on terminating calls to the subscriber and the Call Waiting application is invoked on both Originating and Terminating call attempts to the subscriber (the latter with a lower priority than Call Screening). The Whisper Connect service allows the subscriber to determine who is calling him before deciding whether or not to interrupt his current call in order to talk to them.

The figures below show the main steps in orchestrating the two applications onto a single session to create the combined Whisper Connect service.

Step 1 – Eric calls Dave   Eric initiates a call to Dave as shown in Figure 6.13. Since the example is focused on the orchestration of services for Eric, the session and the service components for Dave are not shown. The INVITE issued by Eric (1) will be received by the P-CSCF which is handling Eric's terminal. The P-CSCF propagates the INVITE (2) to the S-CSCF to which Eric is currently registered. The service profile for Eric (downloaded by the S-CSCF from the HSS at the time when Eric registered his terminal) contains an iFC which indicates that the Call Waiting application should be invoked for the INVITE method with a session case of Originating. The S-CSCF therefore propagates the INVITE (3) to the Call Waiting application Server. As Eric has no currently active sessions, the Call Waiting application simply stores the context of the session and propagates the INVITE (4) back to the S-CSCF. Since there are no further iFC matches within Eric's service profile, the S-CSCF proceeds with call set-up sending the INVITE (5) towards Dave. The session is established.

Step 2 – Amy calls Eric   While Eric is still on the call to Dave, Amy makes a call to Eric. The INVITE (6) from Amy is directed to the S-CSCF to which Eric is currently registered (this would typically be done by the I-CSCF querying the HSS to determine the S-CSCF instance that is handling the destination subscriber, Eric in this case). Upon receiving the INVITE (6) the S-CSCF checks the service profile for Eric and determines that there are two iFC for a terminating INVITE to invoke the Call Screening service and the Call Waiting Service. The Call Screening service has the higher priority and so the S-CSCF invokes this service by sending an INVITE (7) to it. The Call Screening application engages a Media Server (INVITE 8) and connects Amy to it to record a message as shown in Figure 6.14.

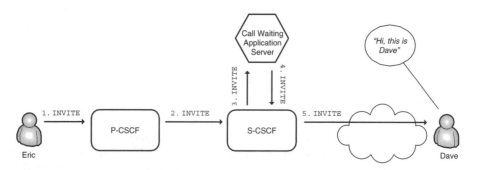

Figure 6.13   IMS Orchestration Example Step 1

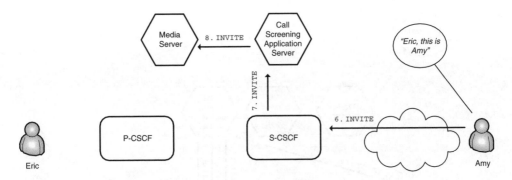

Figure 6.14   IMS Orchestration Example Step 2

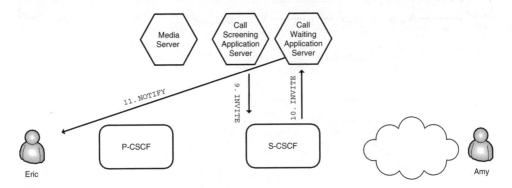

Figure 6.15   IMS Orchestration Example Step 3

Step 3 – Call Screening continues call set-up   Having successfully recorded the message from Amy via the Media Server, the Call Screening application attempts to initiate a call to Eric by propagating the original INVITE (7) received from the S-CSCF. The propagated INVITE (9) is received by the S-CSCF which correlates it with the INVITE sent to the Call Screening application (7) through tag values in the user part of the Route header. The S-CSCF determines that another iFC exists for the terminating INVITE and invokes the Call Waiting application by sending an INVITE (10) to it. The Call Waiting application determines that Eric already has a session in progress and so sends a notification to Eric's terminal. This is shown as a SIP NOTIFY (11) method in Figure 6.15; it is assumed that the terminal has already subscribed to this event, probably at registration time. It should be noted that this is just an example of how call waiting could be implemented; other equally viable mechanisms exist.

Step 4 – Eric accepts the waiting call   As shown in Figure 6.16, Eric accepts the waiting call; again this is through a SIP NOTIFY (12) method (to an earlier established SUBSCRIBE) to the Call Waiting application. The Call Waiting application will then proceed to connect Eric to the new call. Firstly, it puts Eric's call to Dave on hold by sending a re-INVITE with a null SDP to both Eric (INVITE 13, 14 and 15) and Dave (INVITE 16 and 17) and then sending another re-INVITE to Dave for the new call (INVITE 18, 19 and 20); this second INVITE will contain the SDP information

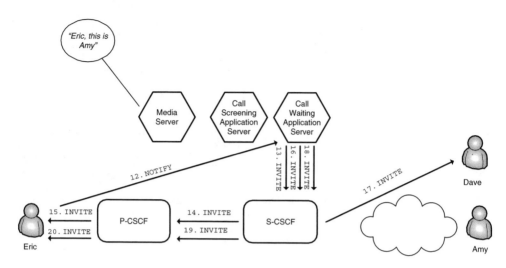

Figure 6.16   IMS Orchestration Example Step 4

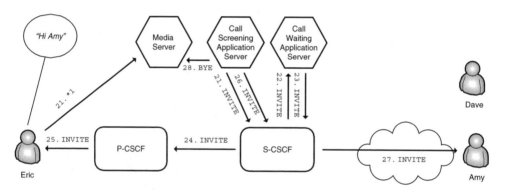

Figure 6.17   IMS Orchestration Example Step 5

for the Media Server used by the Call Screening application (which the Call Waiting application "views" as the calling party). Upon connection with Eric, the Call Screening application will instruct the Media Server to play the message recorded by Amy.

Step 5 – Eric accepts the screened call   As shown in Figure 6.17, Eric decides to accept the call from Amy. This is assumed to be indicated through an in-band DTMF code (*1) which is received by the Media Server (21). The Media Server provides an indication to the Call Screening application that the call has been accepted. The Call Screening application will then connect the parties (Eric and Amy) together, firstly by sending a re-INVITE (21) to Eric in order to establish the media session with Amy's terminal. The re-INVITE will follow the SIP "chain" to Eric passing unaltered through the Call Waiting application Server (22 and 23) and the S-CSCF (twice).

The Call Screening application will then send a re-INVITE (26 and 27) to Amy's terminal to redirect the media streams from the Server to Eric's terminal. Once Eric and Amy

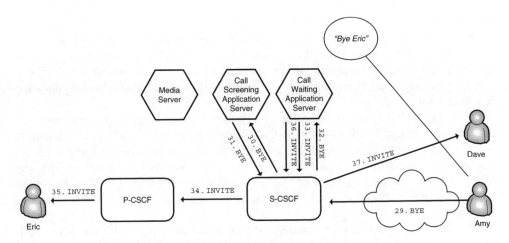

Figure 6.18 IMS Orchestration Example Step 6

are connected, the Call Screening application disconnects the Media Server sending a
BYE (28) to it.

Step 6 – Amy hangs up and Eric is reconnected with Dave Having talked to Eric,
Amy hangs up, causing a SIP BYE (29) to be sent to the S-CSCF (Figure 6.18). The
service profile for Eric does not contain any iFC for the SIP BYE method so the BYE
(30) is propagated along the SIP chain through the Call Screening application (30 and
31) to the Call Waiting application (32). The Call Waiting application receives the BYE
(32) and, since it knows that the original called party (Dave) is "on hold", it attempts
to re-connect the media path to him. Firstly, the Call Waiting application issues a
re-INVITE (33) towards Eric's terminal, which is propagated by the S-CSCF to the
P-CSCF (34) and to Eric's terminal (35). Secondly, the Call Waiting application issues
a re-INVITE (36) towards Dave's terminal to re-establish the media path between
Dave's and Eric's terminals. The re-INVITE (37) is propagated to Dave's terminal,
reconnecting Dave with Eric. The Call Screening application has been removed from
the SIP Chain.

The important principle demonstrated in this example is that the two independent services
of Call Screening and Call Waiting were combined into a higher-level service, "Whisper
Connect", without either service needing to be aware of the other. Since the services
are not aware of each other, they do not need to be modified in any way in order to
be combined.

6.4.2 Conflict Resolution Example – MSF

This example demonstrates the MSF Service Broker resolving service conflicts (sometimes
referred to as *feature interaction*); it makes use of the same two SIP Application Servers
as the previous example (Call Screening and Call Waiting), in addition, to a Voice Mail
application server.

The Call Waiting application is invoked for inbound and outbound call attempts to the
subscriber and Call Screening is invoked on inbound call attempts to the subscriber. Voice
Mail acts on inbound calls and when the subscriber does not answer the call, it connects

the caller to a Media Server which plays an appropriate prompt and allows the caller to record a message (as in the previous example the interaction between Application and Media Servers is not shown in the flows below). The Voice Mail application is triggered on terminating calls to the subscriber, which result in a busy or a ring-no-answer condition. The Call Screening application has the highest priority and the Voice Mail Application has the lowest.

Since Voice Mail connects the calling party to a Media Server to play prompts and record a message, it conflicts with the Call Screening application which also connects the caller to a Media Server since the calling party can only be connected to one Media Server at a time. Without the action of the Service Broker to resolve the conflict, the Voice Mail application would just end up recording the message from the Call Screening application (e.g. "You have a call from Jane Doe, press 1 to accept"), which is clearly not the desired result.

The Call Waiting application can also be considered to conflict with the Voice Mail application, since applying a second call notification (call waiting indication) to a Voice Mail server does not make sense. For this example, we will consider the Call Waiting application to also conflict with Voice Mail.

The figures below show the main steps in orchestrating the three applications onto a session and resolving the conflicts between them.

Step 1 – Amy calls Eric Figure 6.19 shows Amy initiating a call to Eric, resulting in a SIP INVITE (1) being routed to the Call Agent handling Eric's subscription. Eric's profile in the Call Agent indicates that the Service Broker is to be engaged for terminating sessions so the INVITE (2) is propagated to the Service Broker with the addition of a role suffix to the Request-URI to indicate that the Service Broker is being engaged for terminating services. The Service Broker receives the INVITE

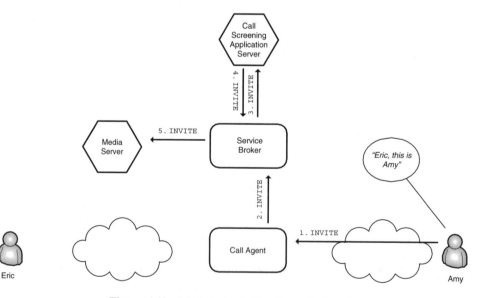

Figure 6.19 MSF Orchestration Example Step 1

(2) and, because the Request-URI indicates that this is a terminating invocation, obtains the terminating subscriber Id (Eric in this case). The service profile for Eric in the Service Broker indicates that two services are invoked on a terminating call attempt, Call Screening and Call Waiting. Call Screening has the highest priority and so it is invoked first by propagating the INVITE (3) to it (since no service has yet been engaged on the session, no conflicts exist). Upon receipt of the INVITE (3), the Call Screening application will connect the caller (Amy) to a Media Server in order to record a message from her. The Service Broker virtualises all Media Server resources in the MSF architecture and so an INVITE (4) is sent to it to request a Media Server resource with the required capabilities (identified in the Request-URI). The Service Broker receives the request for a Media Server resource and resolves it to a physical Media Server instance propagating the INVITE (5) to it. Once the connection to the Media Server is established, the Call Screening application instructs it to prompt and collect a short message from the caller (Amy).

Step 2 – Call Screening continues call set-up   Once the message has been recorded by the caller (Amy), the Call Screening application continues the call set-up by propagating the received INVITE (3) back to the Service Broker as seen in Figure 6.20. The Service Broker correlates this INVITE (6) with the original INVITE (3) through the tag in the user part of the ROUTE header entry. Eric's service profile in the Service Broker indicates that another service, Call Waiting, should be engaged for a terminating call attempt and that no conflict exists between Call Waiting and Call Screening. The Service Broker propagates the INVITE (7) to the Call Waiting application. The Call Waiting application determines that there is currently no active call for Eric so simply stores the call context and propagates the INVITE (8) back to the Service Broker. The Service Broker again correlates the received INVITE (8) with that sent (7) and determines that no further services are to be invoked for this terminating call attempt. The INVITE

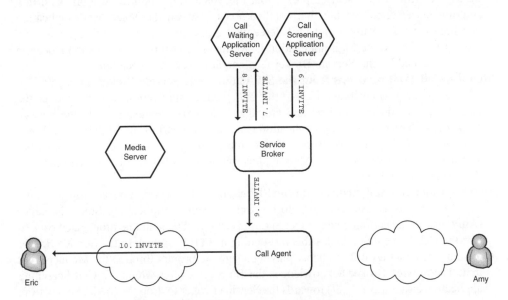

Figure 6.20   IMS Orchestration Example Step 2

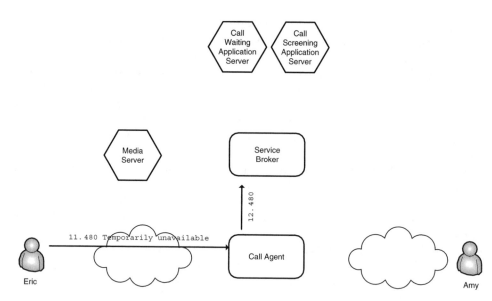

Figure 6.21   MSF Orchestration Example Step 3

(9) is propagated back the Call Agent which continues the call set-up to Eric sending an INVITE (10) to his terminal.

Step 3 – Eric does not answer   Eric does not answer within the defined time period and a 480 Temporarily Unavailable (11) response is returned to the Call Agent, as shown in Figure 6.21, which propagates it (12) to the Service Broker. The Service Broker determines from Eric's service profile that the Voice Mail application is invoked in a ring-no-answer condition but that conflicts exist between the Voice Mail application and both the Call Waiting and Call Screening applications that are already active on the session. The Voice Mail application is not invoked at this point but the invocation event is queued by the Service Broker as a pending service invocation.

Step 4 – Call Waiting drops from the SIP chain   The Service Broker propagates the 480 Temporarily Unavailable (13) to the Call Waiting application, which removes the call context for the call to Eric and propagates the 480 response back to the Service Broker (14). The Call Waiting application is no longer a part of the SIP chain as shown in Figure 6.22. The Service Broker checks the pending service invocation events but still does not invoke the Voice Mail application as a conflict still exists with the Call Screening application.

Step 5 – Call Screening drops from the SIP chain   The Service Broker propagates the 480 Temporarily Unavailable (15) to the Call Screening application. Since the caller (Amy) has already completed the INVITE transaction, 480 cannot be propagated back to her; the Call Screening handles this by sending a BYE (16) back to the Service Broker towards the caller (Amy), probably playing a suitable prompt through the Media Server first. The Service Broker acknowledges the BYE (16), after which the Call Screening application issues a BYE (17) towards the Service Broker to drop the Media Server from the session, and the Service Broker propagates the BYE (18) towards the Media Server.

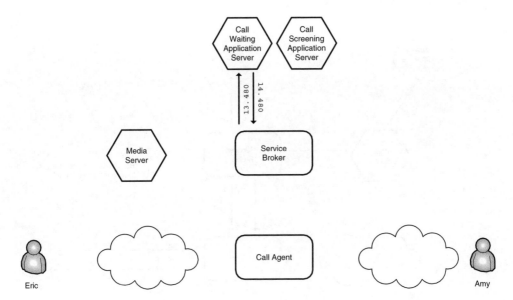

Figure 6.22   MSF Orchestration Example Step 4

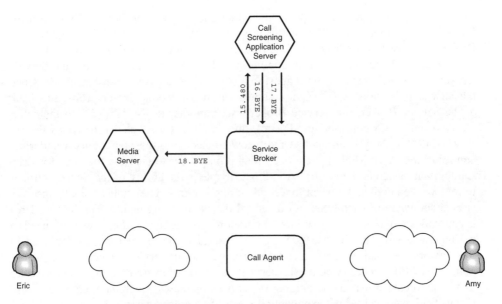

Figure 6.23   MSF Orchestration Example Step 5

The Call Screening application is no longer a part of the SIP chain (see Figure 6.23) and the Media Server has been released.

Step 6 – Voice Mail invoked   The Service Broker, having received the BYE (16), checks the pending service invocation queue. There are no longer any conflicts with the Voice Mail application and so the Service Broker will invoke it as shown in Figure 6.24.

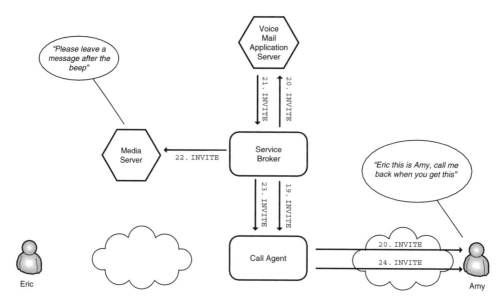

Figure 6.24 MSF Orchestration Example Step 6

Before it does so, however, it puts the calling party (Amy) on hold by issuing a re-INVITE (19) with a null SDP, which is propagated by the Call Agent (20). The Service Broker then invokes the Voice Mail application by sending an INVITE to it (20). The Voice Mail application requests an appropriate Media Server resource from the Service Broker (21), which resolves the request to a physical Media Server instance (22). Once the answer indication (200 OK, not shown) from the Media Server propagates back to the Service Broker, the Service Broker issues another re-INVITE (23) to the caller (Amy) using the Media Server SDP information. The Call Agent propagates the re-INVITE (24) to Amy's terminal. Amy is now connected to the Media Server under the control of the Voice Mail application, which prompts her to record a message for Eric. The important principle demonstrated in this example is the ability of the Service Broker to resolve the conflicts between independent applications. The applications themselves are neither aware of each other nor aware of their conflicting nature. The Service Broker provides a hub model for service integration and interaction. Services only need to be integrated with the Service Broker, not with all of the other services deployed in the network. The hub model scales to a large number of services, whereas a point-to-point integration model (where each service needs to be aware of every other service) inevitably breaks down as the number of services increases, since the integration complexity and overhead grows exponentially with the number of services.

6.5 Service Delivery Platforms

Another model for the delivery of value-added services that is gaining momentum in the industry is the introduction of SDPs. In general terms, an SDP does exactly what its name suggests – it delivers services. However, there is no consistent view in the industry as to what actually constitutes an SDP or indeed what types of service it is intended to deliver.

The Morian Report on SDPs [51] describes an "ideal" SDP as having the following properties:

Providing a complete ecosystem for the rapid deployment, provisioning, execution, management and billing of value-added services.

Supporting the delivery of voice and data services and content in a way that is both network and device independent.

Aggregating different network capabilities and services as well as different sources of content and allowing application developers to access them in a uniform way.

Optionally providing open and secure access to service capabilities for use by external Service Providers and enterprises.

Generally, the SDP is regarded as part of an operator's IT infrastructure rather than an element of the core network. SDPs therefore interwork with many other elements of the operator's IT infrastructure such as CRM, BSS, OSS and AAA in addition to the core network elements such as softswitches and Media Servers.

Although there is no agreed upon definition of what comprises an SDP, they can generically be thought of (as shown in Figure 6.25) comprising four main functional components described below:

Network Abstraction Layer. This component provides a standardised set of interfaces to the different network elements in a technology- and vendor-independent manner. Parlay Gateways often form the Network Abstraction Layer of an SDP although other technologies can form this component such as JAIN servers.

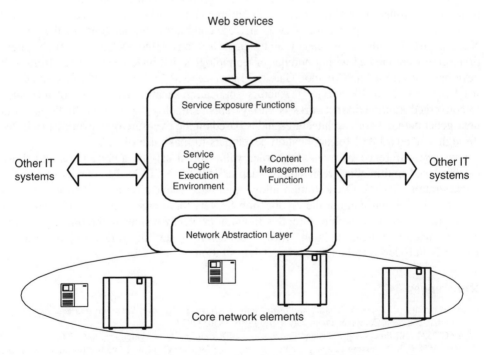

Figure 6.25 Generic Service Delivery Platform

Service Logic Execution Environment. This is a core component of the SDP and provides the delivery and execution environment for a broad range of voice and data services. A number of technologies can form the basis of the SLEE, including J2EE Application Servers. NET Application Servers, a JAIN SLEE or specialised telecom Application Servers. These technologies are often used in combination with, for example, a specialised telecom application server handling the real-time aspects of the service logic and the J2EE application server handling the "business" logic of the service.

Content Management Function. This component is predominantly present in SDPs for Mobile operators and a growing number of fixed-line operators. It provides the provisioning of content to devices, allowing the content to be discovered and delivered to the end devices.

Service Exposure Function. This component exposes the service capabilities of the SDP, usually via Web Services, to third-party Service Providers and enterprises. The Service Exposure Functions allow an operator to open its network through a set of standardised, secure and managed interfaces to applications located in enterprises and other Service Providers.

Summary

Chapter 6 introduces value-added services that are delivered in addition to the core network infrastructure. In traditional networks, such as the PSTN, Value-added services are predominantly delivered through either the standards-based IN model or the more proprietary, but often more flexible, Service Node approach. With the advent of SIP as the primary signalling protocol for converged multimedia networks, a new breed of native SIP application (the SIP Application Server) arrived, combining the flexibility of the Service Node approach with the efficiency and standardisation of the IN model. SIP Application Servers are created with a number of technologies that have been adapted from web development such as the Common Gateway Interface (SIP-CGI) and the far more predominant Servlets (Java SIP Servlets). Another application creation technology of relevance to converged multimedia networks is Parlay. This technology spans both traditional and next-generation network architectures and is based on the exposure of network capabilities through a standard API for application developers to make use of.

One key difference between the earlier value-added service architectures and those of converged multimedia architectures, such as IMS or MSF, is the concept of Service Orchestration. Service Orchestration allows multiple Application Servers to be engaged on the same session without the Application Servers themselves needing to be aware of each other. In the IMS architecture, this function primarily resides in the primary session control function S-CSCF). In the MSF architecture a new network node, the Service Broker is introduced to handle Service Orchestration.

References

1. T. Magedanz, R. Popescu-Zeletin. *Intelligent Networks: Basic Technology, Standards and Evolution*, International Thomson Computer Press, July 1996, ISBN 1850322937.
2. RFC 2824 Call Processing Language Framework and Requirements. www.ietf.org, May 2000.
3. RFC 3880 Call Processing Language (CPL): A Language for User Control of Internet Telephony Services. www.ietf.org, October 2004.

4. RFC 3050 Common Gateway Interface for SIP.www.ietf.org, January 2001.
5. SIP Servlet API Version 1 http://jcp.org/aboutJava/community process/final/jsr116/index.html, February 2003.
6. Open Service Access (OSA); Application Programming Interface (API) Part 3: Framework SCF (Parlay 5) ETSI ES 203 915-3 www.parlay.org, April 2005.
7. Open Service Access (OSA); Application Programming Interface (API) Part 4: Call Control Sub Part 1: Call Control Common Definitions (Parlay 5) ETSI ES 203 915-4-1 www.parlay.org, April 2005.
8. Open Service Access (OSA); Application Programming Interface (API) Part 4: Call Control Sub Part 2: Generic Call Control SCF (Parlay 5) ETSI ES 203 915-4-2 www.parlay.org, April 2005.
9. Service Access (OSA); Application Programming Interface (API) Part 4: Call Control Sub Part 3: Multi-Party Call Control SCF (Parlay 5) ETSI ES 203 915-4-3 www.parlay.org, April 2005.
10. Open Service Access (OSA); Application Programming Interface (API) Part 4: Call Control Sub Part 4: Multi-Media Call Control SCF (Parlay 5) ETSI ES 203 915-4-4 www.parlay.org, April 2005.
11. Open Service Access (OSA); Application Programming Interface (API) Part 4: Call Control Sub Part 5: Conference Call Control SCF (Parlay 5)ETSI ES 203 915-4-5 www.parlay.org, April 2005.
12. Open Service Access (OSA); Application Programming Interface (API) Part 5: User Interaction SCF (Parlay 5)ETSI ES 203 915-5 www.parlay.org, April 2005.
13. Open Service Access (OSA); Application Programming Interface (API) Part 6: Mobility SCF (Parlay 5)ETSI ES 203 915-6 www.parlay.org, April 2005.
14. Open Service Access (OSA); Application Programming Interface (API) Part 7: Terminal Capabilities (Parlay 5)ETSI ES 203 915-7 www.parlay.org, April 2005.
15. Open Service Access (OSA); Application Programming Interface (API) Part 8: Data Session Control SCF (Parlay 5)ETSI ES 203 915-8 www.parlay.org, April 2005.
16. Open Service Access (OSA); Application Programming Interface (API) Part 9: Generic Messaging SCF (Parlay 5)ETSI ES 203 915-9 www.parlay.org, April 2005.
17. Open Service Access (OSA); Application Programming Interface (API) Part 10: Connectivity Manager SCF (Parlay 5)ETSI ES 203 915-10 www.parlay.org, April 2005.
18. Open Service Access (OSA); Application Programming Interface (API) Part 11: Account Management SCF (Parlay 5)ETSI ES 203 915-11 www.parlay.org, April 2005.
19. Open Service Access (OSA); Application Programming Interface (API) Part 12: Charging SCF (Parlay 5)ETSI ES 203 915-12 www.parlay.org, April 2005.
20. Open Service Access (OSA); Application Programming Interface (API) Part 13: Policy Management SCF (Parlay 5)ETSI ES 203 915-13 www.parlay.org, April 2005.
21. Open Service Access (OSA); Application Programming Interface (API) Part 14: Presence and Availability Management SCF (Parlay 5)ETSI ES 203 915-14 www.parlay.org, April 2005.
22. Open Service Access (OSA); Application Programming Interface (API) Part 15: Multi-Media Messaging SCF (Parlay 5)ETSI ES 203 915-15 www.parlay.org, April 2005.
23. Open Service Access (OSA); Parlay X Web Service; Part 1: Common ETSI ES 202 391-1 www.parlay.org, March 2005.
25. Open Service Access (OSA); Parlay X Web Service; Part 3: Third Party Call ETSI ES 202 391-3 www.parlay.org, March 2005.
24. Open Service Access (OSA); Parlay X Web Service; Part 2: Call Notification ETSI ES 202 391-2 www.parlay.org, March 2005.
26. Open Service Access (OSA); Parlay X Web Service; Part 4: Short Messaging ETSI ES 202 391-4 www.parlay.org, March 2005.
27. Open Service Access (OSA); Parlay X Web Service; Part 5: Multi-Media Messaging ETSI ES 202 391-5 www.parlay.org, March 2005.
28. Open Service Access (OSA); Parlay X Web Service; Part 6: Payment ETSI ES 202 391-6 www.parlay.org, March 2005.
29. Open Service Access (OSA); Parlay X Web Service; Part 7: Account Management ETSI ES 202 391-7 www.parlay.org, March 2005.
30. Open Service Access (OSA); Parlay X Web Service; Part 8: Terminal Status ETSI ES 202 391-8 www.parlay.org, March 2005.
31. Open Service Access (OSA); Parlay X Web Service; Part 9: Terminal Location ETSI ES 202 391-9 www.parlay.org, March 2005.

32. Open Service Access (OSA); Parlay X Web Service; Part 10: Call Handling ETSI ES 202 391-10 www.parlay.org, March 2005.
33. Open Service Access (OSA); Parlay X Web Service; Part 11: Audio Call ETSI ES 202 391-11 www.parlay.org, March 2005.
34. Open Service Access (OSA); Parlay X Web Service; Part 12: Multimedia Conference ETSI ES 202 391-12 www.parlay.org, March 2005.
35. Open Service Access (OSA); Parlay X Web Service; Part 13: Address List Management. ETSI ES 202 391-13 www.parlay.org, March 2005.
36. Open Service Access (OSA); Parlay X Web Service; Part 14: Presence ETSI ES 202 391-14 www.parlay.org, March 2005.
37. M. Poikselka, G. Mayer, H. Khartabil, A. Niemi. *The IMS IP Multimedia Concepts and Services in the Mobile Domain*, John Wiley and Sons Ltd, 2004, ISBN 0-470-87113-X.
38. IP Multimedia Subsystem (IMS); Stage 2 3GPP TS 23.228, v6.13.0(2006-03) www.3gpp.org, March 2006.
39. Internet Protocol (IP) Multimedia Call Control Protocol Based on Session Initiation Protocol (SIP) and Session Description Protocol (SDP); Stage 3 3GPP TS 24.229, v6.10.0 (2006-03) www.3gpp.org, March 2006.
52. RFC 3589 Diameter Command Codes for Third Generation Partnership Project (3GPP) Release 5. www.ietf.org, September 2003.
51. The Morina Group. "Service Delivery Platforms and Telecom Web Services and Industry Wide Perspective". June 2004.
40. IP Multimedia (IM) Subsystem Cx and Dx Interfaces; Signalling flows and Message Contents 3GPP TS 29.228, v6.10.0 (2006-03) www.3gpp.org, March 2006.
41. Cx and Dx Interfaces Based on the Diameter Protocol; Protocol Details 3GPP TS 29.229, v6.7.0 (2005-12) www.3gpp.org, December 2006.
42. IP Multimedia Subsystem (IMS) Sh Interface; Signalling Flows and Message Contents. 3GPP TS 29.328, v6.9.0 (2006-03) www.3gpp.org, March 2006.
43. Sh Interface Based on the Diameter Protocol; Protocol Details. 3GPP TS 23.329, v6.6.0 (2005-09) www.3gpp.org, September 2005.
44. Customised Applications for Mobile Network Enhanced Logic (CAMEL); CAMEL Application Part (CAP) Specification for IP Multimedia Subsystems (IMS) 3GPP TS 29.278, v6.1.0 (2005-03) www.3gpp.org, March 2005.
45. 3rd Generation Partnership Project; Technical Specification Group Core Network: IP Multimedia (IM) Session Handling; IM Call Mode; (Release 6) 3GPP TS23.218, v6.3.0(2005-03) www.3gpp.org, March 2005.
46. RFC 3455 Private Header (P-Header) Extensions to the Session Initiation Protocol (SIP) for the 3rd-Generation Partnership Project (3GPP) www.ietf.org, January 2003.
53. RFC 3588 Diameter Base Protocol. www.ietf.org, September 2003.
47. Implementation Agreement for SIP Interface Between Call Agent and Service Broker. MSF-IA-SIP.005-FINAL www.msforum.org, September 2004.
48. Implementation Agreement for SIP Interface Between Service Broker and Application Server. MSF-IA-SIP.006-FINAL www.msforum.org, April 2005.
49. Implementation Agreement for SIP Media Server Interface. MSF-IA-SIP.009-FINAL www.msforum.org, July 2004.
50. RFC 3087 Control of Service Context using SIP Request-URI www.ietf.org, April 2001.
54. MSF Release 2 Architecture. MSF-ARCH-002.00-FINAL www.msforum.org, January 2005.

7

Core Network Architecture

This chapter presents trends in the evolution of the architecture for the core of the converged multimedia network. The motivation for convergence in the core and the requirements that this places on the underlying network infrastructure were presented in Chapter 1. In this chapter, we introduce the network architectures and protocols that enable the core of the network to support both legacy services, which still represent major revenue streams for service providers, new business services such as managed Ethernet and IP Virtual Private Networks (VPNs) and next-generation multimedia and residential broadband services. Multiprotocol Label Switching (MPLS) is widely seen as the most promising candidate technology to allow converged core networks to be built that are optimised for IP-based services. We introduce MPLS, explaining which of its capabilities are important in the core, and why. We then provide a detailed overview of how MPLS enables multiple, service-specific VPNs to be built, each supporting a different carrier service over the common MPLS infrastructure.

7.1 The Convergence Layer: Multiprotocol Label Switching

Historically, many protocols have been touted as the convergence layer for all digital communications services. Protocols in the transport layer, such as Synchronous Digital Hierarchy (SDH) or Synchronous Optical Network (SONET) have, to some extent, provided a single, converged infrastructure over which other services are carried. However, SDH and SONET have focused on providing fixed bandwidth time division multiplexed (TDM) channels. This is inefficient for transporting IP services, which are packet based. In IP, bandwidth is consumed only when there is data to send. However, TDM networks allocate a fixed amount of bandwidth for a service, which may or may not be used. Although this does mean that all services can receive a quantifiable quality of service (QoS), it is not possible to take advantage of the bursty nature of IP traffic to enable statistical multiplexing and hence maximise the usage of the network infrastructure.

Asynchronous Transfer Mode (ATM) goes some way to addressing this, albeit at the network layer. ATM maximises the utilisation of fixed bandwidth channels provided by the transport network while also enabling guaranteed QoS. It does this by transmitting data in fixed length cells that are sent only when an application has data to send. Although

---

Converged Multimedia Networks   Juliet Bates, Chris Gallon, Matthew Bocci, Stuart Walker and Tom Taylor
© 2006 John Wiley & Sons, Ltd

empty cells are sent when an application has no data to send, they are not allocated to a specific connection. This effectively enables statistical multiplexing of traffic over low-bandwidth links. ATM has been very successful in transporting bursty services such as video, data and voice over an SDH infrastructure. In particular, it has proven itself as a key protocol for aggregating consumer broadband access services, such as Asymmetric Digital Subscriber Line (ADSL) between the subscriber and the service provider.

ATM was originally developed with the assumption that it would extend to the end-user application. However, modern networks typically use Ethernet to connect to the end-user; economies of scale have meant that Ethernet LANs cost much less than ATM-based LANs. Furthermore, the growth of the Internet has meant that the bulk of the traffic in a carrier's network is IP. IP uses control protocols such as Transmission Control Protocol (TCP) to provide reliable transmission of data over a variety of underlying layers, which can be ATM, but can also be Ethernet, Frame Relay or many other protocols. This decouples the underlying layer from the application and makes it difficult for that layer to allocate the appropriate amount of network resources. Furthermore, decreases in the cost of providing bandwidth in the core network have negated against the need for the powerful traffic management features of ATM. A further challenge occurs when many IP routers peer across an ATM core network; a full mesh of routing adjacencies must be created between the edge IP routers, which both stresses the IP-routing implementations when the number of routers becomes large (because the number of direct neighbours grows linearly with the number of routers) and means that the number of ATM connections that must be provisioned and managed between the IP routers grows with the square of the number of routing adjacencies. This is the now famous "$O(N)^2$" scaling problem. Therefore, ATM does not always meet the requirements for the converged packet network described above. These considerations have meant that services providers have continued to search for a protocol that meets these requirements.

MPLS has shown promise as a future core network convergence layer and has received substantial attention in the international standardisation and research and development communities in recent years. The history and principles of MPLS are well documented elsewhere [1], and so this chapter will focus instead on how MPLS meets the requirements for a convergence layer protocol.

MPLS forms a part of the set of protocols developed by the Internet Engineering Task Force (IETF) to address specific challenges in the management and performance of IP networks. MPLS was originally intended to enhance packet-forwarding performance in IP core networks. The packet-forwarding decision in IP routers was based on a forwarding table lookup of the IP address contained in the packet header. In MPLS, a label is appended to each packet to form a label-switched path (LSP). This negates the need for a full address lookup; instead, a simple association between the label value and the egress port on which to send the packet to reach the next hop router can be maintained. Packets are assigned a label at the edge of the network by a label edge router (LER) according to a particular forwarding equivalence class (FEC) and are forwarded through the core of the network according to the label value by label-switching routers (LSRs). The association of FEC and label effectively describes how a particular packet should be forwarded and uniquely identifies the LSP on the link to the downstream LSR.

A label-swapping operation by each LSR means that the label value for a given FEC on each link of an LSP will change. Figure 7.1 illustrates the basic MPLS architecture.

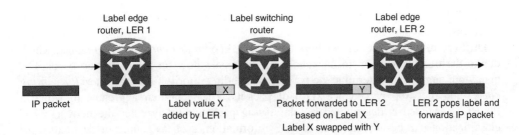

Figure 7.1   MPLS architecture

The MPLS architecture is defined in RFC 3031 [2], while the label stack encoding is described in RFC 3032 [3]. Note that LSPs are unidirectional, and so a pair of LSPs (one in each direction) is required for bidirectional communication between a pair of LSRs. These LSPs may take different paths through the network.

The MPLS label effectively adds a link-connection identifier to IP networks. This also enables enhanced manageability, resilience, traffic engineering and QoS features to be added to those networks. MPLS enables IP flows to be encapsulated in an LSP, or tunnel, that can follow an explicitly traffic-engineered path across the network. Such traffic-engineered LSPs allow resources to be allocated in the nodes through which a tunnel passes. The following sections describe how MPLS achieves this.

7.1.1 Quality of Service in IP Networks

MPLS has become an addition to the IP suite for carriers' core networks. We will therefore first discuss the features that have been developed to support QoS in native IP networks.

IP is a connectionless protocol. Early IP networks forwarded each packet independently from other packets in a given flow, the routing based simply on the globally unique end station address, or IP address, in the packet's header. Each router in the network forwarded packets based on local routing tables, constructed as a result of link state routing protocols such as Open Shortest Path First (OSPF). While such techniques ensure that the reachability of end stations is maximised, it is difficult for the operator to engineer the path taken by flows across the network. This in turn makes it difficult to specifically engineer resources, such as link bandwidths, to expected traffic demands. Therefore, the approach taken to provide QoS was often to over-provision the network. While this is effective in supporting a single QoS guarantee for all services, it does not enable different guarantees to be made to different services, for example, providing separation between real-time services based on the user datagram protocol (UDP) and congestion-aware non-real-time services that use TCP. Such service differentiation is key in generating revenues from premium services.

As a result, a number of tools have been developed to enable QoS to be provided to flows in an IP network. Two methods that have gained prominence are the Integrated Services Framework (IntServ) [4] and the Differentiated Services Framework (DiffServ) [5]. IntServ uses the Resource ReserVation Protocol (RSVP) to allocate resources for Internet traffic flows, providing a number of distinct end-to-end service classes for real-time and guaranteed traffic. However, this requires that routers along the path of a flow must maintain per-flow state information. The amount of state increases with the number of IntServ flows. Therefore, the DiffServ framework has gained traction as a QoS mechanism for IP networks.

DiffServ provides a set of per-hop behaviours, known as *behaviour aggregates*, which define the treatment given to packets of an aggregate flow on a hop-by-hop basis. The behaviour aggregates determine the per-hop scheduling priority (also known as *forwarding class*) and the drop precedence (DP) of packets. The behaviour aggregate to which a flow is associated is determined by classifying packets as they enter the network. The classification can be according to layer 3 information, such as source or destination IP address, or virtually any other factor. Packets can also be marked with "DP," or "colour," depending on their conformance to a traffic contract. This determines the relative priority within a forwarding class with which packets are discarded during network congestion.

Although DiffServ is a useful tool for providing differentiated forwarding behaviour for traffic flows, it does not provide QoS on its own. DiffServ does not provide resource reservation (bandwidth or buffer space) in order to ensure that the network can support the required QoS for a flow. Furthermore, the DiffServ model is primarily applicable only on a single hop; it does not provide a method for signalling QoS requirements along the path of a flow and rejecting a flow if the requested QoS cannot be supported. Instead, the DiffServ PSC (Per-Hop Behaviour Scheduling Class) and DP are indicated using the Type Of Service (ToS) bits in the IP packet's header. See Chapter 8 for more details on the use of DiffServ.

7.1.2 MPLS Traffic Engineering and Traffic Management

MPLS has been extended to incorporate traffic engineering and helpers for traffic management. The PSC and DP for a packet on an LSP can be indicated to the Label-switched Router in one of the two ways. In a label-inferred PSC LSP (L-LSP), the MPLS label implies the PSC; all packets on a given LSP therefore share the same forwarding class. The setting of the experimental (EXP) bits in the MPLS label indicates the DP. Alternatively, an EXP-inferred PSC LSP (E-LSP) can be used, whereby the value of the EXP bits indicates both the PSC and the DP. Each class of service (CoS) can be allocated a share of the bandwidth and can be rate limited. Priority scheduling can ensure, for example, that a higher priority class always receives preferential treatment over a lower priority class. As such, an LSP can be used as a tunnel through the network with an associated QoS and network resource reservation.

The combination of MPLS traffic engineering and DiffServ frameworks provides the ability to compute paths and check the available bandwidth along the path in the network on a per-CoS basis. This is known as *DiffServ aware traffic engineering* [6]. Using this, service providers can deploy and enforce various bandwidth-sharing policies to LSPs of different classes of service throughout the network. Chapter 8 provides more details on MPLS traffic engineering.

7.1.3 Signalling and Routing in MPLS Networks

Signalling and routing are two groups of functions that enable an LSR or LER to determine which next hop LSR or LER to send a packet to, to reach a particular destination router. They effectively assist in building the path through the network for an LSP and automatically configuring the intermediate routers with the LSP labels.

Routing protocols used for MPLS are usually common to the underlying IP network. Routing protocols are well documented elsewhere, so we will only introduce them in the context of their role in the MPLS network. Routers such as LSRs "discover" each other and how they are connected to their peers using interior gateway routing protocols (IGPs) such as OSPF [7] or Intermediate System–Intermediate System (IS-IS) [8]. Large IP networks (such as the Internet) are divided into separate administrative domains (known as *Autonomous Systems (ASs)*), making the overall network highly scalable from a routing perspective. However, IGPs are restricted to one AS (separate IGP instances will run in each AS of a multi-AS network). Routers therefore determine how to reach a destination in a different AS using exterior gateway routing protocols, such as the Border Gateway Protocol (BGP) [9]. Routers at the edge of each AS are known as *Autonomous System Border Routers (ASBRs)* and use a flavour of BGP known as *Exterior BGP (eBGP)* to exchange routes with peer ASBRs in neighbouring ASs. These routes are then advertised to routers within an AS using Interior BGP (iBGP).

MPLS networks also require signalling protocols in addition to the IP-routing protocols to enable LSPs to be established and maintained. Signalling protocols are used to carry the essential attributes of an LSP between neighbouring LSRs. So, although MPLS networks use IP for management, maintenance and operational functions, the signalling traffic often runs on TCP/IP sessions that are segregated from the LSPs they control on separate physical or logical links. MPLS uses signalling protocols for a number of other key functions:

- Enabling neighbouring LSRs or LERs to agree on the MPLS labels to use to send packets to each other.
- Distributing the FECs to LERs to enable them to associate incoming packets to LSPs, and hence tying them to their specific label values and, ultimately, destinations.

The process of signalling thus binds the LSP label to a FEC.

Other uses are also made of signalling protocols, including, for example, carrying traffic parameters describing the bandwidth requirements for an LSP between LSRs and LERs to enable the appropriate resources to be allocated for an LSP by the network. The following sections briefly describe the two signalling protocols used in MPLS: the label distribution protocol (LDP) and an extension to the resource RSVP (known as *RSVP with Traffic Engineering (RSVP-TE)*) that also enables the signalling of MPLS FECs and labels.

Signalling in MPLS is highly flexible, providing a wide range of mechanisms to suit a variety of applications for MPLS. Signalling can be piggybacked on existing protocols, for example BGP or RSVP, but new protocols have also been invented, for example LDP.

Consider a pair of peer LSRs. One is upstream with respect to the flow of the LSP and the other is downstream. The association between FEC and label is signalled in one of the following two ways:

- Downstream on demand. In this mode, the upstream LSR asks the downstream LSR for a label binding to use for a particular FEC.
- Downstream unsolicited. In this mode, downstream LSRs distribute the label bindings to upstream peers without being asked.

The labels distributed by the downstream LSR may be used immediately by the upstream LSR to send traffic for those FECs. If an upstream LSR ceases to need to use a label, it may choose to release the label binding or it may retain the label binding indefinitely in case it needs to use it again. The former mode is known as *conservative label retention* and the latter mode is known as *liberal label retention*.

The most widely used signalling protocol is LDP, standardised in RFC 3036 [10]. This is a simple protocol that exchanges label bindings between peer LSRs, supporting all the modes mentioned earlier. In its basic form, LDP does not support the signalling of other more complex information such as QoS or bandwidth parameters. However, it has been extended to support pseudo wires (PWs) (see Section 7.2.2 below). Also, because LDP allows a downstream LSR to distribute the same label for a FEC to a number of upstream LSRs, it can cause LSPs to merge. This can be advantageous because it minimises the number of labels that the downstream LSR needs to allocate to a FEC, and thus improves the scalability of MPLS. However, the resulting multipoint-to-point structure of the LSP can also be more difficult to traffic engineer than a traditional point-to-point connection.

To address some of these limitations, constraint-based routing was added to LDP. This is known as Constraint-based Routing Label Distribution Protocol (CR-LDP) [11]. This adds traffic-engineering capabilities to LDP. For example, an explicit route for an LSP can be signalled; that is, a list of LSRs that an LSP must transit. It also adds the capability to signal traffic parameters associated with an LSP so that the intervening LSRs can choose how to allocate network resources to the LSP.

However, CR-LDP never quite achieved the popularity of LDP. One possible reason is that it is a hard state signalling protocol; that is, the state information associated with an LSP exists in the LSRs of the network until it is explicitly released, which requires some additional management during network failures. Instead, the IETF chose to focus on extending RSVP, which was already widely deployed in IP networks, to support MPLS labels and FECs. This protocol is RSVP-TE [12]. However, RSVP is a soft state signalling protocol; that is, the state information of the LSP must be periodically refreshed in the LSRs along its path (otherwise it will be aged out) by sending periodic path refresh messages for the duration of an LSP. This can be wasteful of control plane and data path resources in a network with a large number of LSPs. Therefore, methods to reduce the number of path refresh messages have been developed. Optimised versions of RSVP-TE have become the dominantly deployed protocol for traffic-engineered LSPs.

In addition to LDP and RSVP-TE, BGP has also been extended to exchange MPLS labels. BGP has its roots in advertising reachability by distributing routes between multiple edge routers in an IP network. Therefore, it has been extended specifically for multipoint services such as IP VPNs and Ethernet layer 2 VPNs. These are described in more detail later in this chapter.

7.1.4 Protection, Restoration and Service Assurance in MPLS

A converged MPLS network must be able to support services for which there is a strict service level agreement (SLA) that specifies, among other metrics (e.g. latency, jitter, packet loss, etc.), both the maximum amount of time there can be a service outage (or the minimum service availability) and a guaranteed time to repair a fault. These service level guarantees are enabled by guarantees within the network itself. For example, traditional

SDH networks typically guarantee a maximum time of 50 ms to re-establish connectivity following a failure.

In order to conform to such assurances, a network operator needs a set of tools to

(1) indicate if a fault (defect) has occurred;
(2) locate and diagnose the fault;
(3) take remedial action to restore the service as rapidly as possible.

(1) and (2) are functions of the OAM (Operations, Administration and Maintenance) tools, while (3) is a function of mechanisms such as protection switching and fast re-route. This section will discuss how both these sets of mechanisms apply to MPLS networks.

The IP network that underlies MPLS has its own set of OAM tools. These include ICMP (Internet Control Messaging Protocol) Ping and Traceroute. IP networks are connection-less, and therefore the capabilities of these tools are limited to testing the reachability of a destination IP address (using ICMP Ping) or locating the path of a packet in reaching a destination IP address (using traceroute). Recovery from a link or node failure relies on IGP re-convergence; that is, local detection of the failure, flooding of that information through the network and consequent re-computation of the routing tables in each node in the network.

MPLS provides substantial additions to the basic IP maintenance tool set in order to support strict SLAs associated with premium services. In doing so, MPLS comes in different flavours: connectionless flavours in which the MPLS label from different LSPs can merge at each hop and the choice of the next hop depends on the underlying IGP and connection-oriented flavours where labels represent a distinct, and often traffic-engineered, end-to-end path for a specific FEC. MPLS provides a comprehensive set of OAM tools and protection techniques that reflect both these flavours.

We will first consider the OAM tools. These serve a number of functions in an MPLS network. If a defect occurs in the network, defect indication mechanisms must be available that report failures immediately downstream and upstream of an LSR or LER through a return path. There must be mechanisms that provide alarm suppression to prevent alarm storms in higher network layers. OAM tools should also be distributed (for scalability reasons) and automatic. Defect indication mechanisms should be able to achieve rapid notification of defects to the source of the LSP to enable remedial action to occur. Con-nectivity checking and verification tools must be available to provide on-demand tests of unidirectional and bidirectional connectivity. These usually enable defects to be detected over longer timescales (e.g. minutes) than the defect indication mechanisms.

Path Trace tools are needed to allow on-demand control path to data path cross-checking by verifying the flow of packets against the path that the LSP takes through the network. For example, these would enable errors in the label swap operation at an LSR to be detected. Loopback enables specific locations in the network to be selected as points where OAM packets are looped back to the source, enabling the location of defects to be isolated. Finally, performance-monitoring tools may be required to measure delay and packet loss for SLA verification and reporting.

The most basic mechanism for connectivity verification in an MPLS network is LSP Ping, defined by the IETF. This uses messages similar to those used by ICMP Ping. In a defect-free operation, an LSP Ping Echo Request message follows the path of the end-user packets. An MPLS Ping Echo Reply message will confirm that the remote LER is the egress point for traffic for a specified FEC. A defect in the data path inside the MPLS network will cause the LSP Ping Echo Request to timeout. As a simple extension of this, a Trace Route mode of the LSP Ping can be used to localise the node causing the defect. LSP Ping and Trace Route also enable mis-routed user packets and label-swapping errors to be detected, for example, because a LER will receive a packet for a FEC or label that it did not advertise.

In addition to LSP Ping, the IETF is currently developing a set of mechanisms known as *Bidirectional Forwarding Detection (BFD)*. The objective of BFD is to enable very rapid detection of data path failures. It achieves this by using data path messages that require very little processing overhead and can thus be sent in quick succession, providing faster detection of data plane failure such detection on a greater number of LSPs. BFD may also be extended in the future to carry defect indications. In addition, IETF is also defining mechanisms to allow an LSR to test itself, effectively looping back LSP Ping to itself. This is known as *LSR self test*. Both LSP Ping and BFD are designed to function in the context of all flavours of MPLS. The International Telecommunication Union-Telecommunication Standardization Sector (ITU-T) has also defined a set of MPLS OAM, but this is targeted at MPLS deployments where each LSP behaves in a connection-oriented manner.

The ITU-T has also defined a set of OAM mechanisms for MPLS. However, these are targeted at connection-oriented flavours of MPLS. In these, OAM messages flow along trails, represented, for example, by end-to-end LSPs. ITU-T mechanisms include forward and backward defect indications to allow rapid propagation of fault alarms and their suppression at other layers. They also include connectivity verification and a mechanism known as *FEC* Connectivity Verification (*CV*) to test the FEC connectivity where there is no formal binding between the ingress and egress ports of an LSR (this is as opposed to point-to-point connectivity where there is a formal binding). This is intended to enable, for example, mis-merge defects to be detected in multipoint-to-point LSPs established using LDP.

Table 7.1 summarises the OAM mechanisms.

Although OAM enables faults to be detected and localised, an important factor in the design of an MPLS network is how to minimise the occurrence faults in the first place. There are three main requirements. The first of these is to maximise the availability of a network node. This can be achieved through a set of resiliency and redundancy mechanisms that provide failover protection to components in a router, for example the route processor or the switching fabric. Ideally, these should operate in a "hot-standby"

Table 7.1 Summary of MPLS OAM mechanisms

OAM feature	Defect indication	Connectivity verification	Continuity checking	Path trace	Loopback
LSP Ping		X			
LSP Trace Route				X	
LSR Self Test		X		X	
BFD		X	X		X
ITU-T MPLS OAM	X	X			X

mode; that is, the backup component mirrors the protocol state of the protected component, so that if a component fails the protocol software does not need to be restarted and can fail over very rapidly. It is arguable that such "non-stop" routing functionality is more important at the edge of the MPLS network than in its core. This is because it is common to design networks such that the service state is embedded at the edge: it is then less complex to build high-speed cores. LSPs can be protected from core network failures using the fast re-route techniques described below. However, the Provider Edge (PE) may represent a single point of failure that is more complex or expensive to protect using fast re-route, and, therefore, must be more reliable than nodes in the core. It is also more practical to engineer edge nodes for higher reliability using stateful redundancy and non-stop routing because they operate at a lower speed.

The second requirement is that, if a fault does develop in, for example, an individual LSR's control plane that can be cleared by restarting the node or some component, this restart process must minimise the disruption to the IGP of the surrounding routers. For example, if a router fails in an IP network, the surrounding routers will typically detect that failure and remove the links to that router from their routing database. This change will be advertised to their neighbours and thus will propagate through the routing domain. This behaviour can cause instability in a network where LSRs are able to clear the fault themselves by restarting, because links will seem to disappear as the LSR fails and then reappear as the LSR comes back up. Graceful restart is a mechanism that enables the control plane of an LSR to fail over and restart in this manner, without having to completely re-establish state with the surrounding LSRs. The surrounding LSRs prevent this failure being propagated through the network and thus minimise disruption caused by temporary outages.

The third requirement is that, if a long-term fault is detected along the path of an LSP, such as a link or node failure, the LSP must be restored in accordance with the SLA. MPLS, combined with IP, provides a common network layer that can run over almost any underlying transport mechanisms. These transport mechanisms, such as SONET/SDH, or Ethernet, may have their own differing protection mechanisms. However, MPLS enables a common set of mechanisms to be used irrespective of the underlying transport. This enabled both end-to-end and local protection schemes to be engineered over an essentially heterogeneous network.

In non-traffic-engineered MPLS networks, LSPs typically follow routes that are only dictated by the IGP. However, this means that when a link or node fails, and in the absence of other mechanisms, LSP restoration takes as long as IGP re-convergence. This can be many seconds, which is too long to meet the SLAs of many services. MPLS fast re-route claims to enable recovery times that are smaller than the IGP convergence times. Indeed, a 50 ms failover is possible. This is so fast that it can be engineered to be transparent to edge service protection mechanisms; that is, the core network may fail over, but the end-to-end protocol does not notice. MPLS also enables resource-efficient protection, because statistical multiplexing can be used to share backup paths among many protected paths. MPLS also enables granular levels of protection: LSPs assigned different classes of services, or carrying different services, can be protected or not. In addition, route pinning enables backup paths to be engineered and not to change in a transient manner, even if the shortest path first routing algorithm of the IGP re-converges to a different path. Finally,

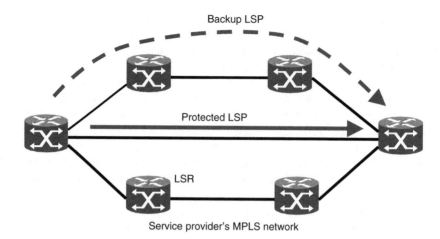

Figure 7.2 Path protection

MPLS fast re-route mechanisms have been developed by the IETF, thus enabling LSRs from different manufacturers to work together in different network environments.

While these capabilities may appear comprehensive, they are not achieved without a degree of network engineering and careful implementation. The remainder of this subsection briefly summarises some of the main MPLS protection techniques.

Simple path protection is illustrated in Figure 7.2. In this technique, a backup LSP protects an end-to-end, or segment, LSP. This may operate in a hot-standby mode, that is, the backup LSP is pre-provisioned so that it can take over as soon as the failure of the protected LSP is detected or in cold standby, where the backup must be established after the failure is detected. The disadvantage of hot standby is that, if each backup LSP protects only one LSP, as much as twice the traffic demand for the protected LSP needs to be allocated in the network. The path of the backup LSP can be calculated so that it avoids all of the links and nodes that the protected LSP passes through. Disjoint paths such as this ensure that there is no single point of failure. The backup path can also be calculated so that other resources can be excluded to ensure, for example, that existing high-priority LSPs are not impacted by the re-routing of other LSPs. One backup LSP can be used to protect each primary LSP (1 + 1 protection) or many primary LSPs (N + 1 protection). N + 1 protection schemes have the advantage that proportionately less bandwidth needs to be consumed in providing backup paths, although planning is required to ensure that single failures do not cause the failure of multiple LSPs protected by the same backup with insufficient allocated bandwidth.

A second form of protection is load balancing. Consider the example illustrated in Figure 7.3. Three LSPs of equal cost are provisioned to handle demand from LER A to LER B. Traffic between A and B is then load balanced across the three LSPs, each carrying 1/3 of the actual load. If one of the LSPs fails, then sufficient resources have already been allocated for the other two to take over the load. The advantage of this approach is that, at most, only 1.5 times the demand from A to B needs to be allocated in the network. Load-balancing hashing algorithms are typically designed so that packets belonging to the same end-to-end service do not become mis-ordered owing to consecutive packets

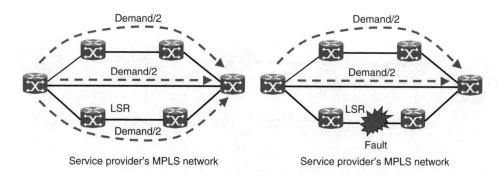

Figure 7.3 Protection using load balancing

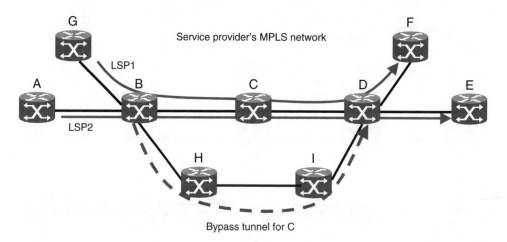

Bypass tunnel for C

Figure 7.4 Local protection: facility bypass

taking different paths through the network, which may have slightly different delay or jitter characteristics; that is, each flow, for example ATM Virtual Channel Connection (VCC) or Frame Relay (FR) Data Link Connection Identifier (DLCI) at the ingress LSR is allocated to a specific LSP.

Local protection is also possible. This can be based on facility bypass or LSP detour. In facility bypass (Figure 7.4), bypass tunnels that extend from a given LSR to the "next next hop" LSR are used to protect LSPs that pass through a bypassed segment.

The advantages of this scheme include the fact that it enables fast re-routing, shared protection of many LSPs is possible and hence fewer backup LSPs may be needed and because it is facility based, specific nodes and links can be protected. However, it is less flexible than the end-to-end mechanisms described earlier, and the impact of both link and node failures may need to be considered when planning the bypass LSPs.

LSP detour–based protection can be specific to a path or a sender. If it is path specific (Figure 7.5), then detours are established from LSRs to the egress LER. Detours merge with each other and with the protected LSP if they share the same next hop outgoing interface on an LSR. The advantage of this scheme is that it enables fast re-routing.

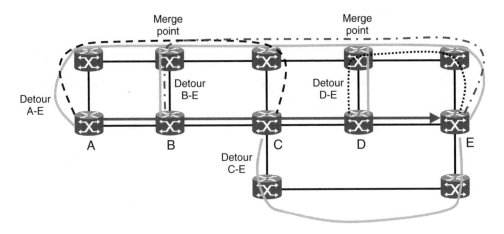

Figure 7.5 Local protection: path specific

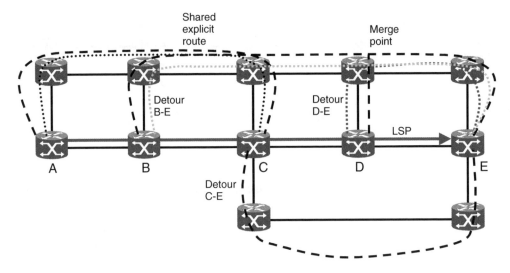

Figure 7.6 Local protection: sender template specific

However, it may require more LSPs that facility bypass and result in less-efficient resource usage.

In the sender template specific case (Figure 7.6), bandwidth can be shared between backup paths, thus making this more resource efficient. Merging occurs when the protected LSP and backup paths at a merge point share the same explicit route to the egress LSR. However, this again results in a larger number of detour LSPs.

While the above fast re-route techniques provide a set of mechanisms that can be used to protect paths within the MPLS network, in order to achieve high levels of reliability, so-called high availability design principles for the LSRs must also be followed. This is important for the Provider routers (P routers) to minimise the number of fast re-route events and for the PEs because it may not always be possible to prevent these from being single points of failure in the network design. Such mechanisms include the following:

- Control or route processor and switch fabric redundancy: These can operate in hot- or warm-standby modes, but, in general, hot standby is preferable so that services are not affected by failovers.
- Redundant common equipment and interfaces: for example, redundant power supplies.
- Configuration redundancy: configuration state is not lost during failures.
- Non-stop routing: The state of protocols and other elements in the control and forwarding planes is maintained during failovers, so that the backup functions can take over those that failed without needing to restart the protocols. Non-stop routing also protects the rest of the network elements from the knowledge of the failure, eliminating the need for neighbouring routers to support a restart of the failed protocols. This reduces CPU load on the neighbouring routers in case of a fault.
- Graceful restart: a set of mechanisms that enable an LSR to restart its control processes without providing massive disturbance to the routing and control protocols in the network.
- Graceful restart helpers: mechanisms typically employed in PE nodes that help neighbouring P and PE routers to restart smoothly.
- In-service upgrades: it should be possible to upgrade elements of an LSR's hardware or software without taking it out of service. This minimises customer service disruption and the cost of the upgrade.

7.2 Virtual Private Networks

A consequence of MPLS is that an LSP effectively creates a tunnel across the IP network. In addition, more than one MPLS label can be stacked on a packet, creating a hierarchy of tunnels. Traffic on a given tunnel is effectively isolated from traffic on other tunnels when it is switched at the LSR. Many virtual overlay networks can thus be created on the MPLS infrastructure, and these are therefore known as *VPNs*. Because the LSP tunnels generally hide the traffic that they are carrying from the LSR[1], they enable the LSP to carry protocols that the LSR cannot natively switch or route. MPLS VPNs are thus an ideal infrastructure to build many different types of virtual networks over a common backbone.

Figure 7.7 shows a general architecture for provider provisioned VPNs defined by the IETF. They are termed *provider provisioned* because the connectivity of the VPN is provisioned between two edges of a service provider's network by that provider, not by the customer or network user.

PE routers reside at the edges of the service provider's MPLS network, while P routers exist within the service provider's network. A packet-switched network (PSN) tunnel extends between those nodes and terminates on them. For an MPLS network, the PSN tunnel is an LSP. However, other tunnel types can be used if the network is not MPLS, for example, Generic Routed Encapsulation (GRE) or Layer 2 Tunnelling Protocol (L2TP).

---

[1] There are some exceptions. For example, LSRs that use equal cost multi-path (ECMP) to load balance traffic across multiple links may look below the MPLS label stack to determine how to load balance e.g. if the LSP carries IP traffic.

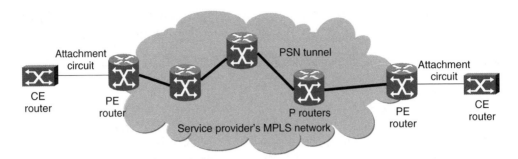

Figure 7.7   MPLS-based virtual private network architecture

Attachment circuits (ACs) connect to the Customer Edge (CE) nodes, which reside at the edge of the customers network. The PEs provide the VPN service using the connectivity across the MPLS network of the PSN tunnel to the routers at the edge of the customer's network (CE routers), which are connected to the provider by ACs. ACs may be ATM, Frame Relay, Ethernet or just about any other kind of connection. As PEs are the first point at which customer traffic enters the provider's network, they have visibility of both the details of the customer traffic (protocol, port, etc.) and the resources and rules of the provider's network. They are therefore the point at which policies such as security or QoS are applied to the customer's traffic.

As noted above, the LSP tunnels hide the type of traffic from the core LSRs in the MPLS network; only the PE devices and CEs need to understand the native service. The IETF defines two types of VPN on the basis of what ToS is provided. Switched services, such as ATM and Frame Relay, and bridged services, such as Ethernet that conventionally belong to layer 2 of the Open Systems Interconnection (OSI) protocol reference model, are carried by layer 2 VPNs. Protocols such as IP that belong to OSI layer 3 are carried by layer 3 VPNs.

7.2.1 Layer 3 Virtual Private Networks

The most common form of VPN is a layer 3 VPN, where layer 3 protocols such as IP are tunnelled through the MPLS network. Such VPNs, typically based on IETF RFC 2547[2] [13], can be used to define multiple service overlays when used with the IP and MPLS traffic engineering and QoS techniques described above and enables the service provider to offer a managed, private, IP network service with specific service level guarantees.

LSPs form a virtual overlay network on a service provider's IP infrastructure corresponding to the services that are to be provided. In addition, MPLS fast re-route and protection techniques can be used to ensure high service availability. RFC 2547 VPNs also allow support for overlapping private customer IP address spaces.

An example VPN architecture is shown in Figure 7.8. CE devices are connected to PE routers at the edge of the service provider's network using IP over almost any access technology. The PEs are fully meshed with multiprotocol (MP) iBGP [14]. This is used to exchange VPN membership and reachability information. Each PE contains a virtual

---

[2] RFC 2547 has been updated to RFC 4364 by the IETF. We retain the term *RFC 2547* here because this is the most familiar reference for BGP IP MPLS VPNs.

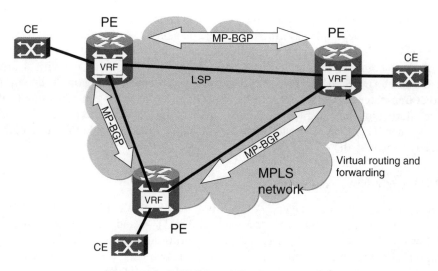

Figure 7.8 MPLS-based IP VPN architecture

routing and forwarding (VRF) instance, one for each VPN that it manages, that provides isolation between the VPNs and also keeps the address space of each VPN private to that VPN. The VRFs are populated with customer routes using a variety of protocols between the CE and PE, for example BGP-4, RIPv2 or OSPF, or they may be statically configured. The customer routers are then advertised between the PEs using MP-BGP, so that they are not visible to the service provider's IGP. In addition, a *route distinguisher* value is used to associate routes that belong to particular VPNs; MP-BGP will only advertise information about a VPN to other PEs that have interfaces with the same route distinguisher value, improving the security and reducing the amount of routing advertisements in the network. When an IP packet is received from a particular CE, the PE consults the forwarding table associated with that site in order to determine how to route the packet.

BGP VPNs are not the only type of layer 3 VPN; Internet Protocol Security (IPSEC) or L2TP can also be used to create VPN service tunnels through the service provider's network. However, the traffic-engineering and QoS capabilities of MPLS have helped make RFC 2547-based layer 3 VPNs a widespread service offering. Nonetheless, the benefit of providing a managed IP VPN service must be balanced against the limitation of such VPNs to IP traffic when many enterprises still require large amounts of non-IP layer 3 protocols (such as Systems Network Architecture (SNA)). These VPNs also require the exchange of customer routes with the service provider, which may not be acceptable to some enterprise customers. Many enterprises also currently build their corporate networks using connectivity provided by a service provider at layer 2, such as ATM or Frame Relay private lines, and the continued ability to provide and grow such services on a converged MPLS network is therefore an important requirement. The remainder of this chapter therefore focuses on new enhancements to the MPLS network to support layer 2 services.

7.2.2 Layer 2 Virtual Private Networks

Layer 2 protocols typically fall into one of the two classifications: switched and bridged. Switched services are those such as ATM and Frame Relay and are typically composed of point-to-point connections (although they may also support point-to-multipoint capabilities).[3] Bridged protocols such as Ethernet operate on a broadcast (or multipoint) paradigm. The IETF layer 2 VPN working group therefore defines two types of layer 2 VPN service. Point-to-point services are enabled on an MPLS network using the Virtual Private Wire Service (VPWS). However, MPLS networks also enable operators to deploy new Ethernet transparent LAN services. In these services, an operator can add value to their service offering by managing the broadcast domain of a wide area Ethernet LAN on behalf of a customer. This type of service provider–managed Ethernet multipoint-to-multipoint service is known as the *Virtual Private LAN Service (VPLS)*.

7.2.2.1 Pseudo wire Architecture

The basic connectivity of a layer 2 VPN is provided by PWs. PWs are defined by the IETF and also in other standards bodies such as the ITU-T and ATM Forum where they are known as *Interworking LSPs*. In their basic form, they allow networks of the same link layer to communicate transparently across a network of a different link layer. In the IETF case, the intermediate network may be MPLS, IP/GRE or L2TPv3 (although MPLS is the most common one), while the attached networks can be ATM, Frame Relay, Ethernet, TDM, Point-to-Point Protocol (PPP) and many other types of layer 1 and layer 2 technologies. Network interworking is performed at the link layer and is typically carried out within the network without the knowledge of the end-user. This enables many existing and new services to be transparently consolidated onto the MPLS network. In addition to point-to-point services, an infrastructure of point-to-point PWs is also used to carry multipoint services across an MPLS or IP/GRE network to create both VPWS and VPLS.

It should be noted at this stage that standards for layer 2 VPNs have been developed in a variety of industry organisations. These organisations have used a variety of terminology to describe the same protocol and functional elements. The remainder of this chapter will use IETF terminology, and an approximate mapping of IETF and MPLS, Frame Relay and ATM Forum (MFA) terms with those used by the ATM Forum and ITU-T is given in Table 7.2.

The basic architecture for PWs [15] is shown in Figure 7.9. It is known as the *PWE3 architecture* (PW emulation edge to edge) because native services are emulated only between the two edges of the service provider's MPLS network. The emulated service is

Table 7.2 Rough equivalence of layer 2 VPN terminology

IETF/MFA term		ATM forum/ITU-T term	
Pseudo wire	PW	Interworking LSP	I-LSP
Packet-switched network tunnel	PSN tunnel	Transport LSP	T-LSP
Provider edge	PE	Interworking network element (physical) or interworking function (logical)	INE IWF

[3] Although ATM also supports point to multipoint (multicast), this is achieved by building a tree of point-to-point VCs.

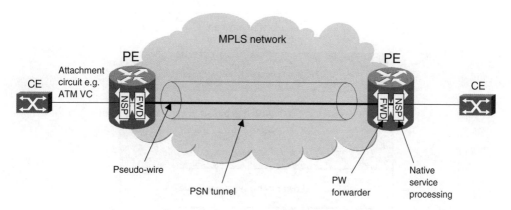

Figure 7.9 PWE3 architecture

not visible in the core P routers, and indeed the end-user may not be aware of it being emulated either, in accordance with the VPN architecture described in Section 7.2.

This consists of an attachment circuit of the native service, for example an ATM Virtual Channel (VC), which is mapped at the PE to a PW that extends through a tunnel to a PE on the other side of the MPLS network. A PW label identifies each PW (PW labels are typically allocated from a PE-wide label space) and there may be many PWs on a PSN tunnel. The role of the PE is to take Protocol Data Units (PDUs), for example ATM cells or Frame Relay frames, encapsulate them in a frame for the MPLS network, add a PW label and then add an outer label for the PSN tunnel, which is an LSP label for an MPLS network. The native attachment circuit terminates in a native service processing (NSP) function in the PE, and encapsulated PDUs are passed to a forwarder on the egress MPLS side of the PE.

Note that a 32-bit control word may also be added between the PW payload and the PW label. The control word is used to carry protocol-specific information, such as the ATM Protocol Type Indicator (PTI) information for certain types of ATM PW, in addition to a sequence number that can be used to check in order delivery of packets (in order delivery is a requirement of many services). There is also a length field that can be used to regenerate the payload if padding has been added by the underlying network.[4] All this information results in a label stack for the PW that is shown in Figure 7.10.

The PW payload can be of an arbitrary length, limited only by the Maximum Transmission Unit (MTU) of the link layer. If the native service PDUs are larger than the MTU, they can be fragmented by the PE on the ingress side of the MPLS network and reassembled on the egress side.

Since LSPs are unidirectional, a pair of MPLS tunnel LSPs is required to provide bidirectional connectivity between any two PEs. These LSPs are established either by provisioning or through MPLS signalling. Since the ACs and the PW are of the same type (i.e. ATM), the resulting layer 2 VPN circuit is termed *homogeneous*.

The exact encapsulation used for different native services was a subject of much controversy during the standardisation of PWs. This was often due to conflicting requirements. For example, with the overhead of adding two labels to the native service, a concern was

---

[4] For example, Ethernet links PAD frames of less than 64 octets.

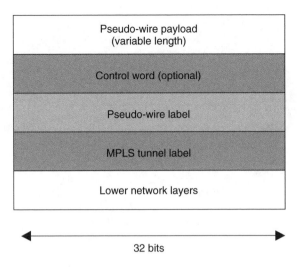

Figure 7.10   Pseudo wire label stack

to try to reduce any other unnecessary overheads in the encapsulation. However, this had to be balanced against the simplicity of the implementation of the PE and the ability to aggregate many native service connections on the PW to scale implementations to support very large numbers of native service circuits.

As an example, consider the specific case of an ATM PW. The ATM Forum [16] and the IETF [17] have specified four encapsulation formats for ATM:

- N-to-one mode
- One-to-one mode
- Service Data Unit (SDU) mode
- PDU mode

N-to-one and one-to-one modes both encapsulate ATM cells in PW frames; one-to-one mode incurs less bandwidth overhead in the process. However, it achieves this by stripping the Virtual Path Identifier (VPI) if the PW carries a Virtual Path Connection (VPC) and both the Virtual Channel Identifier (VCI) and VPI if the PW is a VCC.[5] The N-to-one mode does not do this, enabling a complete ATM port with an arbitrary range of VCI and VPI values to be transported on a PW. However, this incurs more bandwidth overhead.

SDU and PDU modes are used to carry ATM Adaptation Layer Type 5 (AAL5) SDUs or PDUs, respectively. SDUs consume less bandwidth as they do not include the Cyclic Redundancy Check (CRC) or PAD fields, but cannot be fragmented, so ATM OAM cells can become reordered on the PW during the encapsulation process; that is, if an OAM cell arrives among the cells of the SDU during the encapsulation of the SDU, it must be held until all the cells of the SDU have been encapsulated and sent on the PW. The PDU

---

[5] The ATM VCI and VPI values are local to a link and can be regenerated at the egress PE on the basis of the PW context.

mode avoids this by allowing the assembly of the AAL5 frame into a PW frame to halt as soon as an OAM cell arrives. The fragmented frame is sent immediately, followed by the OAM cell and then by the rest of the AAL5 frame in a further PW frame. A detailed analysis of these differences is provided by Bocci and Guillet in [18].

Layer 2 services such as ATM and Frame Relay often have strict absolute or relative QoS guarantees associated with them. Indeed, one of the main contributors to the success of ATM in carrier networks has been its ability to guarantee QoS to user's traffic. This is because it provides a well-understood and standard set of tools to enable QoS. These include the following:

- Standardised service categories and conformance definitions
- Standardised traffic parameters
- Connection admission control
- Traffic policing and traffic shaping.

When such services are migrated to an MPLS network, a translation between the layer 2 service QoS concepts and QoS mechanisms in the MPLS network is required. QoS for PWs is achieved using the mechanisms inherent in the MPLS network, for example, as described in Section 7.1.1, together with traffic management mechanisms including admission control, traffic classification and class-based queuing in the PEs. However, some services also require that timing synchronisation be maintained between the ingress and egress PEs. TDM PWs may need to maintain strict synchronisation for a number of reasons: they are used as scientific timing sources by end-users, to synchronise Private Branch Exchange (PBXs) or for backhaul in a mobile network where synchronisation is required to enable the correct operation of handover of calls between cells. When migrating such services to PWs over an MPLS network, the fact that the MPLS network is packet based may increase the risk of drift and jitter of strictly timed signals in the TDM stream. Therefore, PWs may use mechanisms such as the Real-time Protocol (RTP) to transmit timing information along with the encapsulated TDM data along the PW. This can be used to regenerate the appropriate timing information at the receiving PE.

7.2.2.2 Pseudo wire Control Plane

The control plane for PWs is based on an extension of LDP, which is used in downstream-unsolicited mode. The protocol operates in targeted mode; that is, it operates directly between the PEs at the edge of the MPLS network without the participation of the P routers. To establish a PW, the operator must configure the AC, the FEC and an associated identifier for the PW, together with a set of parameters for the specific AC interface (e.g. the MTU) and the PW type. These are then advertised in a label-mapping message by the downstream PE to the upstream PE for that direction of the PW. PWs are released by issuing a label withdrawal message.

Two FECs are defined that uniquely identify a PW endpoint in a PE:

- PW ID FEC
- Generalised ID FEC.

The PW ID FEC has a number of fields, including the following:

- PW Type: This indicates the native service that the PW carries and how it is encapsulated. At the time of writing, 25 PW types had been defined by the IETF.
- Control Word Bit: This indicates whether a control word is present in the encapsulation.
- Group ID: This can be used to associate a set of PWs together into a group. This could be useful, for example, to issue label withdrawals for many PWs using only one signalling message, thus reducing the signalling overhead.
- PW ID: This is a 32-bit field identifier that, together with the PW type, uniquely identifies the PW between a pair of PEs. The PW ID and the PW type must be the same at both ends of a PW.
- Interface parameters, as described above.

Since the PW type and ID must be unique between a pair of PEs, they must be manually configured at each end in order to prevent the identifier values colliding; that is, the value configured on one PE must not have already been used for another PW on its peer PE. Each PE then launches the label-mapping message to its upstream peer. Although the ACs on both PEs must be configured anyway, such double-sided configuration of the PW ID increases the level of operator intervention required in provisioning a PW. An automatic method for determining how a PW should be identified is desirable, and this is one of the reasons for the development of the generalised ID FEC.

Instead of using a single identifier value for the PW that must be configured to be unique between a pair of PEs, the generalised ID FEC encodes an identifier for the attachment circuit on a PE. This is known as the *attachment identifier* and is composed of an attachment group identifier (AGI) and an attachment individual identifier (AII). As an example, the AGI could identify the VPN, while the AII could be derived from the ATM port, VPI and VCI for the AC. The generalised ID FEC contains both source and target identifiers (the Source Attachment Identifier [SAI] and Target Attachment Identifier [TAI], respectively), the Source Attachment Identifier (SAI) being unique on the source PE and the Target Attachment Identifier (TAI) being unique on the target PE. The pair of SAI and TAI thus uniquely identifies the PW. Furthermore, the source PE can generate the TAI without fear of a collision in the target PE, thus enabling signalling to be initiated from just one end of the PE. Using the generalised ID FEC, it is possible for one of the PEs to act as a master in the signalling process and the other to act as a slave. The master PE sends the label-mapping message for one direction of the PW, containing the FEC, to the slave PE. This PE does not respond with the label mapping for the reverse direction until it has received it. Such behaviour is useful because it reduces the amount of provisioning required to establish the PW. It is also useful for dynamically establishing multi-segment pseudo wires (MS-PWs), which are described in the next section.

In addition to the establishing and releasing PWs, LDP has been further extended to enable the status of the AC or the PW to be signalled between PEs. This enables, for example, a PE to signal to its peer that there is a fault on the corresponding attachment circuit, the PW, or its MPLS interface. The usage of PW status signalling in supporting PW OAM is described later in this chapter.

7.2.2.3 Multi-segment Pseudo wires

PWs were originally envisioned as a method to enable layer 2 services over an IP-based core packet network. However, as the technology has developed, demand has grown to enable convergence in metropolitan area networks as well, which, today, are typically composed of separate ATM, Frame Relay or Ethernet networks. Only a few tunnels carrying highly aggregated services were originally envisioned, and so in the original PWE3 architecture, both the MPLS tunnels and the PWs that they carry extended between only two edges of the provider MPLS network. These are therefore termed *single segment PWs (SS-PWs)*.

Single segment PWs were designed to run only for a single PE-PE hop across the MPLS network. This is manageable when there are only a few PEs around the edge of the core network. However, the deployment of PW-based services between different metro area networks would result in a large increase in the number of PEs.

This means that a full mesh of tunnels is required to achieve PW connectivity between all the PEs around the edge of a service providers network, which grows according to an $O(N)^2$ function as the number, N, of PEs increases. In addition, PWs would have to revert to native service ACs if they were required to cross a boundary between service provider networks and the providers required per-PW control, for example, for reasons such as security, authentication or billing of services or where the packet network changes type, for example MPLS to L2TP.

MS-PWs are a proposed solution to this problem that, at the time of writing, is under development in the IETF. These enable PEs to switch PWs directly between ingress and egress tunnels on a PE, so that only a partial mesh of tunnels and signalling adjacencies is required. Such a PE is termed a *Switching PE (S-PE)*, while the PEs at each end of the MS-PW are called *Terminating PEs (T-PE)*. Each hop of the multi-segment PW is known as a *PW segment*, as shown in Figure 7.11.

In their simplest form, the S-PEs along the path of the MS-PW are statically provisioned; that is, each segment is established as described in the section above. A cross connect is made between the PWs on the ingress and egress sides of an S-PE. Extensions to the PW control protocol are also defined, enabling the S-PEs to be dynamically selected. This requires the signalling protocol to specify the destination PE, in addition to any bandwidth

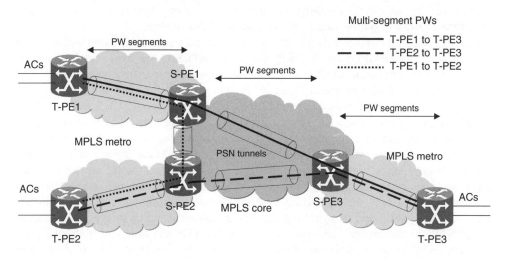

Figure 7.11   Multi-segment pseudo wire example

requirements of the PW so that each S-PE can choose the appropriate next hop S-PE to reach the destination T-PE and select a tunnel that can support the QoS requirements of the PW.

The addition of S-PEs along the path of a PW introduces additional resiliency requirements in comparison with SS-PWs. The tunnel itself can be protected from failures of the P routers using MPLS protection techniques, but the S-PE represents a new point of failure. Protection techniques, similar to those described in Section 7.1.4 but used for the PW itself rather than the tunnel, can be applied to protect MS-PWs from S-PE failures.

7.2.2.4 Multipoint Ethernet Services (VPLS)

Ethernet has become the de facto technology for corporate and residential LANs. Its low cost per bit and flexibility, range of available bandwidths (from 10 Mbps to 1 Gbps to the desktop) and ubiquity on almost every desktop have driven the demand for service providers to offer managed network services tailored to interconnecting Ethernet LANs on remote sites, thus allowing companies to build high-speed corporate intranets.

ATM and Frame Relay services were the first wide area networks to offer widespread support for such services. These provided point-to-point links between Customer Premise Equipment (CPEs) attached to the LAN on each site, encapsulating Ethernet frames according to RFC 2684 [19], for example. Ethernet is a bridged protocol in which frames from one segment are forwarded to the next segment on the basis of a media access control (MAC) address and, optionally, a Virtual Local Area Network (VLAN) tag, contained in the frame header. Ethernet LANs form a part of a broadcast domain in which bridges discover how to reach a particular remote IP address' MAC address by learning which link(s) that end station is heard on. As Ethernet is a layer 2 protocol, it does not rely on layer 3 information (such as IP addressing). As in any layer 2 transport, IP address to MAC address conversion is handled by the source endpoint using the Address Resolution Protocol (ARP). Traditionally, a full mesh of ATM or Frame Relay connections were used to interconnect bridges on remote sites, with the Ethernet spanning tree protocol (STP) used to prevent loops in the resulting network, that is, remove links that did not represent the shortest next hop on the path to the destination MAC address.

The VPLS is a multipoint Ethernet service that provides a network-based emulation of a LAN based on an MPLS (or IP/GRE, but again, MPLS is the most widely deployed variant) backbone network. VPLS-enabled service providers can offer a managed Ethernet LAN service. It can also be used to provide an infrastructure for metro Ethernet because the traffic engineering and protection and restoration mechanisms of MPLS can be applied to each LAN segment. From the service provider's point of view, VPLS offers a clear demarcation for Ethernet services between the customers network and the provider's network domain, while for the customer, there is no need to manage any technologies other than Ethernet. The routing of customer IP traffic is also hidden from the service provider and from other customers of the same provider's network because the service provider has a visibility of only the underlying Ethernet domain. Furthermore, as VPLS utilises the underlying MPLS and IP signalling and routing layer (including loop prevention), the use of STPs in the carrier's network is avoided. As STP was designed with enterprise and campus networks in mind, the stability and scalability issues of STP are avoided in the carrier's network.

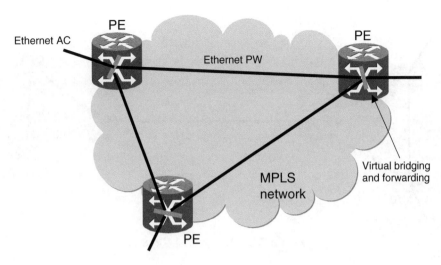

Figure 7.12   Flat VPLS architecture

A simple flat VPLS architecture is shown in Figure 7.12. PEs are interconnected by a full mesh of Ethernet PWs, which are connected to the ACs for each LAN by virtual bridging and forwarding functions in the PEs. Although a shared MPLS infrastructure that supports all the other services that a provider offers is used, from the customer's perspective, it appears as if all their sites are connected to the same private LAN. A key feature of VPLS is that STP is not used to remove redundant links from the LAN. Instead, split horizon forwarding is employed in the PEs. PEs use MAC learning to determine the PW to which a frame may be sent. In split horizon forwarding, broadcast frames are only sent between the AC and the PWs. Frames received on a PW are never sent back on another PW from a PE. This prevents broadcast frames from looping through the full mesh of PWs.

Two methods for the distribution of the PW labels are defined for VPLS. LDP VPLS uses the same LDP-based control protocol as the VPWS described earlier. BGP VPLS uses extensions to BGP to distribute the labels to all the PEs. The LDP method is particularly suited to metropolitan area networks, where the cost of the PEs must be minimised and hence simple LDP is important, but it is also widely deployed for national and international VPLS services. However, BGP may be useful where a service provider is also providing MPLS-based RFC 2547 IP VPNs [13], which already use BGP to distribute VPN routes.

The architecture shown in Figure 7.12 has the disadvantage that the number of PWs and PSN tunnels grows as $O(N)^2$ with the number of PEs. LDP VPLS also defines a hierarchical architecture (Hierarchical Virtual Private LAN Service (H-VPLS)), which is shown in Figure 7.13. This is a hub-and-spoke architecture where smaller, customer-facing PEs (called *MTUs*) are connected to hub PEs using spoke PWs. Hub PEs are interconnected using hub PWs. In this model, multicast replication occurs as close to the destination PE as possible, reducing the amount of bandwidth consumed in the P-PE and PE-MTU portions of the network.

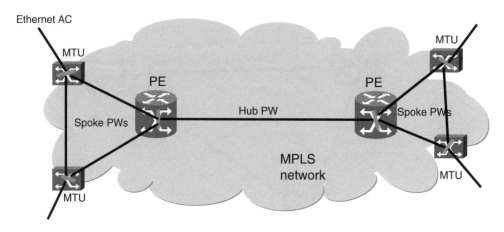

Figure 7.13   Hierarchical VPLS

7.2.2.5 Service Interworking Across MPLS Networks

The services delivered using PWs over the converged MPLS core network described so far share one characteristic in common: the access circuit connected to the CE at each end is the same, that is, it is always Ethernet, ATM or Frame Relay, for example. This is effectively a network interworking architecture because the network layer at the CE is extended edge to edge across the MPLS core and is termed a *homogeneous layer 2 VPN circuit*.

Service interworking, on the other hand, allows CE devices to exchange service layer PDUs transparently across different access link layer technologies. Examples of service layer PDUs include IP and Ethernet datagrams. Interworking is performed at the layer above the link layer, allowing the service providers to offer a service that is almost agnostic of the actual CE to PE link layer. For example, a layer 2 VPN can be offered that allows high-speed Ethernet access to work transparently with established Frame Relay and ATM services. Because the ACs are of different types, the layer 2 VPN circuit is termed *heterogeneous*.

There are three main variants of service interworking that have been developed that use an MPLS core network:

- FR-ATM multi-service interworking
- Ethernet multi-service interworking
- IP multi-service interworking.

These are shown in Figure 7.14.

In FR-ATM multi-service interworking, the link layer of one attachment circuit is extended across the MPLS network as a PW of the same type. If the PW is ATM, PE1 then performs FRF8.2 service interworking [20] between the Frame Relay AC and the ATM link layer of the ATM PW. The ATM PW can be either cell mode, where the FR PDUs are segmented into or reassembled from individual ATM cells on the PW using a segmentation and reassembly function (SAR) in PE1 or, more likely, the PW will be

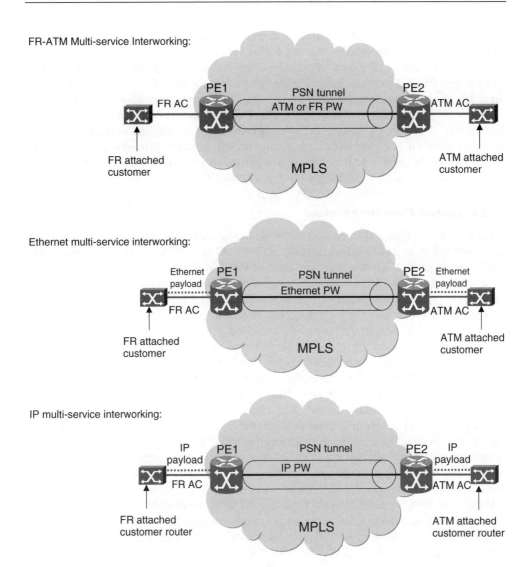

Figure 7.14   Service interworking variants

SDU mode, where SAR in not needed in PE1 and the complete FR PDU can be carried in a frame on the PW. In this case, the SAR function exists in PE2. If a Frame Relay PW is used as an alternative, the service interworking function resides in PE2.

The second variant is Ethernet multi-service interworking, which can be used to extend an Ethernet service to sites attached via Ethernet, ATM and Frame Relay circuits. Here, a bridged Ethernet payload is carried by the ATM and Frame Relay ACs. The Ethernet PDU is encapsulated in an Ethernet PW, which is extended across the MPLS network between PE1 and PE2, thereby extending the Ethernet service layer end to end. This could also be used to offer Ethernet VPN service in areas where only ATM or FR access is available.

In this case, sites that can be connected to native Ethernet bearers are thus connected, and other VPN sites are connected using PWs over ATM or FR.

The final variant is IP interworking. Here, the end-to-end service is IP and the ACs carry routed PDUs instead of bridged Ethernet PDUs. An IP PW is used, which has the same encapsulation as a regular LSP. The point of this service is to allow an IP-optimised service that operates at layer 2 from the point of view of the service provider. This means that the service provider's network is not exposed to the routing domains of the customer's networks, which may be viewed as being more private from the customer's point of view. It also allows the service provider to offer a partially managed service (rather than fully managed, in the case of BGP-MPLS layer 3 VPNs).

7.2.2.6 Control Plane Interworking

A number of strategies can be envisaged for the deployment of an MPLS core network and the replacement of legacy infrastructures, such as ATM or TDM networks. Perhaps the most radical one is a wholesale replacement of the legacy networks, deploying MPLS as close to the edge of the service provider as possible. In this scenario, the connection to the customer is typically a statically provisioned ATM, FR or TDM connection, Ethernet port or VLAN. However, more conservative strategies can be envisaged. In these strategies, the service provider initially deploys only a limited region of MPLS for layer 2 services, maintaining a significant legacy of ATM or Frame Relay aggregation networks. These legacy networks commonly have complex control plane mechanisms that are used to dynamically establish, release and maintain services in the layer 2 networks. This must continue to be supported as portions of the network migrate to MPLS.

Control plane interworking is therefore required between the disparate layer 2 networks and the MPLS network, in order to enable dynamic end-to-end connectivity to be established. Without control plane interworking, network management or manual intervention would always be required to establish a connection or recover from a fault at the point in the network where the two signalling domains meet. This presents a challenge because of differences in the deployment and capabilities of the ATM, Frame Relay, Ethernet and MPLS control planes. Consider ATM as an example. ATM has extensive control plane functionality. An important capability of existing ATM networks is ATM-switched services, based on Soft Permanent Virtual Connections (SPVCs) and Switched Virtual Connections (SVC). SVCs (and their virtual path equivalents, Switched Virtual Path (SVPs)) enable end-users to dynamically signal end-to-end ATM connections, while SPVCs (and Soft Permanent Virtual Path (SPVPs)) appear as a permanent connection to the end-user; the intermediate hops on the connection between the first and last ATM switches in the service provider's network are established through signalling, rather than by network management.

SPVCs are used to simplify the provisioning of ATM services and support dynamic traffic engineering and faster restoration in the event of network failure, because routes are typically calculated by the source ATM switches and signalling typically progresses through the network more rapidly than configuration instructions from a network management system; the network management does not act as a bottleneck when large numbers of connections must be re-routed around a network failure. This means that ATM SPVCs can be used to provide a higher availability (i.e. less outage time in the case of a failure) than ATM Permanenct Virtual Circuit (PVCs). This capability benefits non-ATM services, which can also be carried using ATM SPVCs, such as Frame Relay or Ethernet, as well as TDM voice, to non-ATM endpoints. These other services therefore also drive the deployment of ATM SPVCs. By transparently supporting ATM-switched services over MPLS, existing provisioning tools and operational procedures can be used. It is therefore important to provide functionally equivalent capabilities when ATM-based services are converged onto a core MPLS network. Control plane interworking is one way to deliver this functional equivalence during the migration period when a service provider has significant regions of native switched ATM network infrastructure that must be carried over newer MPLS network regions. The ATM Forum Private Network–Network Interface (PNNI) is the most common ATM-signalling and routing protocol that operates between nodes within a service provider's ATM network. The ATM User-network Interface (UNI) is used to allow signalling between the user and the service provider's network or between two interconnected service provider's networks, and the ATM Inter-Network Interface (AINI), for signalling between nodes where it is undesirable for dynamic routing information to be exchanged. This may be the case for reasons of privacy or where static routes are exclusively used for ATM connections. Note that all these ATM control protocols share a similar set of signalling messages. Therefore, although control plane interworking is described below in the context of PNNI, these techniques can also be adapted for use with other ATM control protocols. It is primarily a question of whether routing information needs to be interworked in addition to the signalling messages.

There are two main deployment scenarios for ATM-switched services when migrating to a converged MPLS core network:

i. Extending ATM services between endpoints that support ATM signalling and, optionally, routing.
ii. Extending ATM services between endpoints where one or more do not support ATM signalling and routing.

In (i), ATM signalling and routing are tunnelled transparently through the intervening MPLS network, so that the ATM networks on either side are not aware of its presence. There are a number of possible ways to do this, but for the sake of brevity, we will discuss two here. These are ATM virtual trunks and Extended PNNI (also known as ATM-MPLS Signalling Interworking).

ATM virtual trunks are illustrated in Figure 7.15. Here, all ATM connections on the link between the ATM switch at the edge of the ATM network and the node at the edge of the MPLS network that share a contiguous range of VPI values are associated with a virtual trunk. Connections in the ATM networks can be switched onto this trunk by switching their VPIs in, for example, ATM Edge 1. All the connections on a specified virtual trunk are then multiplexed onto an ATM PW using N-to-one encapsulation that can carry cells from many ATM connections on the same PW. The connections are then de-multiplexed from the PW at the egress MPLS edge. The ATM control channels, for example the signalling channel and Routing Control Channel (RCC) for PNNI, would also be multiplexed onto a PW in a similar manner, allowing ATM signalling and routing traffic to be carried transparently across the MPLS network. The PW is usually established using the PW control protocol on the basis of the Targeted LDP as described previously.

One objective of virtual trunks is to minimise changes to the existing functionality of the ATM and the MPLS networks. In particular, it is not necessary to implement ATM

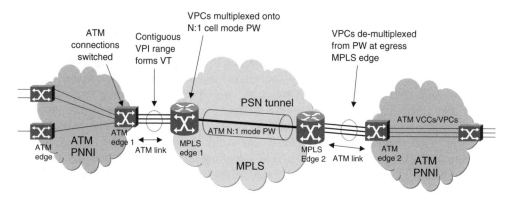

Figure 7.15   ATM virtual trunk

signalling or routing on the node at the edge of the MPLS network or MPLS on the ATM edge node. The ATM and MPLS edge functions can therefore reside on separate nodes. This also reduces the minimum number of PWs that must be provisioned. However, the unrecoverable failure of a single PW will cause all the ATM connections carried to fail and so be re-routed. Furthermore, because many ATM connections with different QoS requirements are mapped to a single PW, that PW must support the highest QoS of all its ATM connections. This may result in less-efficient multiplexing of ATM connections into the PW. Each ATM virtual trunk must also be statically provisioned; that is, virtual trunks and their associated PWs are not dynamically established and released as needed by the ATM networks.

Extended PNNI is shown in Figure 7.16. This takes a different approach to virtual trunks in that ATM PWs are established and released dynamically as ATM connections are set up and released. The PSN tunnel represents a PNNI link, which is extended transparently across the MPLS core network between two attached ATM PNNI networks. For PNNI to establish connections on a link, an RCC and a signalling channel are required, and

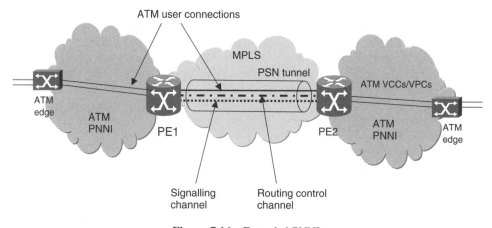

Figure 7.16   Extended PNNI

each of these is associated with a dedicated virtual channel. The RCC exchanges routing information (e.g. routing tables and network topology state updates) between PNNI nodes at either end of the link, while the signalling channel carries ATM-signalling messages.

Rather than using targeted LDP to signal the PW labels, Extended PNNI specifies extensions to PNNI to enable it to negotiate the values of PW labels and their mapping to the corresponding VCI and VPI of the ATM connections, as well as the PW encapsulation mode. While this requires a PNNI protocol stack to be implemented on the PEs, no changes are needed to the existing edge ATM networks or the MPLS core network. However, both the PNNI protocol software and the PW control software (which would still be used to establish non-ATM PWs) must be able to access a shared pool of PW labels. However, there is no direct protocol interworking between PNNI and LDP.

Extended PNNI can be used to implement multi-segment ATM PWs, addressing a problem similar to that described previously. Only a partial mesh of PSN tunnels is required across the MPLS network, with PNNI being used to route connections via transit PEs at the edge of the MPLS core, maintaining the same scaling properties and resilience as current ATM networks. Today, PNNI is deployed to rapidly re-route connections from a primary route on which a link fails to an alternative route, thus maximising the availability of the ATM service. Extended PNNI continues to provide protection in this manner. The high call arrival rates expected on the alternative route have no impact on the MPLS core since they are tunnelled over the Transport LSP (T-LSP). Extended PNNI thus minimises the load on core routers, maximising the network availability and minimising the impact on other services.

Deployment scenario (ii) is primarily aimed at cases where ATM services provided to users of an existing ATM network must be extended to users attached to an MPLS network. For example, this can be useful where a service provider is extending the footprint of its services to an area where it does not have an ATM network, but where an IP/MPLS network has been deployed for IP services. The objective is to enable end-to-end ATM services, with characteristics such as high failure resiliency (through rapid re-routing) and mostly automated set-up, while avoiding the need to implement and manage PNNI on the PEs attached to the MPLS network. This scenario could be addressed using SPVC-PWE3 interworking (also known as layer 2 mediation), which was still undergoing development in the MFA Forum at the time of writing this book.

One version of SPVC-PWE3 interworking is illustrated in Figure 7.17. PNNI signalling and routing are supported on PE1, while PE2 only needs to support the PW control protocol. PE1 effectively acts as a gateway between the ATM and MPLS networks. PNNI routing is terminated on PE1, which is configured with the externally reachable addresses at PE2 and advertises these to the ATM network. PNNI-signalling messages are interworked with the PW control protocol at PE1; an ATM set-up message arriving at PE1 causes LDP label-mapping messages to exchange PW labels between PE1 and PE2. An ATM release message arriving at PE1 would trigger a label release for the PW. If PE1 fails, then the ATM SPVC would be released by PNNI and re-routed via a different gateway PE (if one exists), with a corresponding new PW set up between PE2 and the new gateway, in order re-establish the end-to-end ATM connection.

Although SPVC-PWE3 interworking allows ATM-switched services to be extended to interworking with non-PNNI-enabled PEs, the PW control protocol message cannot carry information such as ATM connection traffic parameters. Therefore, it is preferable

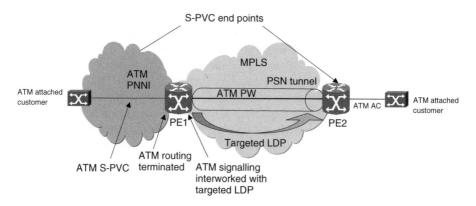

Figure 7.17   SPVC-PWE3 interworking

if the ATM connections originate in the ATM network rather than on PE2; that is, if the connections originate on PE2, PE1 will not know what ATM traffic parameters is included in the PNNI set-up message that it sends to its attached ATM network. Furthermore, ATM connections that originate in the ATM network cannot extend beyond PE2, because PE2 does not know what traffic parameters to associate with the connection.

7.2.2.7 Fault Management and OAM for Pseudo wires

Convergence on MPLS has some significant implications on OAM and particularly the way that faults are detected and managed in the network. The introduction of PWs introduces new functionality that can be the source of additional faults that must be detected and diagnosed. In addition, fault notifications or alarms must be translated at technology boundaries, that is, between ACs and PWs.

A converged network must provide and coordinate OAM at the relevant levels in the IP/MPLS network, namely, at the PSN tunnel level, the PW level and the service level. Both proactive and reactive OAM mechanisms must be provided, which are independent at all levels; that is, mechanisms must automatically detect faults when they occur and report them, and tools must be provided that enable an operator to localise and diagnose that fault once detected. Independence means that faults at specific levels should not cause alarms at other levels; this helps in isolating the fault and prevents storms of potentially false fault notifications.

OAM at the PSN tunnel level includes those mechanisms described in Section 7.1.4 for MPLS networks. At the service level, this includes mechanisms that are native to the emulated service, such as ATM OAM and Ethernet OAM. For VPLS, service OAM mechanisms have also been defined that enable the connectivity and configuration of the VPLS to be checked, for example mechanisms to test that a specific MAC address is known by a remote PE (MAC ping) and to trace the topology of the VPLS (MAC trace).

PWs have required the development of a new set of OAM mechanisms, which include the following:

- PW status signalling
- Virtual Circuit Connectivity Verification (VCCV).

PW status signalling has been described in depth earlier and allows out-of-band notifications regarding the status of a PW or its ACs to be passed between adjacent PEs. This is termed an *out-of-band OAM mechanism* because messages are passed using the control plane rather than the data plane for a PW. Such an out-of-band OAM is of limited use in testing the data path of a PW as it can take a different route through the network to the user traffic on a PW and so it is used only for status indications. VCCV, on the other hand, is an in-band mechanism that allows OAM messages to flow along the data path of a PW. This means that the messages "fate share" with the user traffic on the PW; that is, they follow exactly the same path, and if a fault develops it will affect the OAM and the user data equally. This increases the chances of detecting a fault. VCCV works by establishing an in-band channel on a PW, identified by a special PW control word. Other OAM tools can then be run inside that VCCV channel, such as Ping or BFD.

The distinctions between homogeneous and heterogeneous layer 2 VPN circuits, which were introduced previously in the context of network and service interworking, are important for OAM. The first factor to consider is the locations where defects can occur. Figure 7.18 shows a simplified defect location model for a PE for both homogeneous and heterogeneous layer 2 VPN circuits that span a single PSN tunnel. Three main architectural components are shown: the layer 2 interface that terminates the attachment circuit (e.g. ATM, FR or Ethernet), the PW termination and the MPLS interface on which the PSN tunnel terminates. An IP layer also exists that provides edge routing on an MPLS-network-facing layer 2 interface. The OAM model considers that defects occur on the attachment circuit in the layer 2 network, on the PE AC interface, on the PE MPLS and in the MPLS network.

In the homogeneous case, the AC link layer is extended transparently across the MPLS network in a PW because both ACs and the PW are of the same type. OAM for the native service is carried in-band on the PW and is used for both AC and PW defect indications. This is useful for ATM, which has a comprehensive set of in-band OAM tools, but is not

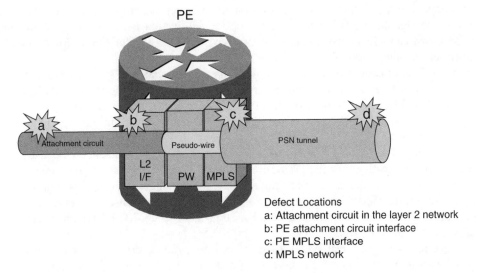

Figure 7.18 Defect locations

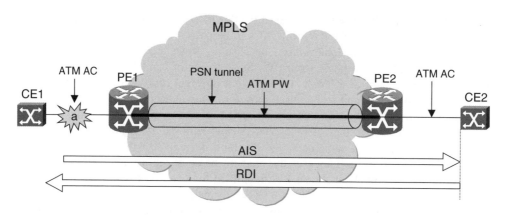

Figure 7.19   Defect notification propagation for a homogeneous layer 2 VPN circuit

possible for Frame Relay, which relies on out-of-band defect indications carried using the link management interface (LMI) over the AC.

Figure 7.19 shows an example of defect handling for the homogeneous ATM case. The OAM segment runs from CE1 to CE2. A defect occurs through a failure of the attachment circuit at (a) in the ATM network. This will result in an ATM Alarm Indication Signal (AIS) OAM cell being sent downstream to PE1. This OAM cell is encapsulated along with other ATM traffic on the PW and sent across the PW to PE2, where it is forwarded on the ATM AC to CE2 to indicate failure of the ATM circuit. CE2 then acknowledges with an ATM Reverse Defect Indicator (RDI), since this is the terminating node in the OAM segment.

In the heterogeneous case, the AC link layer is terminated at the PE because at least one of the ACs and the PW are of different types. If the type of the PW does not support native service OAM, for example if it is Frame Relay, then the OAM always terminates at the AC endpoint in a PE. In these cases, a PW-specific defect indication, such as out-of-band PW status signalling is used to carry the defect notification between the PEs. In-band ATM OAM can still be used where the PW type is ATM (e.g. for Frame Relay – ATM multi-service interworking).

Figure 7.20 illustrates how defect notifications are propagated in an ATM-Frame Relay heterogeneous layer 2 VPN circuit, where a Frame Relay PW is used. An ATM AIS is sent to PE1 following a failure of the ATM network at (a). However, since there is no equivalent in-band Frame Relay OAM alarm, a PW status signalling message indicating a receive fault on PE1's AC interface is sent between PE1 and PE2. This is translated into a Frame Relay Local Management Interface (LMI) status indication on the ATM AC to CE2, indicating that the Frame Relay circuit is down. Note that since there is no forward and reverse defect state in Frame Relay, only a single indication is sent to CE2. PE2 responds to the PW status message to PE1 indicating that it has received a defect indication on its receive direction, which is translated to an ATM RDI on the reverse direction of the ATM AC.

There are two additional points to note regarding the descriptions above. Firstly, a defect indication loop must be created such that the far end of the circuit acknowledges defects in the forward direction with an RDI. This means that the source of the circuit can

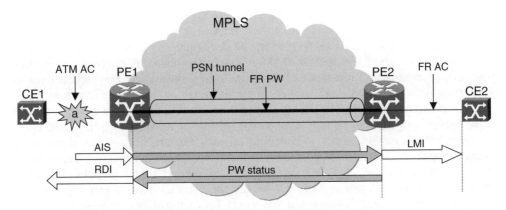

Figure 7.20   Defect notification propagation for a heterogeneous layer 2 VPN

tell that there is a fault somewhere in its transmit direction and, thus, can also tell when the fault has cleared. Secondly, the translations of the defect indications occur at technology boundaries in the PEs. In some cases, such as the Frame Relay case, a PE effectively proxies for a bidirectional defect indication mechanism. This results in a modified defect loop.

OAM for PWs is a complex topic, and the reader is directed to literature such as Ref. 21 for further information.

7.3 Summary

This chapter has introduced an architecture for the converged network core. This architecture is based on MPLS, which while being originally developed for enhanced forwarding and traffic engineering for IP, has been enhanced to support layer 2 services such as ATM, Frame Relay and Ethernet. LERs add a shim label to the network packet, effectively associating that packet with an LSP. This enables the path of the LSP to be engineered. IP traffic management mechanisms, such as DiffServ, can be used with traffic-engineered LSPs to enable them to support end-user QoS guarantees. Signalling mechanisms such as LDP and RSVP-TE are used to establish and maintain the LSPs. In addition, OAM and protection and restoration techniques have been developed that enable faults to be detected, localised and diagnosed, and for LSPs to be re-routed automatically around failures.

These mechanisms provide a basis for the convergence of both legacy and next-generation services onto the MPLS core. MPLS-based VPNs are emerging as a key enabling technology for both layer 3 and layer 2 services. Layer 3 VPNs are an established and well-understood technique for delivering managed IP services from a carrier's MPLS network. However, layer 2 VPNs are emerging as a key convergence enabler because they allow services that have traditionally been supported over legacy broadband networks, such as ATM, to also be delivered by the MPLS network. This chapter has therefore concentrated on providing a detailed description of layer 2 VPNs. These are based on an underlying infrastructure of PWs and support both point-to-point services, such as ATM, Frame Relay and Ethernet private lines, as well as multipoint services, such as VPLS. Service interworking using PWs also allows end-users to connect to layer 2 VPNs using dissimilar access technologies, increasing the flexibility when compared with traditional layer 2 services. However, the development of the PW layer has also required a special set of PW-specific OAM and signalling mechanisms. Such capabilities will become even more important as PWs grow from simple single hop architectures to supporting native PW switching between multiple metro and core domains or between multiple providers. Such MS-PW functionality is required to scale layer 2 VPN services if the MPLS network is to replace current deployments of ATM-based infrastructure.

Layer 2 VPNs enable service providers to broaden the range of services that can be delivered by the MPLS network, potentially reducing operational expenditure and enabling them to take advantage of a new infrastructure to support future services. However, there are a number of factors that will impact the viability of core convergence. For example, the cost of organisational and operational restructuring and the ability to migrate existing users from the legacy networks to the new converged network will affect the economic justification for moving to the new infrastructure. In addition, while more mature technologies such as layer 3 VPNs, single hop PWs and VPLS are becoming widespread, some of the newer techniques such as MS-PWs are currently under development and will take time to appear in the service provider's operations. All these factors mean that the decision to converge on a single core network should be taken on a case-by-case basis according to an individual service provider's business.

References

1. B.S. Davie, Y. Rekhter. *"MPLS: Technology and Applications"*. Morgan Kaufman, ISBN: 1558606564, May 2000.
2. E. Rosen, A. Viswanathan, R. Callon, *et al*. IETF RFC 3031. "Multiprotocol Label Switching Architecture". January 2001.
3. E. Rosen, D. Tappan, G. Fedorkow, Y. Rekhter, D. Farinacci, T. Li, A. Conta, *et al*. IETF RFC 3032. "MPLS Label Stack Encoding". January 2001.
4. R. Braden, D. Clark, S. Shenker. IETF RFC 1633. "Integrated Services in the Internet Architecture: An Overview". June 1994.
5. S. Blake, D. Black, M. Carlson, E. Davies, Z. Wang, W. Weiss. IETF RFC 2475. "An Architecture for Differentiated Service". December 1998.
6. F. Le Faucheur. IETF RFC 4124. "Protocol Extensions for Support of Diffserv-aware MPLS Traffic Engineering" June 2005.
7. J. Moy. IETF RFC 1583. "OSPF Version 2". March 1994.
8. R. Callon. IETF RFC 1195. "OSI ISIS for IP and Dual Environments". December 1990.
9. Y. Rekhter, T. Li, *et al*. IETF RFC 1771. "A Border Gateway Protocol 4". March 1995.
10. L. Andersson, P. Doolan, N. Feldman, A. Fredette, B. Thomas, *et al*. IETF RFC 3036. "LDP Specification". January 2001.
11. B. Jamoussi, L. Andersson, R. Callon, R. Dantu, L. Wu, P. Doolan, T. Worster, N. Feldman, A. Fredette, M. Girish, E. Gray, J. Heinanen, T. Kilty, A. Malis, *et al*. IETF RFC 3212. "Constraint-Based LSP Setup using LDP". January 2002.
12. D. Awduche, L. Berger, D. Gan, T. Li, V. Srinivasan, G. Swallow, *et al*. IETF RFC 3209. "RSVP-TE: Extensions to RSVP for LSP Tunnels". December 2001.
13. E. Rosen, Y. Rekhter. IETF RFC 2547. "BGP/MPLS VPNs". March 1999.
14. T. Bates, Y. Rekhter, R. Chandra, D. Katz. IETF RFC 2858. "Multiprotocol Extensions for BGP-4". June 2000.
15. S. Bryant, P. Pate, *et al*. IETF RFC 3985. "Architecture for Pseudo Wire Emulation Edge to Edge". March 2005.
16. ATM Forum Specification. AF-AIC-0178.001 "ATM-MPLS Network Interworking, Version 2.0". August 2003.

17. L. Martini, M. Bocci, J. Jayakumar, N. El Aawar, G. Koleyni, J. Brayley, *et al*. Internet Draft; draft-ietf-pwe3-atm-encap-10.txt. "Encapsulation Methods for the Transport of ATM over MPLS Networks". Work in progress. September 2005.
18. M. Bocci, J. Guillet. *"ATM in MPLS Based Converged Core Data Networks"*. IEEE communications Magazine, January 2003.
19. D. Grossman. IETF RFC 2684. "Multiprotocol Encapsulation over ATM Adaptation Layer 5". September 1999.
20. Frame Relay Forum, Frame Relay/ATM PVC Service Interworking Implementation Agreement, September 2004.
21. M. Aïssaoui, D. Watkinson, M. Bocci. "OA&M In a Converged IP/MPLS Network". Alcatel Telecommunications Review, December 2004.

8

Guaranteeing Quality of Service in the NGN

8.1 Introduction

This chapter describes how resources are managed in the next generation network (NGN) so that Quality of Service (QoS) can be achieved to enable service differentiation. We describe in detail how QoS can be provided in the NGN for a number of services including PSTN replacement, residential video services (i.e. broadcast TV and video on demand) and business services. The chapter describes the QoS solution at all levels, from the routers and switches to the bandwidth managers and the application servers as follows.

The capabilities of the underlying packet network elements are discussed, along with how they classify and prioritise packets, and how the use of Multiprotocol Label Switching (MPLS) and IP transport has challenged the implementation of QoS. This includes consideration of how the core network can be traffic engineered using MPLS to provide a suitable platform for the NGN.

The requirements that must be met for the PSTN to be safely migrated to the NGN are discussed and a number of options for satisfying the QoS needs of such a service are examined. The MultiService Forum (MSF) solution for bandwidth management is covered in detail because it is a physical instantiation of the TISPAN-defined RACS architecture for bandwidth management and provides an insight into how bandwidth managers can provide rigorous QoS. The protocols used, including Diameter and SIP precondition signalling, are discussed and solutions for multi-domain QoS and IMS roaming support are examined.

A major driver for QoS in broadband services is the so-called triple play services which provide video, voice and Internet in a bundled package. These services have specific QoS requirements that are different again from the PSTN and have their own impact on network architecture. This chapter also looks at these services and how their needs for QoS can be met.

The needs of business services are examined; these are very different from residential or PSTN services and have their own set of solutions. This chapter considers one of the more pressing problems faced by VPN providers of how best to extend QoS guarantees across network boundaries in the MPLS BGP VPN architecture.

Converged Multimedia Networks   Juliet Bates, Chris Gallon, Matthew Bocci, Stuart Walker and Tom Taylor
© 2006 John Wiley & Sons, Ltd

Finally, a major but neglected issue of QoS is how to ensure that the network continues to function in the event of an overload or denial of service (DoS) event that impacts not the transport component but the application servers. The techniques for dealing with such problems are discussed and some possible and evolving solutions for protecting the NGN are described.

It should be noted that at the time of writing, QoS solutions were still evolving rapidly. Some of the solutions described in this book have been widely deployed, others are in their infancy and still more may never achieve volume deployment because either alternative solutions emerge or the problem being addressed goes away.

8.2 Defining QoS

QoS is a network performance concept whose definition varies depending on the perspective. From a user's perspective, it is the set of treatments given to the traffic belonging to a service that determines the perceived quality of that service. If the QoS is insufficient, then the application that relies on that service may not operate satisfactorily. For example, video services require a low packet loss, but they are typically not interactive and so packet delay and delay variation can be higher. Voice services, on the other hand, require low delay and delay variation because they are interactive, but can usually withstand a limited number of packets containing voice samples being lost by the network.

From a network perspective, QoS is characterised by a number of quantifiable attributes. The following are examples of the QoS commitments that a packet network can give:

- Within a call, connection or flow, the packet loss ratio, end-to-end delay and delay variation are the main parameters. Other parameters may include factors such as the probability of mis-ordering packets. Such parameters can be explicitly specified in the service level agreement, giving rise to "hard" QoS commitments. Alternatively, "soft" QoS commitments may be given that imply priority of a user's traffic over other users during periods of congestions or "low" levels of delay and loss.
- At the call level, the time taken to establish or release the call, the probability of a call attempt being successfully completed by the network and the probability of a call that is in progress being dropped are important.

A network represents a fixed resource (in terms of bandwidth/delay and processing capacity) and although it is possible to design a network so that there is never any contention for resources and all services receive the best possible level of QoS, this is not always desirable. Service differentiation is a key way in which service providers can maximise the revenue-generating potential of the network. Users must be encouraged to pay for the level of QoS that they receive, and premium services must be given an appropriate level of QoS compared with low-cost services. In addition, emergency services may need to be given priority over other services, for example, allowing emergency calls to cause other non-emergency calls to be dropped or degraded if network resources are insufficient. However, the complexity required to manage and enforce differentiated levels of QoS with sufficient granularity must be weighed against the cost of simply over-provisioning the network. In recent years, the cost of bandwidth and processing capacity has dropped significantly, reducing the pressure on network designers to implement potentially complex QoS mechanisms.

A differentiated level of QoS is given to services by sharing the network resources among users with service awareness; that is, the network knows what level of QoS is committed for a service and treats the packets accordingly. Mechanisms operate at two levels to enable sharing of network resources according to the commitment given by the network:

- The network maintains a model of the available resources and decides if it can admit traffic while achieving QoS commitments required for the services. This is typically known as *admission control*.
- The network treats a flow of packets in the data path in order to achieve the commitments given to that flow. This treatment generally consists of prioritising packets relative to those of other flows in order to achieve particular loss or delay commitments when they are contending for the same resource, for example, transmission opportunities on a link.

For these mechanisms to achieve a quantifiable level of QoS, the behaviour of each traffic source of a service must be understood, or characterised. The network then knows how to allocate resources to the source, and the QoS commitment therefore takes the form of "if you send at and up to a given rate, then I commit to giving you a certain packet loss/delay/delay variation." In order to enforce this traffic contract and to prevent users from taking more than their fair share of network resources and negatively impacting the QoS of other users' services, traffic policing or shaping mechanisms can be used at the edge of the network. These drop or mark packets that arrive in excess of their committed rate or in the case of shaping, delay packets that arrive too quickly after a previous packet, so that the traffic flow conforms to the contract.

The challenge in providing QoS in the NGN is to deliver the appropriate levels of QoS while minimising the complexity of the solution, ensuring that it can scale to all users and balancing this with the cost and risk of over-provisioning.

8.3 QoS in IP Networks

Chapter 7 described the use of MPLS as a convergence layer in the core of the NGN. However, MPLS has been largely designed as an addition to the IP protocol suite for carriers' core networks; that is, MPLS reuses many of the protocols and mechanisms of IP for its operation and management. Chapter 7 also briefly discussed IP and MPLS QoS mechanisms. In the following section, we discuss these mechanisms in more detail, before demonstrating how they enable QoS for MPLS-based services.

As explained in Chapter 7, IP is a connectionless protocol in the sense that a router makes a decision on the next hop to send a packet solely on the basis of information in the packet header and independently from other packets belonging to a particular flow. Early IP networks forwarded each packet simply on the basis of the globally unique end station address, or IP address, in the packets header. Each router in the network builds its forwarding tables, which are used to determine the next hop for a packet, based on local routing tables. These are constructed as a result of link state routing protocols such as Open Shortest Path First (OSPF) and IS–IS, or exterior gateway protocols such as the Boarder Gateway Protocol (BGP) v4. OSPF and IS–IS are interior gateway routing protocols because they enable routers to determine their peers within an Autonomous

System (AS), together with the link and node topology of the AS. Protocols such as BGP enable routers to determine how to reach the gateway to the next AS. Interior BGP (iBGP) runs within an AS, while Exterior BGP (eBGP) runs between boarder routers on adjacent ASs. These routing protocols give information only on the reachability of end stations. For example, if a link fails, the routers connected to it will flood information about this through the network, enabling the other routers to re-calculate their routing tables to avoid the failed link.

While such techniques ensure that the reachability of end stations is maximised, it is difficult for the operator to engineer the path taken by flows across the network. This in turn makes it difficult to specifically engineer resources, such as link bandwidths, to expected traffic demands. Although IGPs such as OSPF have been extended to enable traffic-engineering information, for example the amount of available bandwidth on a link to be flooded (this is known as *OSPF-TE*), the approach taken to providing QoS was often to over-provision the network. While this is effective in supporting a single QoS guarantee for all services, it does not enable different guarantees to be made to different services.

An additional consideration that impacts QoS in IP networks is the control protocol that is used for the traffic. The following two control protocols are common:

- Transmission Control Protocol (TCP)
- User Datagram Protocol (UDP).

In TCP, a feedback loop exists in that the source can detect when a packet is successfully delivered because it receives a TCP acknowledgement message. If an acknowledgement is not received within a specific time window of the transmission of a packet, the source assumes that the packet has been lost and will retransmit the packet. Too many lost packets cause the source to back off and wait for a random time before retransmitting the packet. This TCP back off and retransmit behaviour causes TCP sources to adjust their rate so as not to cause congestion in the network. However, while it does mean that the delivery of information can be guaranteed, delay is added to the flow of packets. Therefore, TCP is typically used for services that are not delay sensitive, but do require low loss, such as email or file transfer protocol (FTP).

In UDP, the application does not receive feedback from the network as to the state of congestion. IP packets sent using UDP are simply discarded during congestion. There is no explicit feedback loop provided by the IP network and sources typically just continue sending packets with no back off and retransmit behaviour. While this means that the transmission of a packet cannot be guaranteed, it also means that packets that do get through do not incur additional delay or delay variation (jitter) because of the source behaviour. UDP is therefore commonly used for real-time interactive services where loss is less of a concern, such as voice over IP (VoIP).

A number of mechanisms have been developed to enable classification and prioritisation of packets in an IP network. These mechanisms can be used, together with the appropriate reservation mechanisms, to enable QoS to be achieved. These include the following:

- Integrated Services Framework (IntServ) [1].
- Differentiated Services Framework (DiffServ) [2].

IntServ was the first of the two to be defined by the IETF. The IntServ framework consists of a number of components that, when combined, can provide per-flow QoS. These include the following:

- Packet scheduling. Packets that require a low delay receive scheduling priority over "best effort (BE)" packets.
- Classification of packets, for example, according to source or destination IP address, into traffic classes for appropriate QoS treatment by the network.
- Admission control to ensure that there are sufficient resources to deliver the requested QoS to admitted and existing flows.

IntServ enables a number of classes of service to be provided, in addition to basic BE. These include the Guaranteed Service (RFC 2212 [3]) that provides firm quantifiable bounds on delay and bandwidth and the Controlled Load Service (RFC 2211[4]) that does not provide hard guarantees, but aims to give a very high percentage of packets the same treatment that they would receive (in terms of queuing delay) from an unloaded router.

IntServ uses the Resource Reservation Protocol (RSVP) to signal the traffic parameters (and hence indicate the amount of resources to be reserved) to nodes in the network, using the *flowspec* object. It also enables filters to be specified to indicate which subset of packets receives the reserved resources.

Although RSVP is a soft state signalling protocol, IntServ does require that routers along the path of a flow must maintain some per-flow state information. The amount of state increases with the number of IntServ flows, which may lead to very large amounts of state information in the core of an IP network.

DiffServ is an alternative QoS architecture, which aims to minimise the amount of per-flow state in the network. It does this by implementing classification and conditioning functions (such as policing and shaping) only at the edges of a network domain, by marking the per-hop treatments that nodes should give to packets in the 6-bit DiffServ field of each IP packet and by applying those treatments to traffic aggregates. Processing in core routers is kept very simple, with only packet queuing and scheduling treatment per class. DiffServ (the architecture is specified in RFC 2475 [2]) has thus gained traction as the preferred QoS mechanism for IP networks. The architecture consists of a number of functional elements:

- A set of per-hop forwarding behaviours
- Packet classification
- Traffic conditioning such as metering, marking, shaping and policing.

The per-hop forwarding behaviours, which associate with behaviour aggregates, determine the per-hop scheduling priority (also known as *forwarding class*) and the drop precedence of packets. There are three main groups of Per-hop behaviour Scheduling Classes (PSCs):

- Expedited Forwarding (EF, described in RFC3246 [5]) is intended for services that require a low loss, low delay and low jitter.

- Assured Forwarding (AF, described in RFC 2597 [6]) is intended for low loss services, but for which no specific delay guarantees are intended. There are four AF classes, each of which provides three levels of drop precedence (DP) for packets during congestion. Note that although RFC 2597 does not mandate specific levels of priority between AF classes, typical implementations use AF4 as the highest and AF1 as the lowest.
- BE or Default Forwarding (DF).

Note that the order of packets within each PSC must be maintained, and thus form part of the same ordered aggregate.

A typical node architecture consists of a classifier, followed by priority-based queuing, where EF is serviced in strict priority over AF, and EF and AF are queues over BE.

Table 8.1 summarises the DiffServ classes and applicable drop precedence.

The behaviour aggregate to which a flow is associated is determined by classifying packets as they enter the network. Classification can in accordance with layer 3 information, such as source or destination IP address, or other factors, such as layer 2 information, for example destination MAC address. Packets can also be marked with "drop precedence," or "colour," depending on their conformance to a traffic contract. This determines the relative priority within a forwarding class with which packets are discarded during network congestion. A number of marking mechanisms have been defined for IP, and these may be used depending on the end-to-end service characteristics and the forwarding class. Note that these may also be used with IntServ but are described here due to the relative popularity of DiffServ.

RFC 2697 [7] describes a single-rate three-colour marker. Metering is based on three traffic parameters: Committed Information Rate (CIR), Committed Burst Size (CBS) and Excess Burst Size (EBS). The CBS and EBS are effectively equivalent to the thresholds of two leaky buckets at rate CIR. Traffic is marked "green" (which corresponds to the lowest DP) if it does not exceed the CBS, "yellow" if it is between the CBS and the EBS and "red"

Table 8.1 DiffServ classes

Class	Drop Precedence
Expedited Forwarding (EF)	Not applicable
Assured Forwarding Class 4 (AF4)	Low (AF41)
	Medium (AF42)
	High (AF43)
Assured Forwarding Class 3 (AF3)	Low (AF31)
	Medium (AF32)
	High (AF33)
Assured Forwarding Class 2 (AF2)	Low (AF21)
	Medium (AF22)
	High (AF23)
Assured Forwarding Class 1 (AF1)	Low (AF11)
	Medium (AF12)
	High (AF13)
Default Forwarding (DF)	Not applicable

if it exceeds the EBS (corresponding to the highest DP). This marker is useful because it enables a service to be policed on the basis of only the burst size, not the peak rate.

RFC 2698 [8] describes a two-rate three-colour marker. The main difference from the single-rate three-colour marker is that a second rate, the Peak Information Rate (PIR), which is higher than the CIR is introduced. Traffic that exceeds the PIR is marked "red," while traffic between the PIR and CIR is marked "yellow," and that below the CIR is marked "green." This enables a separate peak rate as well as a committed rate to be enforced on a service.

Although DiffServ is a useful tool for providing differentiated forwarding behaviour for traffic flows, it does not on its own provide QoS. Although IntServ can be used over DiffServ domains (see RFC 2998 [9]), DiffServ in itself does not provide resource reservation (bandwidth or buffer space) in order to ensure that the network can support the required QoS for an end-to-end flow. Furthermore, the DiffServ model is primarily applicable only on a single hop; it does not in itself provide a method for signalling QoS requirements along the path of a flow and rejecting a flow if the requested QoS cannot be supported, although later work in the IETF has addressed per-domain behaviours for DiffServ [10].

Instead, DiffServ tends to be used as a component in an end-to-end set of QoS mechanisms in an IP network. The remainder of this chapter describes the overall QoS architectures for multimedia services that often take advantage of mechanisms in the underlying IP network, such as DiffServ.

8.4 Traffic Engineering in the MPLS Core

In the "traditional" BE Internet, congestion was mitigated through a combination of network engineering and congestion-aware traffic control algorithms, such as TCP. However, network engineering with the objective of minimising congestion (and hence providing sufficient QoS to support premium services) can result in significant over-provisioning, and thus under-utilisation of the network. It can also be difficult to predict a priori where congestion will occur, either during normal network operation or during failure conditions. Network engineering tends to be a long-term activity involving the provisioning of links and nodes in the network to account for long-term predictions of traffic trends.

Network engineering therefore needs to be augmented with Traffic engineering (TE). TE is concerned with the characterisation and control of the traffic, so that the specific performance objectives of applications can be met. A major goal is to facilitate efficient and reliable network operations while simultaneously optimising network resource utilisation and traffic performance. The requirements for MPLS TE are detailed in RFC 2702 [11]. Motivations for TE include support for resilience through enabling diverse routes, or explicitly routed primary and backup LSPs, and to efficiently support QoS by avoiding congestion without having to over-provision every link. TE enables the operator to route traffic along paths where there is sufficient bandwidth and where the QoS requirements can be met. TE is applicable over short- to medium-term timescales, and may be online or offline. In online TE, the network nodes calculate routes through the network on the basis of dynamically flooded information about changes in topology and bandwidth utilisation. While this may be able to react quickly to such changes, operators may require more centralised control over routing policies, for example to avoid traffic transiting through politically undesirable geographical regions. Offline TE uses a centralised tool that gathers statistics about the topology and utilisation of the network and generates optimised routes through the network, allowing capacity to be allocated in a centralised, coordinated manner. However, offline TE may be slower to react to sudden changes in topology or demand, for example during network failures.

Existing IP interior gateway routing protocols (IGPs), such as OSPF and IS–IS, only allow a simple link-cost calculation of the best path across a network. TE extensions have therefore been added to enable traffic-engineering metrics to be advertised. OSPF-TE and ISIS-TE include link metrics such as bandwidth, unreserved bandwidth, available bandwidth, delay, delay-jitter, loss probability, administrative weight and economic cost. These are then used by the ingress LSR to build a TE database, which is then used by a constraint-based routing algorithm to generate explicit routes for LSPs. The path can thus take into account bandwidth reservation, include or exclude specific links or nodes and utilise loose or strict routing; that is, portions of the path of the LSP can be specified as a list of specific LSRs or can be left up to local routing decisions. The explicit route is then passed to the signalling mechanism to enable the LSP to be established.

As discussed in Chapter 7, RSVP-TE has become the predominant signalling mechanism for explicitly routed, traffic-engineered LSPs. RSVP-TE uses UDP as the transport mechanism for its signalling messages, which means that TCP sessions need not be maintained. RSVP-TE establishes LSPs using two signalling messages as follows: First a *Path* message is sent downstream from the ingress LER to the egress LER. The Path message contains a label request, causing downstream LSRs to allocate a label for that direction of the LSP, together with an Explicit Route Object (ERO), a Route Record Object (RRO) and a Flow_Spec object. The ERO is generated by the ingress LER on the basis of its traffic-engineering database, which specifies the path through the network that the Path message must to be forwarded on. The Path message therefore effectively specifies the route of the LSP, although sections of this may be strict, that is, the exact next hop LSRs are specified, or loose, that is, a local routing decision can be made to the strict next hop listed in the ERO. The Flow_spec contains the requested traffic parameters.

As the Path message is sent downstream, the state is established for the LSP in each intermediate LSR, and the IP hop address is added to the RRO. When the Path message reaches the egress LER, a *Resv* message is generated and sent back upstream along the route of the LSP. This causes connection admission control (CAC) to reserve resources in the intermediate LSRs. RSVP-TE allows significant flexibility in the way that resources are allocated, which is extremely useful in setting up the path protection mechanisms described in Chapter 7. This is achieved through the use of a filter in the RESV message, which enables intermediate LSRs to make separate reservations for all upstream sources, a single shared reservation for all upstream sources or a shared reservation for a subset of the upstream sources. It also sends the allocated label from the downstream to the upstream LSR.

RSVP-TE is a soft state signalling protocol because it requires a session to be explicitly refreshed in order to maintain it. Once an LSP is established, periodic sending of Path and Resv messages causes a state to be maintained; if a path refresh is not received by an LSR within a certain time, then the state is deleted. This has the advantage that if the underlying routing changes in the network, the next Path message automatically creates the LSP state along the new route, without having to tear down the LSP and re-establish it. However, this path refresh activity can result in a large processing overhead in the LSRs when there are a large number of LSPs. Although some steps have been taken to reduce this through path refresh reduction extensions to RSVP, this has been one factor in limiting the deployment of RSVP-TE LSPs. Ensuring that each RSVP-TE LSP carries a very large number of aggregated traffic flows can mitigate this. For example, it is possible to tunnel large numbers of inner LSPs over a small number of highly aggregated outer RSVP-TE LSPs to reduce the number of traffic-engineered LSPs that must be managed by the core of the network.

MPLS has been extended to incorporate the traffic-engineering functionality in the data path, in addition to the control plane traffic-engineering extensions described above. It does this by using the MPLS label value, together with a set of three experimental (EXP) bits in the MPLS shim header to infer the treatment given to a packet by an LSR on one of the following two ways:

- Label-inferred PSC LSP (L-LSP)
- EXP-inferred PSC LSP (E-LSP).

In an L-LSP, the MPLS label implies the PSC; all packets on a given LSP therefore share the same forwarding class. The setting of the EXP bits in the MPLS label indicates the DP. Such LSPs are also sometimes called *class-based LSPs* because all packets are treated as if they are of the same scheduling class. This also means that, if traffic of different forwarding classes is multiplexed over this LSP, it will receive the same treatment as the "best" class; that is, the LSP provides no differentiation in scheduling priority between high and lower DiffServ classes, for example EF and BE. Therefore, oversubscribed BE traffic can cause congestion to EF traffic if they are multiplexed onto the same L-LSP. The solution is to either only multiplex traffic of the same DiffServ class onto a given L-LSP or to allocate enough resources to the L-LSP so that congestion never happens, which is much less bandwidth efficient.

For an E-LSP, the EXP bits indicate both the PSC and the DP for the packet. This means that each packet of an LSP can receive a DiffServ treatment that is appropriate to its forwarding class and DP. E-LSPs can thus be used to implement both class-based and class-multiplexed LSPs. Each forwarding class can be allocated a share of the bandwidth on the LSP and can also be rate limited. The LSP effectively acts a virtual tunnel through the network, which is able to provide QoS differentiation in its own right. Note, however, that since there are only three EXP bits, only a subset of the available DiffServ classes can be indicated on the LSP. The mapping between the DiffServ class and the EXP bit values is therefore usually configurable on an LSR.

The combination of MPLS TE and DiffServ frameworks provides the ability to compute paths and check available bandwidth along the path in the network on a per-class of service basis. This is known as *DiffServ-aware traffic engineering*, and allows granular class-based optimisation of network resources by mapping traffic associated with particular Forwarding Equivalence Classes (FECs) and DiffServ classifications to particular LSPs. Class-specific constraints can be implemented on LSPs, and so can class-specific policies such as re-routing or preemption priority. The capabilities of DiffServ-aware TE are therefore very useful for converged networks where many services with multiple QoS requirements must be supported over the same core network without over-provisioning the network. They are also useful where portions of the network transit a third-party network provider. Maximising the intelligent use of such rented bandwidth is important to minimise costs. However, a counter to this is the relative complexity of implementing and managing such fine-grained QoS in the core of the network.

8.5 Video Services

Competition from mobile operators and low-cost or free Internet voice services is squeezing revenues from residential voice services, which, historically, have represented the bulk traditional fixed-line operators' revenue. These operators are therefore looking to compensate by generating revenue from new residential services. One way to do this is to utilise the new broadband infrastructure to augment existing voice and data (high-speed Internet) services with video, in the form of video on demand and broadcast television. Such "triple play" offerings take advantage of the bandwidth available from new broadband access technologies, such as Digital Subscriber Line (DSL) and Fibre to the Home (FTTH) to allow these operators to compete with both cable access service providers, many of whom are also looking to expand into voice and data services, and direct broadcast satellite TV providers. Triple play services are delivered from a common access and aggregation network infrastructure in order to minimise operational and capital expenditures.

The following are some key attributes of video services that impact the way the resources are managed and QoS is delivered:

- A huge increase in the bandwidth per subscriber: Future networks will need to deliver multiple standard definition and high definition TV channels, as well as high-speed Internet and voice. This results in a requirement for upwards of 20 Mbps per household, possibly reaching 100 Mbps in the future.
- Video services are mostly asymmetric in nature: the majority of the bandwidth requirement is in the downstream direction; that is, from network to subscriber. Traffic in the upstream direction is mostly control traffic.
- A high degree of personalisation of services: Triple play requires QoS differentiation and protection of guaranteed services such as video and voice, from oversubscribed "BE" high-speed Internet services.

In order to understand the resource management mechanisms required to support such high bandwidth demands together with QoS differentiation, consider the network architecture shown in Figure 8.1.

The following key elements should be noted:

- A residential gateway or network termination equipment (NTE) containing a DSL modem. This may be integrated into a TV set top box and is connected via a DSL-enabled telephone line to a Digital Subscriber Line Access Multiplexer (DSLAM) residing in the service provider's local exchange.
- An aggregation network. Owing to the demand for high bandwidth at cost points that are attractive for a residential service, this is increasingly seen as an Ethernet network. MPLS may also be used in the aggregation to enable TE and resiliency, as described in Chapter 7 of this book. In that case, Virtual Private LAN Service (VPLS)

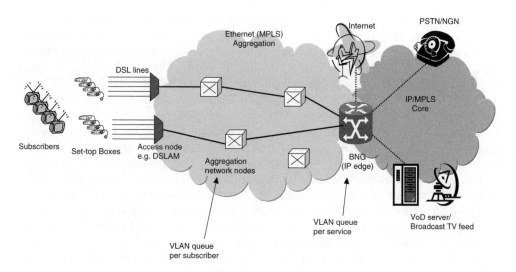

Figure 8.1   Example video delivery network architecture

or Virtual Leased Lines (VLLs) are used to support the delivery of Ethernet-based services. In order to scale the network while supporting QoS differentiation, nodes in the aggregation network support a high degree of subscriber awareness. This is described in more detail in the following paragraphs.

- One or more Broadband Network Gateways (BNG) or service-aware IP edge routers. These terminate the subscriber sessions and hand off traffic from either video servers (in which case a service-aware IP edge router can be used) or Internet service providers (in which case, a BRAS – Broadband Remote Access Server – can be used). The BRAS and service edge router function may be combined in the same node or may be separated, allowing cost optimisation of nodes for their specific function.

Two types of TV services are commonly provided. These are broadcast TV and video on demand. Broadcast TV services rely on multicast replication between the video server and the subscriber to deliver traffic from specific TV channels to multiple simultaneous subscribers. In order to minimise the utilisation of bandwidth in the aggregation network, multicast replication should happen as close to the subscriber as possible, usually in the aggregation network or in the access node or DSLAM. Video on demand is a point-to-point service, with a video stream being delivered directly via an Ethernet VLAN from the video server to the requesting subscriber. Video servers should be placed as close to the subscriber as possible to prevent resources being wasted in backhauling large amounts of video-on-demand traffic across the core of the network.

Network engineering may be used to ensure that sufficient are resources available to handle the expected peak demand for video services alongside other services delivered by the converged network. In addition, CAC mechanisms, such as bandwidth manager architectures as described elsewhere in this chapter, may be used to reserve resources for video services, on an on-demand basis, as well as to ensure that sufficient resources are reserved to continue to maintain QoS when traffic is re-routed during network failure events.

In addition to network engineering and admission control schemes, QoS mechanisms must exist in the data path, and the key factor here is the trade-off between scalability, due to the large number of residential subscribers and services that must be supported on each network interface, and the level of granularity and QoS policy control. In a typical triple play deployment, services for each subscriber may be delivered (at the access to the network) on a unique Ethernet VLAN. QoS differentiation can be indicated using either the Ethernet P bits or the DiffServ code point of each packet.

In the downstream direction, per-subscriber queuing is provided in the access and aggregation nodes, with real-time video and voice services receiving priority over BE high-speed Internet services. This allows per-subscriber QoS policies to be implemented. Interfaces on these nodes only carry traffic from a limited number of subscribers, which reduces the number of queues required. High-speed Internet can be rate limited on a per-subscriber basis in these nodes to prevent congestion during peak periods. Only per-service priority queuing is therefore required at the BNG, which reduces the requirement to support large numbers of queues at this point.

As mentioned above, there is very little video traffic in the upstream direction. That which does exist is mostly composed of control messages, for example Internet Group Multicast Protocol (IGMP). Therefore, there is no need to provide per-service/per-subscriber queueing in the upstream direction, Instead, only a simple real-time and non-real-time priority queueing is needed in the access and aggregation nodes.

8.6 Business VPN Services

Business data services such as ATM, Frame Relay or Ethernet private lines, VPLS and IP VPNs tend to have holding times that are orders of magnitude longer than session-based services such as consumer multimedia or voice. Dynamic signalling across user-network interfaces is typically not necessary. Instead, connectivity is established under longer-term contracts or management agreements. However, dynamic signalling of QoS requirements and traffic parameters is required in the provider's network.

Figure 8.2 shows an example of how QoS can be provided to business services in a converged MPLS core network. The example shows aggregation networks, which are ATM or Ethernet here, connected across the MPLS core using a pseudo-wire (PW) on an LSP tunnel. Sufficient resources are reserved against the underlying network for all of the PWs on the LSP tunnel by assigning a specified bandwidth to the traffic-engineered LSP tunnel, using RSVP-TE to establish the LSP through the MPLS core.

Within the attached layer 2 aggregation networks, established ATM QoS mechanisms, or provisioning mechanisms combined with prioritisation using the P bits of the Ethernet header are used to support the QoS requirements. At the ingress to the MPLS core, classification of packets according to layer 2 markings such as the Ethernet P bits, or the ATM service category and cell loss priority, enables a DiffServ class of service and DP to be assigned to the packet. The ATM Forum provides a set of recommended mappings for ATM over MPLS in [12]. Note that this classification can also be based on existing DiffServ code point markings applied to traffic in an IP VPN.

Ingress policing can be used to ensure that traffic entering the MPLS network conforms to the traffic contract. The policing parameters for each user connection at PE1 should be the same as that at the ingress to the access network so that conforming traffic from the customer is not policed.

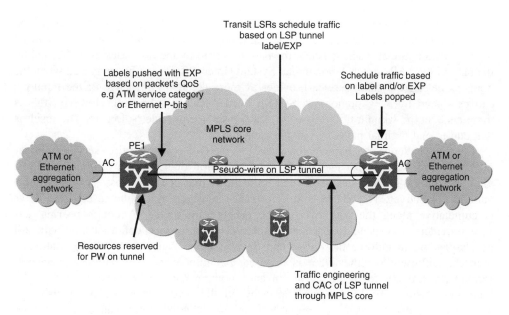

Figure 8.2 Business service QoS example

Queuing and scheduling mechanisms in the PE arbitrate among competing connections to ensure that each connection receives its appropriate share of the allocated bandwidth to achieve its committed QoS. In the control plane, admission control can be applied for the PWs against the assigned LSP tunnel bandwidth. A process known as *Egress remarking* then assigns a value to the EXP bits of the outer LSP tunnel on the basis of the class of service and DP assigned to the PW packet.

In the core LSRs, appropriate queuing and scheduling treatment is given to packets on the PW according to the outer LSP type. If this is a label-inferred LSP (L-LSP), then the DiffServ treatment is inferred from the label of the LSP, while the EXP bits infer the DP of the packet. However, if it is an EXP-inferred LSP (E-LSP), then both the priority and DP are based on the EXP bit settings.

In the egress PE, scheduling is based on the EXP and/or tunnel label values, as in the core LSRs. The outer labels are then popped and the PW mapped back to the appropriate ATM service category or other layer 2 connection QoS class.

8.7 Extending QoS for VPN Services across Multiple Providers

The above QoS architectures are primarily targeted at the delivery of services from a single service provider's network. The advantage is that one consistent set of policies and mechanisms can be applied to a service to ensure that it achieves its QoS commitments. Furthermore, the contract between the customer (or user) and the service provider only requires one service provider to manage the delivery of QoS commitments for the service. However, few service providers' networks can reach all the possible users for a service. Global services therefore require that all the providers along the path of a service agree on and deliver the appropriate QoS commitments. This raises a number of challenges when one tries to apply the above QoS model across multiple providers' domains.

The first of these challenges includes the ability to reserve aggregate bandwidth and select optimal paths for the LSPs across multiple domains. Since end-user flows are carried over aggregate tunnels such as LSPs, the path followed by the flow packets is dictated by the path of the LSPs. Inter-Autonomous System (Inter-AS)TE can be envisaged when the domains belong the same service provider as more detailed knowledge of the topology can be shared by the Autonomous System Border Routers (ASBRs). However, this is more challenging when traffic crosses to another service provider's domain. The result is that end-to-end optimal paths may not be selected.

A further challenge is achieving consistency of the QoS across multiple service provider domains. Each service provider has its own definition of network-level classes of service and the QoS objectives associated with them, that is packet loss and delay/delay variation objectives. The end-to-end performance achieved by individual end-user flows is cumulative along the path over which packets are routed. Although peering service providers may need to harmonise their class of service definitions, it will still be challenging to enforce the end-to-end Service Level Agreements (SLA) contracted with the end-user. Specifically, it is required that monitoring of QoS metrics be performed pro-actively for each service in each domain and that cumulative metrics be maintained using a method agreed upon by all the involved service providers. This way, a report can be made to subscribers on the achieved performance on a regular basis.

Bandwidth manager architectures bring additional challenges: communication and coordination among the bandwidth managers of the multiple domains are required in order to admit an end-user flow; that is, each bandwidth manager along the path of a flow must admit the flow to its domain, which requires that they must be aware of the QoS requirements and bandwidth requested for each flow, the neighbouring domain from which the request originates and the next domain along the path.

One proposal for a multi-provider QoS architecture is to reuse the concept of VPNs to create virtual ASs that only transport traffic from a particular service class and only peer with other virtual ASs that transport traffic from the same service class. Each such Virtual AS is, in fact, a VPN. In this way, a network of VPNs is created, which is a virtual Internet for a particular service class. All routes that are advertised within such a virtual Internet are guaranteed to be reachable via an unbroken chain of virtual ASs that support the same service class. Such a virtual Internet can gradually grow with similar dynamics and scale as the current BE Internet and without disturbing the BE Internet. The service class of such a virtual Internet can only be guaranteed if there is no congestion or packet loss in this virtual Internet. A bandwidth manager can carry out the allocation of resources to flows, as described earlier in this chapter, where bandwidth managers from peering virtual ASs signal to one another to ask for resources in the neighbouring network.

Figure 8.3 illustrates an example of how peering VPNs can provide QoS between separate domains. There are no end-to-end LSPs within the network. Each VPN has its own local set of LSPs. The end-to-end path between two QoS-enabled edge points is a concatenation of LSPs. The concatenation points are IP routing points. Traffic from LSPs that terminate on a particular concatenation point is re-distributed over a new set of LSPs by performing a routing look-up.

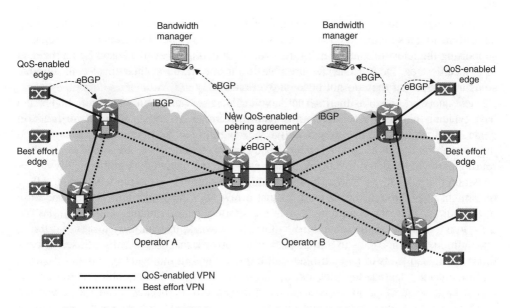

Figure 8.3   Peering of VPNs to create an inter-domain QoS-enabled network

8.8 QoS and the PSTN

The nature of the existing TDM-based PSTN is that it is a carefully dimensioned network with finite bandwidth available for carrying calls and with relatively constrained routing behaviour (i.e. there are a limited number of routes a call can take through the network and re-routing is strictly controlled to limit the options available for a call to be passed through the network). Where more dynamic routing is used within the PSTN, it typically uses mechanisms such as trunk reservation to ensure that a voice trunk will only accept re-routed calls if it operates below a certain capacity. When a subscriber places a call in this network, the local exchange performs a resource check when selecting the outgoing route for the call so as to ensure that sufficient bandwidth is available for the call; if this is not the case, then the call fails and the subscriber will hear the network busy tone played back.

In addition to protecting against exhaustion of the traffic-carrying capacity, the UK PSTN also protects against processor capacity exhaustion. This may happen in the event of a focussed network overload caused, for example, by a television phone vote or a telephone phone-in, or it may happen because some external event causes a lot of people to make phone calls in an area within a short space of time (typically caused by large-scale accidents or an event such as a terrorist attack). In this case, many more calls are accepted into the network towards a given destination than can be handled by that destination and because customers are unable to get through to their required number, they often re-try the call. This can lead to extremely high call rates far beyond those that the network switches are capable of handling and can, in the absence of protection mechanisms, lead to a complete failure of a switch as it spends all its time rejecting calls and is unable to perform the maintenance required to carry the existing calls. In order to protect against this type of overload, PSTN switches are able to invoke call-gapping algorithms to choke focussed overload at source exchanges before it gets to the exchange supporting the televoting number. In the event of a mass call event caused by an incident in a specific area, local exchanges are able to throttle traffic at their local concentrators so as to ensure that they do not become overloaded.

Packet networks such as the Internet have evolved very differently from the PSTN and have typically not had the same controls against overload. Today's packet networks are, in general, BE networks in that they effectively carry traffic to the limit of their capacity and then discard at the point at which their capacity is exceeded. They utilise highly dynamic routing protocols to determine the optimum path through the network and typically do not constrain the paths that a flow may take through the network; so in the worst case, a flow that might pass through the network in just a few hops may take tens of hops because in the current network condition, that is the shortest path available. Unlike the PSTN, they do not perform admission control within the network; instead, they just accept traffic unconditionally. This leads to a highly efficient network that can be fully utilised at peak times but it also leads to uncontrolled traffic discard and in the event of a major network failure, this may degrade the network for all users for a significant period of time. Because of the large scale of packet networks, the bursty nature of packet-based traffic and the relatively tolerant applications being run over them (e.g. file transfers or web surfing), this approach has worked very well to date; and in extreme circumstances, it has been shown to outperform PSTN networks that do not have the full set of overload controls in place.

Recently, however, packet networks have begun supporting QoS; initially, this was primarily driven by the business market and the need to guarantee customers set bandwidths over packet cores. There are a number of technologies that have been adopted to provide this capability, the most important of which are DiffServ-based admission control over connectionless IP networks and MPLS traffic-engineered networks. These technologies can also be used by voice and video over IP providers to ensure that their real-time voice and video services get better treatment when the network is heavily loaded than the less time-critical applications such as web surfing and file sharing.

As networks start to converge and network operators look to migrate their PSTN networks onto packet-based infrastructure, considerable work is being undertaken by the industry to understand what QoS mechanisms are actually required in the NGN and how they might be implemented. There are four main philosophies that might be followed by a network operator wishing to implement a network to support both traditional BE web surfing services and time-critical peer-to-peer services, such as voice and video calling.

1. Ignore the problem. The assumption is that the network works well almost all the time and in the unlikely event of unusual calling patterns or network failures, the quality will suffer.
2. Over-provision the network. The assumption is that the cost and complexity of implementing rigorous QoS mechanisms is such that it is cheaper to buy more capacity than you think you will ever need (both bandwidth and call-processing capacity) and run a very lightly loaded network. In this case, call-processing congestion cannot be encountered because the Call Agents are capable of supporting the end-users regardless of calling frequency, and the network will only ever become congested owing to a lack of bandwidth in the case of multiple unrelated failure scenarios that are so unlikely that they can be discounted.

3. Provide a simple DiffServ-based QoS solution and some network over-provisioning. The assumption is that it is not cost-effective to completely over-provision the network to protect against all failure scenarios. However, if end-user traffic is limited at the edge of the network according to pre-calculated traffic loads and if time-critical traffic is prioritised over non-time-critical traffic, then acceptable performance will probably be achieved in all but the most extreme failure cases. Call-processing overload can be prevented either by over-provisioning or by providing some means of limiting call set-up attempts sent to the call agents.
4. Provide a DiffServ-based QoS solution with rigorous CAC. The assumption is that while the network can be over-provisioned to a degree, it is either uneconomic to protect against failure by over-provisioning or the services being supported are sufficiently critical that some degree of guarantee must be met even in unlikely failure cases. The network cannot become congested because CAC is performed for each peer-to-peer voice/video session and in the event of network failures, mechanisms are provided to terminate in-progress sessions in such a way as to give priority to business-critical sessions.

The solution chosen by any given network operator depends very much on the services that they wish to support and the SLAs that they have with their customers. For example, a low-cost or free voice service is likely to adopt the first model and ignore QoS as their customers can always use a traditional land line or a mobile phone if they have a critical call to make. A low-cost additional line service might seek to differentiate itself by providing some QoS for business or premium subscribers using DiffServ while providing no guarantees for low-value residential customers. However, for network operators looking to migrate their PSTN onto a packet network, they are likely to have significant regulatory requirements with respect to handling emergency calls and governmental/key worker calls. This means that they are likely to have to either massively over-provision the network or implement a full bandwidth–managed network solution with rigorous CAC.

8.9 QoS Architectures for PSTN Services

This section describes at a high level how QoS may be realised for PSTN services in packet networks. It looks at three different solutions of varying complexity, scalability and robustness. These solutions are as follows:

1. A simple DiffServ-based QoS solution.
2. A simple CAC solution that uses the session border controllers (SBCs) at the edge of the network to provide QoS.
3. A bandwidth manager–based QoS solution.

Each solution has its advantages and disadvantages and the decision as to which solution fits a given network will depend on commercial factors, service obligations and network sizing.

8.9.1 A Simple DiffServ-based QoS Solution

In this deployment, the network operator combines over-provisioning with a simple DiffServ implementation to ensure that both the quantity of traffic admitted to the network

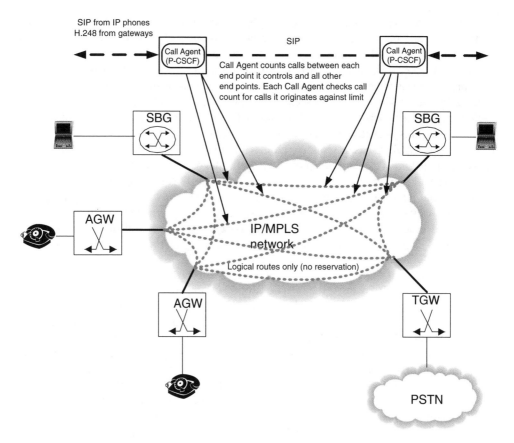

Figure 8.4   Simple DiffServ QoS with call counting

is controlled and the network has sufficient capacity to handle this traffic even in failure scenarios. This is combined with a simple call-counting mechanism in the Call Agents as shown in Figure 8.4.

In this architecture, the network supports DiffServ-based QoS and polices traffic at the boundary of the network. This means that at a given access or trunking gateway, only a set capacity of traffic will be accepted into the network. Any excess traffic beyond this limit will be dropped to prevent mis-configuration or overload of one network element destabilising the entire network. Similarly, at any packet interconnect with peer networks, DiffServ policers are configured on the interface to limit the traffic sent into the network to a maximum, as agreed between the two operators as part of a commercial interconnect agreement.

However, just limiting the total volume of traffic accepted into the network, while being effective in ensuring the stability of the core network, is less helpful from the point of view of guaranteeing the peer-to-peer voice service. When the number of calls entering the network exceeds the limit for traffic at that ingress point, the policer in the network element will start to drop packets to bring the total flow within the contract. Unfortunately, this action cannot distinguish between individual sessions; so instead of dropping all packets for a session, it will drop some packets from all sessions. The impact of this is that all voice services will suffer some packet loss which will effectively degrade the voice quality of every subscriber whose media flow passes through the ingress point. In the worst case, the packet loss will reach such a point that sufficient packets will be dropped from all the sessions passing through the ingress point such that every session will be rendered unintelligible. Clearly, what is required is some sort of mechanism to limit the number of active sessions accepted at the ingress point, and in this network architecture, the Call Agent can perform a simple call-counting function.

Each Call Agent active in the network must understand a range of traffic sources and traffic sinks that it will either receive calls from or terminate calls on. When it receives a request from a call, it performs the call routing and set-up signalling, and it must examine the media end point information (carried within the SDP of the SIP signalling). From this signalling, the Call Agent can identify the two media end points of the call, which might correspond to two access gateways or an access gateway and an interconnecting SBC. The Call Agent is provisioned with the total bandwidth (or if a single codec is used for all calls, the number of calls) that it can accept between the two end points. This can be thought of as creating a virtual tunnel of fixed size between the two end points (it is virtual because it is not necessary for the network to actually reserve this bandwidth); these are the dashed lines shown in Figure 8.4. When a call arrives, the Call Agent checks the bandwidth that it has already admitted between the two end points and adds the new call's bandwidth to the total; if it is within the configured limit for the virtual tunnel, then the call is accepted; if not, the call is rejected. In this way, calls are only admitted up to the limit of the ingress points capacity, thus ensuring that whole sessions are rejected when the network hits capacity and preventing already established sessions from being impacted by packet loss. This relatively simple solution can also be enhanced through the use of bandwidth reservation on the virtual tunnels. For example, when a tunnel hits 90% of capacity, a Call Agent might reject any calls except emergency or governmental calls; as these are typically a small proportion of the total calls admitted, these will receive a very high grade of service compared to normal calls as there will always be some spare capacity for them.

In order to engineer the network to support such a solution, the network operator must plan the network on the basis of predicted and observed traffic patterns between each media end point, that is traffic sources and traffic sinks, and dimension the network according to this plan while allowing sufficient capacity to handle network failures. They must configure the policers at the end point such that no ingress point allows more than the planned maximum traffic into the network and then determine which Call Agents are responsible for handling traffic from each source and set-up and configure the virtual tunnels against which they will perform CAC. Although this may sound like a cumbersome task, it is in fact a process that the PSTN has followed for many years with considerable success and so the overhead on the network operator is significantly less than might be expected.

Note that although the virtual tunnels represent a single Call Agent's view of network resource, there may be multiple Call Agents and even multiple services sharing the same network resource. This implies some degree of inefficiency as network operators have to virtually partition the resources between each Call Agent when they assign the capacity limits to the Call Agent.

There are a number of advantages of this approach to QoS:

- It is relatively simple to engineer and can be realised without requiring the emerging bandwidth management technologies.
- Multiple virtual tunnels share the same underlying network infrastructure so that the solution can exist in multi-service environments as each service can be allocated its own slice of bandwidth independently of the others.
- In the absence of severe network failures, it provides a means of guaranteeing voice traffic priority.
- Additional capabilities can be supported by the Call Agent to ensure that certain types of call (emergency call) and certain types of users (governmental or key workers) are prioritised.

However, this type of solution does not provide a complete guarantee of QoS in all circumstances and it suffers from a number of disadvantages.

- The actual resources available in the network are not tracked; so if a severe failure takes place, such that the bandwidth reserved for the service by DiffServ-based network engineering is no longer available, the call control layer is not informed and so will continue to allocate calls as though its full bandwidth was available. This will result in uncontrolled discard of voice traffic and, hence, potentially a complete breakdown in the voice service even though the network retains some useful capacity.
- Unless the network architecture is such that there are relatively few traffic sources and sinks and they support a very large number of individual users, the allocation of virtual tunnels between source and sinks on a per–Call Agent basis is inefficient. It is possible that some architectures which have a few very large access gateways and only limited points of interconnect may meet this criteria. However, in many cases, the number of sources and sinks will render this approach uneconomic. This is because the classic Erlang B formula which is used to dimension peer-to-peer telephony networks is efficient when used to dimension very large capacity routes but very inefficient when used to dimension small capacity routes. For many real networks, this solution will produce a very large number of long thin tunnels that stretch edge to edge through the network each of which must be reserved a set amount of virtual bandwidth. In order to guarantee an acceptable grade of service for traffic, significantly more bandwidth must be allocated to these tunnels than is likely to actually be used at any one time. While this may be acceptable in some networks, especially as voice is a relatively low bit rate service, as the network scales and the bit rate increases from around 120 kbps for voice to upwards of 500 kbps for video streams, this may become prohibitively expensive.
- Although it does not require the use of emerging bandwidth management technology, this solution requires the Call Agent to support relatively sophisticated QoS features; in particular, it must support the concept of virtual routes and must be configured to determine the source and sink points of the traffic from the IP addresses seen in the call set-up signalling. As an alternative to requiring the Call Agent to perform these functions, some SBCs offer similar capabilities, tracking bandwidth through the network for individual routes. Although this provides a solution for a network operator where the Call Agent cannot support QoS, it does not offer significant additional advantages

and access gateways (likely to be end points of most subscribers in a PSTN replacement service) do not typically support such facilities.

8.9.2 A Session Border Controller–based Solution with Explicit Reservations

A variation of the simple DiffServ-based solution described above would be to use SBCs to physically reserve bandwidth for the virtual tunnels through the network using MPLS TE or RSVP, as shown in Figure 8.5.

In this type of deployment, the network is dimensioned as before for the majority of services; but for critical peer-to-peer services, such as voice, the SBC instantiates the virtual tunnels by making a real reservation in the packet network, that is, it creates edge-to-edge tunnels for the service.

When the SBC receives each call, it identifies the tunnel that it will use and performs CAC on the tunnel by either accepting or rejecting the call on the basis of the tunnel occupancy. If for some reason, the network should suffer a failure causing the tunnel to

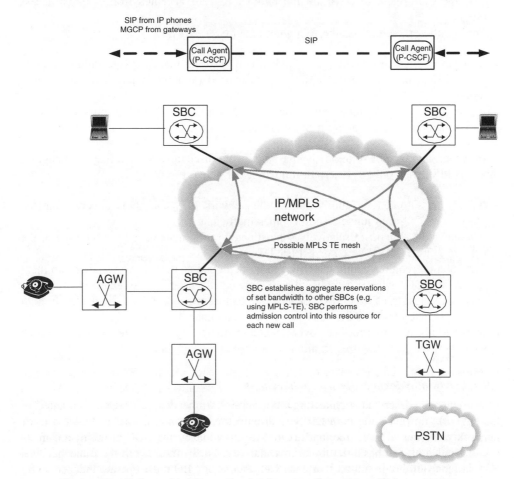

Figure 8.5   QoS provided by session border controllers with explicit reservations

lose some of the bandwidth available to it, the SBC can change the CAC parameters and hence the number of calls it can accept before it considers the resource to be used up. In the worst case, the SBC may even release low priority calls until the bandwidth used by the service falls within that available for the tunnel; this allows it to preserve critical calls at the expense of less important calls. The benefit of this approach is that in severe failure cases, the network will still be able to carry some peer-to-peer calls as long as at least some connectivity remains. This offers a significant improvement over the simple DiffServ-based solution described previously.

Another benefit of the solution is that it is slightly more efficient than the DiffServ-based solution because many different access gateways may share a SBC; hence, it provides a first point of concentration reducing the number of long thin tunnels that must be supported. However, in order to achieve this, it must be able to track resources available on the access links or be able to rely on a highly resilient network architecture whereby a link to an access gateway is either available or unavailable (in which case it operates at either full capacity or no capacity). This can be combined with call counting by the Call Agent to constrain the number of calls accepted from a given access point.

This solution therefore has the following benefits.

- It makes explicit reservations and tracks them ensuring that in severe failure cases at least some calls will always be able to be carried (as long as physical connectivity remains).
- It provides some additional concentration and therefore reduces the number of long thin tunnels required, making it more efficient than the DiffServ-based solution.

However, this type of solution also has significant disadvantages that may well suggest that it is of interest only in a certain limited set of network scenarios.

- It adds only limited scalability and as the network grows, it starts experiencing the problem of long thin tunnels and becomes inefficient.
- It requires the support of relatively sophisticated traffic management in the core of the network, either RSVP based or MPLS TE.
- The solution requires additional SBCs to be deployed as there is no usual requirement to deploy SBCs where the subscriber is connected via an access gateway; in effect, every call must transit the SBC for this architecture to work. Given the relatively high costs associated with SBCs, this may not be economically viable.
- In order to be able to prioritise government and emergency calls in the event of resource shortages, the SBC requires significant call control intelligence.

8.9.3 Bandwidth Manager–based Architectures

The previously described architectures can support some degree of QoS guarantee in the network; however, the pure DiffServ architecture is limited in that it does not track network resources in real time, and explicit reservation by the SBC is inefficient in its use of bandwidth and has strict limits in scalability. Furthermore, both these mechanisms also suffer from the limitation that each Call Agent or SBC must operate independently of other Call Agents or SBCs.

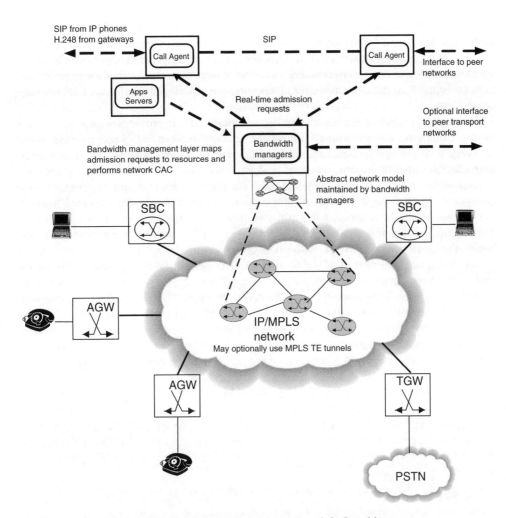

Figure 8.6   Bandwidth management–based QoS architecture

In order to support a true multi-service approach to QoS, the telecommunications indus-try has proposed the use of a resource control function that is able to receive requests from multiple application servers (such as Call Agents, but also multimedia servers such as video on demand servers) and based on knowledge of the underlying network and its current capacity to accept or reject the requests. This entity is defined in the MSF archi-tecture as a *bandwidth manager* and in the ETSI TISPAN architecture as the *Resource Admission Control Function (RACF)*. A high-level view of a bandwidth manager–based QoS architecture is shown in Figure 8.6.

The bandwidth managers are responsible for tracking resources in the network, receiv-ing requests for bandwidth from the application servers and performing CAC on the basis of the capacity available in the network. The bandwidth management layer is responsible for understanding the topology of the network such that it can determine the resources required to support a reservation between two end points and the capacity that is currently spare on those resources. In the event of a network failure, the bandwidth management layer must be able to determine the impact of the failure and the resulting traffic patterns and terminate any reservations that it determines may no longer be supported. The bandwidth management layer may consider the relative importance of individual reservations, thus ensuring that critical services and sessions are prioritised in the case of resource contention.

The use of bandwidth managers has the advantage that it provides a guaranteed QoS mechanism based on admission control and the real-time state of the network while allowing multiple applications and services to request bandwidth from it, thus permitting a flexible real-time partitioning of resources between the various services active in the network. This does not remove the need for the network operator to dimension the network for each service and ensure that capacity is available for the expected loads. It does, however, allow additional flexibility in how the total bandwidth in the network is distributed between the services in real time and depending on the network state, and hence allows a far more efficient allocation of bandwidth.

The actual implementation of the bandwidth management layer depends greatly on the type of network in which it is employed, its total size, the types of services supported by the network and the underlying technology on which the network is implemented. However, a major benefit of this solution is that the network is abstracted from the application servers that are using it to provide service. This allows the application layer vendors to ignore the underlying network complexity, keeping the solution focussed on the service layer and allowing the network operator to change network technologies without impacting the application servers running on the network.

Furthermore, an additional benefit of a bandwidth manager–based solution is that it can be scaled to very large networks indeed, if the bandwidth management layer is suitably architected. Typically, this is achieved by avoiding bottlenecks such that the application layer can send requests to multiple bandwidth managers, and these bandwidth managers may themselves consult additional bandwidth managers when determining whether to accept or reject a reservation.

In summary, the benefits of a bandwidth manager–based architecture are as follows:

- It is potentially very efficient in its use of network bandwidth by allowing multiple services and application servers to share a common resource pool.
- It tracks resources in real time in the network, thus allowing critical sessions to be carried between two end points as long as basic connectivity remains.
- It can prioritise mission-critical sessions in the event of a large-scale failure.
- It allows the application servers to have minimal understanding of the underlying network.
- It can scale to very large network topologies.

However, there are some disadvantages to the solution which mean that it is not the automatic choice for all network architectures.

- In large networks, it may add to session set-up delay.
- It adds another layer of application servers to the network and hence incurs additional processing cost.

- The bandwidth management layer adds complexity to the network design and must be understood and managed by the network operator.

One of the characteristics of the bandwidth management–based solution is that in the event of network failures, bandwidth mangers must perform significant processing tasks to analyse the impact of the failure and identify which resources must be released by the application layer. This means that the network operator must purchase processing capacity to handle failure cases rather than the more benign steady state and this must be accounted for when considering the processing cost of a bandwidth manager–based solution.

8.10 The MSF Architecture for Bandwidth Management

The MSF bandwidth management architecture was designed primarily with the aim of supporting mission-critical peer-to-peer services such as PSTN voice. These services, as discussed previously, are usually regulated and the regulator may place strict requirements on network performance and behaviour. For example, they may dictate how it will behave in the event of overload with regard to the handling of calls to the emergency services or how it handles calls involving certain classes of user, for example key workers, security services or governmental users. The MSF architecture for bandwidth management is like all MSF network architectures designed around physically deployable boxes rather than being a functional architecture like the ETSI RACS architecture [13] that is described in Chapter 5. It is possible to see the MSF architecture for bandwidth management as a physical instantiation of the RACS architecture, if the functions of the ETSI Service Policy Decision Function (SPDF) are considered to have been split between the MSF bandwidth manager and the Call Agent and/or S-SBG. The original bandwidth management architecture in the MSF was based on its release 2 (MSF R2) architecture which was targeted at the fixed network. This architecture is described in Refs 14 and 15. The MSF release 3 (MSF R3) architecture adds support for the 3GPP (3rd Generation Partnership Project) IMS components and defines the interfaces and network elements of a truly converged network. Because of the explosion of interfaces that IMS brings to the network, the component interfaces of the architecture have been re-labelled during the transition from the MSF R2 architecture to the MSF R3 architecture, although from a bandwidth management perspective the interfaces remain fundamentally unchanged. The diagram in Figure 8.7 shows the MSF architecture for bandwidth management as defined in the MSF R3 architecture [16]. For each interface in the architecture, it lists the signalling protocols that may be used. For interfaces to the bandwidth manager, it provides the MSF interface name for both the MSF R2 architecture and the MSF R3 architecture.

The first point to note is that in this architecture the SBC is replaced with a decomposed Session Border Gateway (D-SBG) as described in Chapter 3. In the MSF architecture, the bandwidth manager receives requests for bandwidth from Call Agents or S-SBGs (possibly acting as P-CSCFs). For a fixed network call from a subscriber connected by a gateway (access or trunk) terminating on a similar gateway, the request for bandwidth will be from the Call Agents. For a call from an IP phone or IMS terminal that enters the network via a D-SBG and is either terminating within the network or passing to another network, the bandwidth reservations will be from the S-SBGs. Note that Figure 8.7 does not show an inter-S-SBG interface for simplicity; however, two S-SBGs can communicate using SIP.

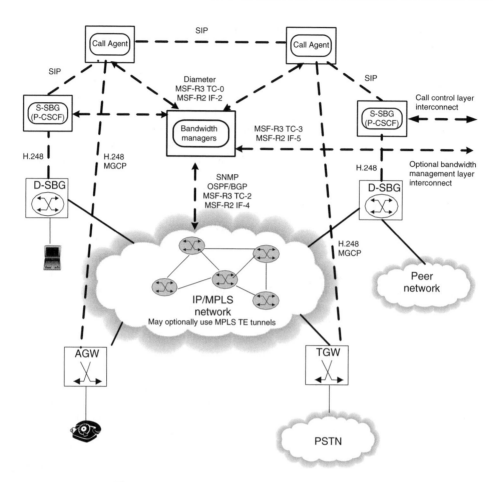

Figure 8.7 MSF bandwidth management architecture

To understand how this architecture works, consider the simple case of a call originating on an access gateway and terminating on a locally connected IP phone that is reached via an SBG. The calling subscriber's access gateway is controlled by a Call Agent; this is the originating Call Agent for this example case. The call set-up proceeds as for a standard originating call over an H.248 interface (see Chapter 4), to the point where the Call Agent can determine (by number analysis) where the call will terminate. In this case, the Call Agent will send SIP signalling (see Sections 8.10.5 and 8.10.6) to the terminating Call Agent (i.e. the one hosting the subscriber) which will determine that the subscriber is reached by a particular SBG and forward the signalling to the correct S-SBG. At some point in this process, prior to ringing the subscriber, the call control layer determines that it must request the bandwidth manager to reserve capacity between the network interface of the calling party's access gateway and the called party's network facing interface on the D-SBG.

The bandwidth manager examines its network model and determines the path through the network that the call will follow (on the basis of the IP addresses); it may send additional requests to peer or subordinate bandwidth managers to reserve the bandwidth. If it determines that bandwidth is available, the request for bandwidth will be accepted and the call can proceed. If there is insufficient bandwidth, the request from the call control layer will be rejected, which will ultimately result in the calling party's access gateway being instructed by the originating Call Agent to apply a busy tone. Using the same principles, when the call is terminated by the call control layer, those network elements that reserved bandwidth from the bandwidth managers must instruct the bandwidth manager to release the reservation.

The bandwidth reservation requests are made over the IF-2/TC-0 interface using Diameter; see Section 8.10.7. These bandwidth requests may be on a per-session basis, or a Call Agent may request bandwidth in chunks (i.e. to support multiple calls). However, the role of the bandwidth manager is the same, that is, it performs CAC in real time on the basis of the state of the network.

There are a number of models for making this reservation. One option is that the terminating Call Agent (or S-SBG) can make a single reservation for all bearers required by the call. The terminating Call Agent is the correct one to do this because it is this entity that determines that the call has reached a state where a bearer set-up is required. This is the most efficient approach in terms of signalling; however, in some call scenarios, this requires significant call-handling intelligence in the bandwidth manager and a simpler approach is for both the originating and terminating Call Agent to reserve some of the bandwidth required for the call.

Although the bandwidth management requests may be received on a per-session basis, the bandwidth manager maps these requests to its network model which contains information about pre-reserved bandwidth aggregates that are available to carry traffic in the network. Current implementations make aggregate reservations in the network using the Operational Support Systems (OSS) to configure the routing elements and the information is either configured in the bandwidth manager or discovered by auditing relevant MIBs[1] and snooping the routing protocol signalling. Since these aggregate resources are pre-reserved, the bandwidth manager need not make per-session reservations for bandwidth in the network, avoiding the scaling issues associated with the IETF IntServ architecture. The bandwidth manager does, however, require an interface to the network layer to determine the current state of its resource (and possibly in future implementations to request additional bandwidth aggregates). The MSF has identified this interface as IF-4/TC-2 and this is discussed in Section 8.10.2

In addition to the interfaces to call control and the underlying network, the bandwidth manager also has interfaces to other bandwidth managers (IF-5/TC-3). This interface may be used internally within the network to scale the bandwidth management solution, see Section 8.10.1, or to provide transport level network interconnect, see Sections 8.10.4 and 8.7.

---

[1] MIB stands for Management Information Base. It is a structure set of objects that can be managed on any given network element. These objects may be read/write (i.e. they can be set or read by the management interface) or read-only (i.e. they can only be read by the management interface). A given network element may implement many MIBs because the IETF defines MIBs on the basis of functionality. So, for example, a router may support an MPLS MIB and a DiffServ MIB. The MPLS MIB would be used to get or set values pertaining to MPLS support on the router and the DiffServ MIB would be used to get or set values that apply to DiffServ support on the router.

8.10.1 The Bandwidth Management Layer and Scaling the Network

The MSF bandwidth management layer of the network purposefully avoids bottlenecks by ensuring that no individual bandwidth manager is required to understand the state of all of the reservations in the network. Therefore, it is possible to scale the network to an arbitrarily large size by simply adding additional bandwidth managers. To facilitate this, the MSF solution envisages creating a bandwidth management layer which contains multiple bandwidth managers that support both hierarchy and peering. This bandwidth management architecture is shown in Figure 8.8.

The bandwidth management layer has an element of hierarchy in that the Call Agents request bandwidth from top-level bandwidth managers. There are multiple top-level bandwidth managers, each of which has a view of the network topology and handles a set of defined Call Agents. Below the top-level bandwidth managers are multiple lower-level bandwidth managers, these elements are responsible for reserving bandwidth for a portion of the underlying network and for tracking the resources available within those sub-networks. These lower-level bandwidth managers require an understanding of the topology of the network and the state of the resources over which they allocate bandwidth.

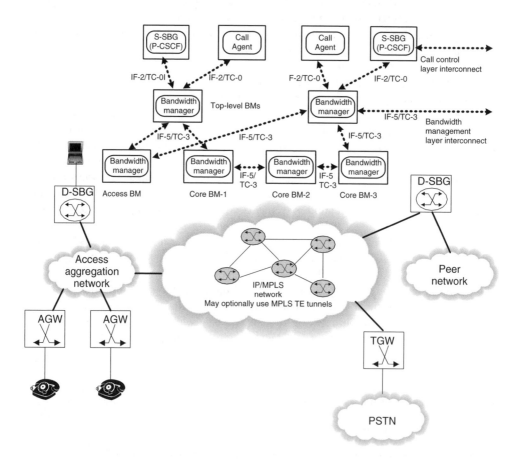

Figure 8.8   Large-scale bandwidth management

This hierarchy allows the network to be scaled by adding additional top-level bandwidth managers to cope with additional Call Agents and to add lower-level bandwidth managers to divide the tracking of resources in the underlying network into manageable chunks. A top-level bandwidth manager is required to understand the topology of the underlying network only to the degree that it is able to determine the lower-level bandwidth managers responsible for handling resources for the path that the call will follow. These resources are then reserved by the top-level bandwidth manager from, potentially many, lower-level bandwidth managers.

It is possible for lower-level bandwidth managers to peer with other lower-level bandwidth managers in order to optimise the reservation of bandwidth through the network by caching requests from its peer lower layer bandwidth managers. This allows bandwidth to be reserved through multiple sections of the network by a single lower layer bandwidth manager without requiring queries at the bandwidth managers responsible for handling those network sections. The decision as to whether to peer lower layer bandwidth managers or to remain with a strictly hierarchical solution is a design optimisation and depends on the particular characteristics of the network and the services being supported.

8.10.2 Interactions with the Underlying Network

The bandwidth manager provides CAC functions for the network or for that portion of the network that it is responsible for. In order to do this effectively, it must maintain some sort of model of the resources of the underlying network and sufficient knowledge of the topology and routing of the network to be able to determine the resources that will be required to pass a media flow between the two end points of a given reservation. The bandwidth manager will identify a route through the network for a flow and reserve resources for the flow against its map of the underlying network. If at any point in the reservation a bandwidth manager determines that accepting the flow would overcommit its available network resource, then it must either reject the reservation for the flow or identify a lower priority flow to bump to make space for the new flow. Typically, in the PSTN network, this bumping of flows or preemption is not performed because it is considered better to reject a new call than disrupt an existing call; however, this may not be the case for other sorts of networks.

The network model maintained by the bandwidth manager must be sufficiently complex to enable it to map a reservation accurately to the set of resources that must be reserved but it should ideally be abstracted as far as possible to prevent the bandwidth manager from understanding every entity in the network. The network model must also be updated if either the routing in the network changes, effectively re-mapping reservations to resources, or some failure alters the quantity of resources available to the bandwidth manager to reserve bandwidth against. In the MSF bandwidth management architecture, this interface has been identified as the IF-4/TC-2 interface, but the protocols that it maps to and the network model that must be maintained depend very much on the underlying network technology.

If the bandwidth manager reserves bandwidth against a core IP network, either IPv4 or IPv6, then it must understand the physical topology of the network and be able to map the current routing of traffic against this topology. If the routing in the network changes, then the bandwidth manager must re-map its reservations to potentially a different set of resources (which correspond to physical links). The information required from the network in this case is obtained from two sources operating on very different timescales. The operator's management systems (known as the *Operational Support Systems or OSS*) provide information about the bandwidth associated with each link and how the links interconnect the IP layer entities, and this information changes very slowly. The actual routing in place in the network can be determined by snooping (i.e. participating in) the routing protocols active in the network (e.g. OSPF or BGP) which may change very quickly. The bandwidth manager must be able to rapidly re-calculate the links used for any reservation on the basis of changes in the routing protocols. This is a challenging problem in a large network; however, commercial experience suggests that such solutions are viable.

If the underlying network technology is MPLS based, then the bandwidth manager can abstract the network by making use of MPLS-TE capabilities. In effect, this means that its knowledge of the network topology may be limited to the MPLS topology (i.e. the tunnel end points) and the routing information required to identify which tunnels would be used to carry a flow between two interfaces at the network edge. One of the benefits of MPLS is that if the tunnels are fast re-route protected, the bandwidth manager only needs to be aware of a network-level failure when both the primary and the backup tunnel have failed (or the backup tunnels are overcommitted). This means that the network can cope with fairly severe failures without impacting the services running over it.

However, as for the native IP case, the bandwidth manager in an MPLS network requires real-time routing and tunnel state updates from the network, via the IF-4/TC-2 interface, to be able to perform CAC and maintain an accurate network model. For example, the bandwidth manager must be informed by the network of a failure at the MPLS layer that cannot be re-routed. This might be by a control protocol, OSS notification or by a mechanism such as an SNMP[2] trap. If such an event occurs, then the bandwidth manager must adjust its reservations accordingly, see Section 8.10.8.

This begs the question what protocol should this interface use, and currently there is no consensus in the industry as to the protocol to be used by the MSF IF-4/TC-2 interface. A pragmatic approach to the problem is to use SNMP to audit key MIBs on the routers and combine this information with routing protocol snooping. Such an approach has been shown to work, but requires customisation of the bandwidth manager interface for every deployment. Work is now underway in the MSF that will formalise this approach and propose a standard SNMP-based interface that can be used between routers and bandwidth managers for an MPLS network.

If the network operator adopts an MPLS-TE-based network, another question that must be asked is what tunnel architecture should be adopted to support mission-critical services such as the PSTN. There are two choices that can be made, one is to provide a full mesh of edge-to-edge MPLS traffic engineered tunnels and the other is to place some hierarchy into the solution by connecting the edge nodes together via a fully meshed set MPLS core routers deep in the core of the network. Each approach has its advantages and disadvantages and as always, the choice will be dictated by the network operator's circumstances.

---

[2] SNMP stands for Simple Network Management Protocol. It is used to allow the network element manager to get and set attributes on the MIBs implemented by the network element. An SNMP trap is a lightweight alarming mechanism that is triggered (i.e. sent) when a specific event has taken place, for example an interface failure.

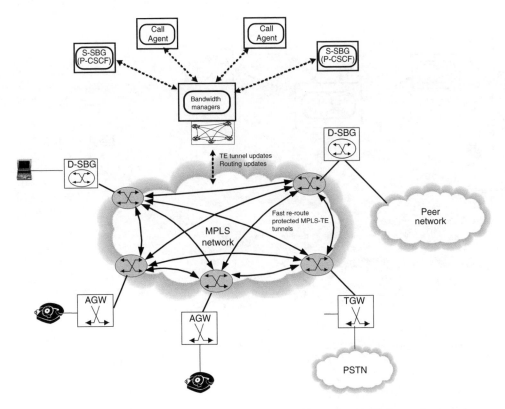

Figure 8.9   Bandwidth management with edge-to-edge MPLS tunnels

The simplest approach from a MPLS-TE viewpoint is to connect the edges of the network together with the full edge-to-edge mesh of MPLS tunnels as shown in Figure 8.9.

One of the problems with the edge-to-edge model of bandwidth reservation is that it does not scale to very large networks as it suffers from the well-known "long thin pipes problem". If a network is made up of a large number of long thin pipes in a mesh edge to edge, then Erlang B tells us that the bandwidth that must be reserved to guarantee a sufficient grade of service is very large relative to the number of calls being carried. The PSTN overcame this problem by using transit switches to aggregate traffic onto a few large trunks through the core. MPLS-based bandwidth management solution can obtain the same benefits without adding transit switches. This can be achieved by stitching together a succession of short fat pipes (MPLS tunnels) to form an edge-to-edge route as shown in Figure 8.10. Note that a real-world implementation would generally connect each edge router to two core routers for resilience; however, these secondary tunnels are omitted from the diagram for simplicity.

In this type of network, the bandwidth managers maintain a map of the tunnels that they are responsible for performing admission control into. The bandwidth managers will map a reservation between two end points on the basis of their knowledge of the current network routing which can be obtained by snooping the routing protocols. This solution gives optimum bandwidth efficiency, but as always it is not a one-size-fits-all problem.

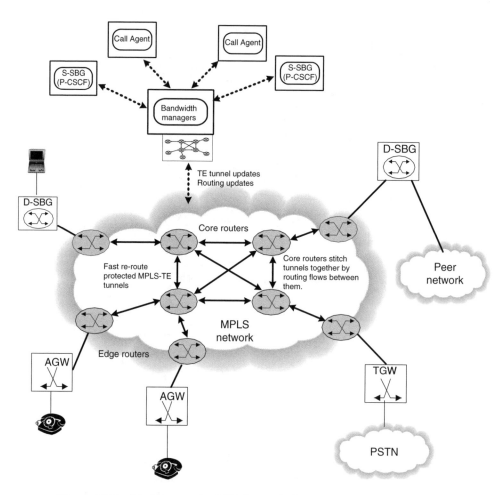

Figure 8.10 Stitching together MPLS tunnels for greater bandwidth efficiency

If a network operator has a relatively small network, perhaps 80 or so large edge routers (or D-SBGs), then the edge-to-edge model will probably be sufficient. However, if the number of routers increases significantly beyond this, into the hundreds (and in some networks the thousands), then the savings offered by tunnel stitching may become very significant. For example, a 200-router network would need over 39 thousand tunnels for an edge-to-edge mesh (MPLS tunnels being uni-directional). However, if 20 core routers were added, the same network could be built with just 1180 tunnels (allowing for a full mesh between the core routers). Although this solution looks similar to the PSTN transit network architecture described in Chapter 2, the introduction of packet technology has changed the rules in that now many more edge nodes can be supported before a full mesh is impractical. Furthermore, the addition of a second tier of routers to form a core does not require call control intelligence in the way that the PSTN transit nodes did.

In addition to designing a suitable tunnel architecture, the network operator must also consider a number of other issues if they intend to deploy this type of bandwidth management solution; these issues include how they might evolve from an existing architecture and how they ensure that their routing layer converges quickly in the event of a failure. For more information about how bandwidth managers interact with underlying networks and design considerations, see Ref. 17.

8.10.3 Handling Network Interconnect

The MSF architecture is fundamentally designed around PSTN type peer-to-peer services and it envisages a deployment whereby individual operators own their network and deploy a bandwidth management layer to allocate resources over it in response to requests from their own Call Agents and application servers. Where a call originates on one operator's network but terminates on another, it must pass over a point of interconnect to another network operator. Because the bandwidth manager is such a central part of the operator's network infrastructure, the MSF solution does not recommend allowing peer networks to request bandwidth for a call directly from the peer operator's bandwidth management layer. Instead, the architecture recommends that interconnect between networks takes place at the application layer. This is the same approach that has served the PSTN well for many years and crucially allows the network operator to take advantage of existing commercial frameworks.

If a call is required to pass through more than one operator's network, then it must be passed over a defined point of interconnect. These interconnect points are typically policed by SBGs (or SBCs) to ensure the security of the network, and it is usual for a commercial agreement to be in place that allows the operator that terminates the call to gain some revenue from the operator on whose network the call originated. Similarly, business models exist that allow for transit networks, which carry a call between the originating and terminating networks but do not themselves terminate the calls. Not all operators, however, allow their networks to be used as transit networks and so may prevent such traffic from entering their network.

In the MSF architecture, this call control layer interconnect is typically performed by S-SBGs. The interconnect point is shown as part of the MSF architecture in Figure 8.7 and is labelled the "Call Control layer interconnect". The signalling session border gateway (S-SBG) at the interconnect point behaves as a SIP Back-to-Back User Agent which, as discussed in Chapter 3, allows it to hide the topology of its own network from the peering network. The call appears to originate on this "network edge" SBG and is passed into the peer network where it is processed by another S-SBG (again acting as a SIP Back-to-Back User Agent). The peer network is able to determine from the SIP signalling the originating point of the call (which, for the media, is the D-SBG at the point of interconnect), and the terminating end point of the call in its network (which may be a subscriber's terminal or another point of interconnect to a third network). It is then able to reserve resources between these two end points by requesting bandwidth from the bandwidth management layer as for the intra network case.

In addition to the commercial benefits of this approach to interconnect, it also has a number of technical benefits. It enhances network security by preventing third parties from being able to directly access a network's bandwidth managers, thus preventing potential DoS attacks (such as reserving large amounts of bandwidth arbitrarily). It also ensures that the topology of a network is hidden from third parties as the S-SBG can hide the SIP headers that provide information about how the call was internally routed within the operator's network; the IP addresses that are seen by the peer network are restricted to those IP addresses that have been allocated to the interconnecting SBGs. It also aids scalability because passing the routing and forwarding decision back to the call control and application servers allows route analysis to be done again, thus allowing bandwidth to be allocated in short hops. In this model, bandwidth is allocated between the originating terminal and the interconnect point, between two interconnect points and between the second interconnect point and the terminating subscriber; the alternative end-to-end QoS reservation would be over a single long hop traversing one or more entire networks. However, as always, there is a trade-off, and in this case it is that the operator incurs the additional costs of the interconnecting SBGs and the additional SIP-processing costs at the intermediate Call Agents.

8.10.4 An Alternative Approach to Network Interconnect

The MSF architecture also supports network interconnect at the bandwidth management layer by extending the inter-bandwidth manager interface over the network interconnect boundary. This interface is optional in the MSF architecture and is shown in Figure 8.7 as the "Bandwidth Management layer interconnect" interface. In this type of interconnect, one network requests bandwidth directly from its peer network's bandwidth managers. The reservation would be made between a point on the far side of the peer network, corresponding to the known location of the user, and the boundary point between the two networks. Because this request is for a specific bandwidth pipe, rather than for a call or session set-up, the IP addresses between which the pipe is to be set up must be known and in order for this to be effective, both networks must have a common view of the IP addresses used for the session. If the network operator is using SBGs to hide the details of their internal SIP routing infrastructure, then it is possible that the S-SBG that they are peering with is in a third-party network and not the adjacent network. An example of this type of interconnect is where two networks have a SIP level interconnect with each other but connect via a third-party transport network. This is shown in Figure 8.11.

This type of interconnect would typically be used in the case where a transport network operator offers a basic bandwidth service, that is, acts as a bit pipe for other networks' applications. Many traditional telco operators are reluctant to offer this type of service because they perceive this as moving down the value chain, that is, they are no longer able to gain revenue by charging for call-routing services but they must still pay the premium costs of operating a guaranteed QoS. However, in some cases, this is an attractive model, as is the case of 3G mobile networks which interconnect for packet data using third-party-provided IP Exchange networks that provide a set of tunnels linking the many diverse mobile operators. Today, this type of interconnect provides a strictly BE service; however, it is possible to envisage that such a commercial model might evolve to offer guaranteed bandwidth interconnects between both fixed and mobile operators.

8.10.5 Signalling QoS Requirements

In order to be able to invoke QoS for a given session, it must be possible to signal the requirement for QoS to the underlying network. In the MSF solution signalling, QoS requires the interaction of two distinct signalling protocols, SIP which is used for call

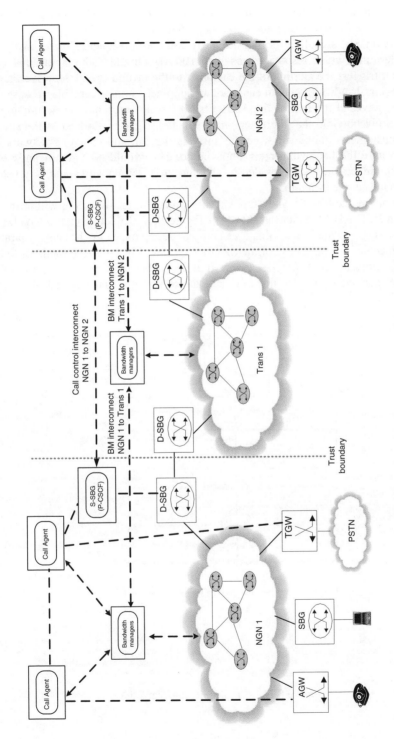

Figure 8.11 Interconnect via a third-party transport network

control signalling and Diameter which is emerging (in ETSI TISPAN and 3GPP) as the protocol of choice for requesting and authorising network bandwidth.

An issue with SIP signalling is that the basic call set-up sequence sends an INVITE to the end terminal which responds back with 100 Trying and 180 Ringing that is, the INVITE itself triggers the alerting of the customer to the incoming call. However, for services that require bandwidth guarantees, such an approach is not acceptable because when the INVITE arrives at the terminating subscriber, no reservation has yet been completed and it is entirely possible that the reservation will fail. This would lead to the situation whereby a customer's phone rings, he/she answers the call and then gets network busy signalled because the bandwidth reservation failed. This would be a significantly worse customer experience than that provided by the current TDM networks and so a solution had to be found to overcome this.

The solution adopted by the industry is to use precondition signalling within the SIP, as defined in RFC 3312 [18] and is shown in Figure 8.12. Effectively, a precondition is placed on the session which states that it can only be set up once QoS has been guaranteed for it. This mechanism makes use of the SIP PRACK and SIP UPDATE methods to carry information within the SDP for the session indicating that the originator of the session requires QoS guarantees and for tracking the state of that guarantee at each stage in the

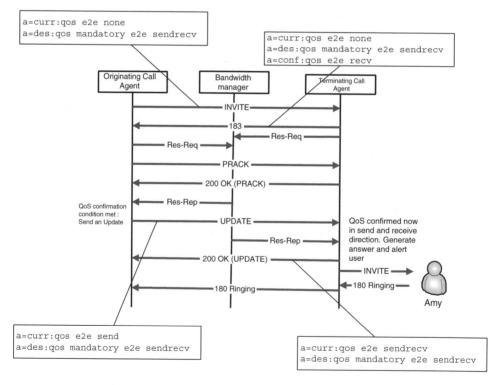

Figure 8.12 SIP precondition signalling

call set-up. The customer is only alerted by the end terminal when it determines that an end-to-end guarantee of QoS has been provided by the network.

Another issue with SIP signalling is that it supports an offer–answer mechanism which means that the actual destination that the call is established towards, and the required bandwidth, is only known once the terminating entity has responded to the offer. Furthermore, SIP signalling is very promiscuous in that it is possible that large amounts of signalling will be generated to multiple destinations that can only ever result in a single call to a single destination being set up. This is particularly true for services that have been built using SIP forking (as described in Chapter 2), and clearly reserving bandwidth towards every possible destination would not be an effective use of network resources.

Both these issues lead to a model whereby SIP signalling requesting QoS preconditions is sent from the originating Call Agent to the terminating Call Agent. Once the terminating Call Agent has determined that it wishes to establish a call, it responds back towards the originating Call Agent again, indicating that preconditions are required. If the call's media path passes through multiple networks, then each network must in turn reserve the resources and update the preconditions starting with the network that originated the call and finishing with the network that terminated the call. Once the terminating Call Agent sees from the SIP signalling that the preconditions have been met, it alerts the end-user, sends a 180 Ringing message and the call proceeds as for a normal SIP-based call set-up. A detailed example of this mechanism is provided in Section 8.10.6

Although the Call Agents use SIP to signal preconditions and request QoS guarantees from their peer Call Agents, it is not suitable for reserving resources from the bandwidth managers. In the Global MSF Interoperability 2004 event (GMI2004), a SIP Implementation Agreement for the interface from the Call Agent to the bandwidth manager was defined as a stopgap, because it was feared that the early MSF-preferred protocol for resource reservation, NRCP [19], would not be sufficiently mature to allow interoperability testing. As work progressed in ETSI TISPAN on the RACS interfaces, the authors of NRCP worked to ensure that its reservation model was supported in the enhancements being made to the Diameter protocol that was emerging as the protocol of choice for resource reservation. This ensured that the ETSI RACS functional architecture and the MSF bandwidth management physical architectures were broadly compatible. For the latest release of its bandwidth management architecture, the MSF has therefore adopted the ETSI Gq' interface ("Gq prime") which uses Diameter as defined in the following specifications: the MSF diameter implementation agreement [20], the ETSI TISPAN Gq' specification [21] and the Diameter base protocol [22].

8.10.6 Signalling QoS with SIP Preconditions

RFC 3312 [18] allows end-to-end QoS to be reserved and the status of the reservation communicated within the SIP domain. The mechanism chosen to do this was to define additional attributes within the SDP carried by the SIP and to use the SIP PRACK and UPDATE messages to pass this information. RFC 3312 defines the following SDP attributes.

Attribute Name	SDP encoding	SDP parameters
Current status	"a=curr"	Precondition type Status type Direction tag
Desired status	"a=des"	Precondition type Strength tag Status type Direction tag
Confirm status	"a=conf"	Precondition type Status type Direction tag

RFC 3312 [2] also defines the following parameters.

Parameter	SDP encoding	Values
Strength tag	Strength-tag	"mandatory" "optional" "none" "failure" "unknown"
Precondition type Status type	precondition-type status-type	"qos" "e2e" "local" "remote"
Direction tag	Direction-tag	"none" "send" "recv" "sendrecv"

The current status attribute describes the current status of QoS reservation for the SIP-initiated session and the desired status describes the required status of QoS reservation for the SIP-initiated session. A strength tag of mandatory would ensure that the session initiation could not proceed until the QoS preconditions had been met (i.e. when the current status equals the desired status).

So, for a bi-directional call that requires QoS guarantees before alerting the customer, the following attributes would be sent in the SDP.

a=curr:qos e2e none
a=des:qos mandatory e2e sendrecv

This can be seen in the example flow shown in Figure 8.12 which shows QoS reservation in a single network where each Call Agent is responsible for reserving resources in the send direction only.

The initial INVITE contains an SDP offer which describes the media streams being requested and also contains attributes for QoS preconditions. When the terminating Call Agent receives the INVITE, it does not immediately alert because the QoS preconditions have not been met. If the end-user terminal also supports QoS preconditions, then the message may be forwarded to the end terminal to allow it to determine which codec to accept; however, in many cases, the terminating Call Agent may make the selection of codec on behalf of the end-user according to their service preferences or network policy. In this case, the terminating Call Agent selects a codec and generates an SDP answer; as part of this answer, it requests confirmation from the originating Call Agent when resources have been reserved on the bearer path towards it and informs the originating Call Agent that it has no current reservation for the SIP session. The confirmation is required here (as indicated in the a=conf attribute) because in this network model the Call Agents can only reserve resources in their send directions (i.e. towards the other party in the call) and are not aware of the status of reservations from the other party towards them. Alternative implementations may chose to make the terminating Call Agent responsible for requesting reservations in both directions in which case the confirmation request would not be necessary.

Once the terminating Call Agent has sent the answer to the offer, it requests bandwidth towards the calling party from its local bandwidth manager using a protocol such as Diameter. The originating Call Agent receives the 183 with the SDP answer, and it looks at the c line of the SDP to determine the IP address the far end wishes to receive the media on and it reserves bandwidth between the originating media end point and the known terminating media end point.

When the originating Call Agent receives confirmation from the bandwidth manager that resource has been made available in the network to carry the call, it sends a new SDP offer towards the terminating Call Agent. This is triggered because it has reserved resource in its send direction (which is the terminating Call Agents receive direction), and the terminating Call Agent had requested confirmation when this action had been completed. This results in an UPDATE message being sent containing the new offer with the current QoS attribute set to "e2e send".

The terminating Call Agent also receives an indication from its bandwidth manager that resources have been made available between its media end point and the calling party's media end point and so is aware that resources have been guaranteed in its send direction. When it receives the Update from the originating Call Agent, it can see from the current QoS attribute that resources have also been reserved in its receive direction. This means that the original QoS preconditions, as defined by the desired QoS attribute of mandatory end-to-end send/receive QoS, have now been met and it generates an answer to the offer reflecting the new status. It sends this answer to the originating Call Agent in the 200 OK to the update. Because the mandatory desired QoS preconditions have now been met, the terminating Call Agent can alert the user without fear that the call will fail post-ringing owing to network resource exhaustion and the call set-up can proceed.

This model of signalling QoS reservations can also be extended over network boundaries with each network being responsible for reserving bandwidth within its domain. In this case, each network may wish to reserve resources fully in each direction before indicating to the peer network that reservation has taken place in the UPDATE message. Interworking functions at the edge of the two peering networks (e.g. at an S-SBG) may be required to ensure that SIP signalling propagates the QoS status in each network correctly, see Section 8.10.10.

8.10.7 Bandwidth Reservation Using the Diameter Protocol

Although SIP signalling propagates the status of the QoS reservations to call control, the MSF model assumes that a bandwidth management function in each network will actually reserve the resources on behalf of the SIP-based call control layer. In order to provide this reservation, ETSI and the MSF have proposed the Diameter protocol because of its widespread use with 3G IMS networks.

The Diameter protocol [22] was originally proposed by the IETF to fix a number of shortcomings with the Radius protocol which is widely used to perform authorisation and accounting functions within data networks. Although this background might not seem to make Diameter the obvious choice for a NGN protocol aimed at bandwidth management, it was adopted by the 3rd Generation Partnership Project (3GPP) for the Gq interface between an application server and a Policy Decision Function (PDF) in its IMS architecture (see Chapter 5). One of the reasons for the choice is that Diameter protocol can be easily extended by defining additional applications for it. This is achieved in Diameter by explicitly defining the application being invoked in the Diameter signalling so that both entities understand the required capabilities. To define a new application, an organisation must apply for a vendor identity from IANA[3] and can then use a separate vendor-specific application identifier for each application that it has for the Diameter protocol.

ETSI TISPAN has extended the 3GPP Gq interface for the fixed network and this has become the Gq prime interface (Gq') [21]. In the ETSI RACS architecture, this interface is between the Application Function and the SPDF; see Chapter 5 for more information on RACS. The MSF bandwidth management architecture maps closely to the ETSI RACS architecture and, therefore, the MSF also uses the Gq' interface protocol between the Call Agent and/or the S-SBG (P-CSCF) and the bandwidth manager. However, where the ETSI architecture assumes that the SPDF is responsible for controlling gates[4] in the SBGs, the MSF architecture assumes that this is a function of the S-SBG (P-CSCF) using the H.248 protocol to perform this gate control as defined in Ref. 23. This is a more natural fit between the functions of a bandwidth manager (which does not contain SBG control functionality) and an S-SBG (which most definitely does contain SBG control functionality). In reality, many vendors may seek to bundle the functions of the S-SBG into their Call Agent and the S-SBG is itself evolving to support 3GPP P-CSCF functionality.

The impact of this on the MSF's use of Diameter is that the Call Agent uses it to request bandwidth through the network from the bandwidth manager, it does not use the additional capabilities supported by the full Gq' flavour of Diameter to reserve gates and obtain NAT bindings. The MSF has produced an implementation agreement describing this use of Diameter [20]. It should be noted that for other services that do not use such

---

[3] IANA stands for Internet Assigned Numbers Authority. It is responsible for allocating a wide range of numbers and codepoints for the Internet. Part of this responsibility is to allocate vendor identifiers for the Diameter protocol. Because these are administered by IANA, it can be guaranteed that when an organisation is issued with a vendor Id, it can be confident that the same number will not be used by other organisations.

[4] A gate, which may also be known as a *pinhole*, is an opening into the network at a D-SBG. A gate has a packet classifier and a policer associated with it. Only packets that match the classifier are admitted and then only if they are within the policing (i.e. bandwidth) parameters for the gate, other packets are dropped.

a feature-rich control plane, it may be more appropriate for the bandwidth manager to configure gates in the SBG on behalf of these application servers and it would do so with Diameter.

In order to use a bandwidth manager, a Call Agent or S-SBG (P-CSCF) must first establish a secure connection to it in the shape of a TCP or Stream Control Transmission Protocol (SCTP) connection running over a secure transport, either an IPsec tunnel or a TLS connection. This is achieved in Diameter by the Call Agent sending a Capabilities Exchange Request message to the bandwidth manager. This contains information about the Call Agent's realm and IP address and also the particular Diameter application being invoked. For a Gq' interface, this would be achieved by including in the header two different Supported-Vendor-Id AVPs[5], one set to 13019 (ETSI) and the other to 10145 (3GPP), in addition to the Auth-Application-Id AVP set to 16777222 (which indicates the 3GPP Gq vendor-specific application) and a Vendor-Specific-Application-Id set to 10145 (3GPP). This sequence is shown in Figure 8.13.

Once the capabilities exchange has been successfully completed, the bandwidth manager is able to provide resource reservation services for the Call Agent within its network domain. A typical reservation sequence is shown in Figure 8.14.

The "Authenticate and Authorize Request" (AAR) message sent by the Call Agent (S-SBG/P-CSCF) to the bandwidth manager must contain a Media Component Description AVP. This performs a function within Diameter that is identical to that of the embedded SDP seen in SIP and H.248 in that it defines the bandwidth characteristics of the media stream being requested, thus allowing the bandwidth manager to determine how much resource and between which network, ingress and egress points are being requested. An example of an AAR message in an MSF bandwidth management application is shown in Table 8.2 (note that unlike SIP and SDP, Diameter is not a text-based protocol and so the table represents the values of fields rather than their exact encoding).

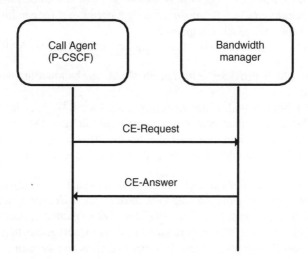

Figure 8.13 Diameter capabilities exchange

---

[5] AVP stands for Attribute Value Pair. It is effectively a Diameter protocol information element, the encoding and meaning of which are defined in the Diameter protocol.

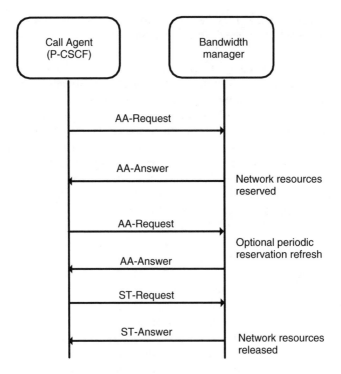

Figure 8.14   Resource reservation using Diameter

This example shows a bi-directional reservation; however, at the time of writing, the proposal in the MSF was that reservations would be made uni-directionally. This would be achieved by including only a permit-in or only a permit-out (depending on reservation direction) AVP in the media sub-component.

There are a number of other AVPs that might be included in the AAR as follows:

- RR-Bandwidth which provides information about the bandwidth required for RTCP receiver reports.
- RS-Bandwidth which provides information required for RTCP sender reports.
- AF-Charging-Identifier which provides a value to aid in charging correlation.
- SIP-Forking-Indication which informs the bandwidth manager that SIP forking has taken place and that it should adjust its CAC for the request accordingly.

The use of the Authorization-Lifetime AVP determines that the bandwidth manager will apply soft state reservation management. This means that reservations will expire if they are not refreshed in a given period. This has the advantage that it removes the need to audit the bandwidth manager after failure cases to detect reservations that are in existence even though the call that required them has matured (this can happen if communication between the Call Agent and bandwidth manager is lost for some reason). However, it might also mean that reservations could be erroneously released because communication between the Call Agent and the bandwidth manager is lost, preventing reservation refresh from taking place.

Table 8.2 Diameter AAR message in the MSF application

Diameter message fields	Comments

`<Diameter Header: Version =1,`
`Length, Command code= 265, REQ,`
`PXY>`
	The Diameter protocol has a fixed header format defined in Ref. 22. The key fields are the command code of 265 which indicates an AAR message and the command flag REQ which indicates a request and PXY which indicates that the message can be forwarded by a proxy.

`<Session-Id="cs23.`
`net1.bigtelconetwork.com;`
`1634322;23347">`
	A globally unique identifier that identifies the session. The bandwidth manager uses this to relate requests to existing reservations (if the session is known) or to identify requests for new reservations (if the session Id is unknown).

`{Auth-Application-Id=16777222}`	Identifies the Diameter application as the Gq interface.
`Origin-Host="cs23.` `net1.bigtelconetwork.com"`	The domain name of the host requesting the reservation.
`Origin-Realm="net1.` `bigtelconetwork.com"`	The administrative realm that the requesting host resides in.
`Destination-Realm="net1.` `bigtelconetwork.com"`	The destination realm that the request is addressed to. Because in this example, the Call Agent and the bandwidth manager reside in the same administrative area this is the same as the origin realm.
`[Media-Component-Description =`	The start of the Media Component Description Grouped AVP which contains a number of media sub-components each of which defines the ingress, egress and bandwidth of a media stream.

`[Media-Component-Number=0]`
`[Media-Sub-Component =`
`{Flow-Number =0}`
	Start of the Media sub-component.
	The first and only flow description contained in this media sub-component.

`{Flow-Description = "permit in` `10.20.10.3 24000"}`	The origin of the media flow in the network, that is, the point at which it is expected to appear. In this case, it is the network facing side of the D-SBG at the edge of the operator's network that the customer attaches to the network through (see Chapter 3 for the D-SBG configuration). It might also be the network facing side of an access or trunking gateway. The port number for the RTP stream is included but is not necessary for the bandwidth manager as it is not itself configuring pinholes but just reserving bandwidth between interfaces.

(continued overleaf)

Table 8.2 (*continued*)

Diameter message fields	Comments
{Flow-Description="permit out 10.20.200.136 1956"}	The destination of the media flow in the network, this might be the network facing side of a D-SBG through which the session will exit the network. It might also be the network facing side of an access or trunking gateway. Again the port number is not really required by the bandwidth manager but the syntax of the protocol requires it to be sent.
{Flow-Status = ENABLED}	The Flow-Status AVP informs the bandwidth manager whether the flow for the media sub-component should be opened for user traffic in the uplink, downlink or in both directions. A setting of ENABLED means that it should be opened in both directions.
	In the MSF architecture, where the bandwidth manager is just reserving capacity, the flow status will always be set to ENABLED to avoid the need to modify it later with a separate message. However, if the bandwidth manager were also configuring pinholes, then the media would be reserved but access to the resource would be barred to the customer until the appropriate point in the call set-up. In this case, settings of DISABLED, ENABLE-UPLINK or ENABLE-DOWNLINK would be used to disable a flow, enable it into the network or enable it out of the network.
{Flow-Usage = NO_INFORMATION(0)}	This is the default value of the Flow-Usage AVP, if required; however, this can be used to explicitly identify RTCP flows. In this example, however, the RTCP overhead is allowed for in the bandwidth reservation and is not separately reserved.
{Max-Requested-Bandwidth-UL=120000}	The bandwidth reserved in the direction from the in address to the out address in bits per second. This bandwidth covers the IP, UDP and RTP overheads but does not allow for the lower layers. A value of 120000 is approximately what is required for a G.711 codec with a packetisation time of 10 ms.
{Max-Requested-Bandwidth-DL=120000}	This is the bandwidth reserved in the direction from the out address to the in address. This is a symmetric reservation to cover a voice service. It is equally possible for the bandwidth manager to receive requests in only one direction, which would result in two separate reservations for a call.

Table 8.2   (*continued*)

Diameter message fields	Comments
]	Indicates the end of the media sub-component description.
[AF-Application- Identifier="PSTN-call"]	Identifies the application that invokes the service; this can be used to provide additional information to the bandwidth manager (that cannot otherwise be signalled in DIAMETER) about the QoS that should be applied to the reservation. This would be used to activate a pre-defined QoS profile on the bandwidth manager
[Media-Type=Audio(0)]	Indicates that the media is audio. The bandwidth manager may use this information to determine the required traffic class of the reservation (if it supports multiple types of traffic).
{Flow-Status = ENABLED }	Describes the flow status for the media component that applies to any included media sub-components that have not had this value explicitly set. In this case, this value is overridden by that in the media sub-component.
[Reservation-Priority = DEFAULT (0)]	Describes the reservation priority for the media component that applies to any included media sub-components that have not had this value explicitly set. In this case, this value is overridden by that in the media sub-component.
]	The end of the media component grouped AVP.
[Authorization-Lifetime = 450]	Indicates the time period that the reservation is valid for in seconds. This is an optional AVP but if it is included, then the reservation will timeout if it is not refreshed by the Call Agent within this time plus a grace period allocated by the bandwidth manager.

The AA-Answer would typically contain the header, the session Id and the host, the realms plus a result code (in this case, the result code would be "DIAMETER_SUCCESS" but in a failure case, it might be "INSUFFICIENT_RESOURCES"). If the Authorization-Lifetime AVP was sent in the AA-Request, then the actual lifetime granted by the bandwidth manager would be returned in the same AVP in the AA-Answer along with an Auth-Grace-Period AVP which informs the Call Agent (P-CSCF) of the number of seconds after the reservation lifetime that the bandwidth manager will clear the reservation.

Reservation clear down is performed by the Call Agent by sending a Session Termination Request (ST-Request) which would contain the header, the host Id, the realms and a cause for the release. The Session Termination Answer provides an acknowledgement that the termination was successful.

Additional Diameter messages are used to allow the bandwidth manager to inform the Call Agent that a sessions resource has been lost for whatever reason, that is, dropped owing to a network failure. This is achieved by sending an Abort-Session-Request (ASR) containing an abort cause, and the Call Agent would acknowledge this with an Abort-Session-Answer (ASA) message.

8.10.8 Challenging Cases and Responding to Network Failures

Conceptually, the role of the bandwidth manager in receiving requests for reservations, mapping the reservations to the underlying network and either authorising or rejecting the request looks straightforward enough. However, there are two cases that significantly complicate the problem.

1. Handling network failure scenarios
2. Supporting SIP forking and complex call handling.

Network failure scenarios are problematic for the bandwidth manager because at the point at which it detects resource contention, it must perform a number of tasks very quickly.

• It must re-calculate the resources available to it on the basis of the updated routing information from the network or information about the failure of an MPLS tunnel.
• It must determine if it now has reservations for more bandwidth than it can guarantee, given the new state of the network. If it has more reserved bandwidth than available bandwidth, it must rapidly identify which reservations will be released (probably on the basis of the reservation priority).
• It must identify and inform the network element that requested the reservation (either a Call Agent or another bandwidth manager) that the reservation is to be released.

This is a non-trivial task because if the failure happens in the core of the network, a significant number of individual reservations might need to be analysed and released. Furthermore, in order to effectively release a reservation, a bandwidth manager must notify the Call Agent so that the Call Agent can close the pinholes and signal the end of the call to the end-user. Until the pinholes are closed, the traffic will continue to enter the network even though the bandwidth manager has released the reservation. This means that too much bandwidth is being admitted to the network and all sessions may be impacted and may suffer packet loss. Therefore, the bandwidth manager must inform the Call Agents or S-SBGs which reservations have been released quickly and the Call Agent or S-SBG must close the pinholes rapidly. However, in doing this, the bandwidth manager must take care not to flood the Call Agent (or top-level bandwidth managers) with too many signalling messages, thus triggering control plane overload, see Section 8.11.

Fortunately, there are compromises that can be made to assist the network operator. The first decision to be taken is the likelihood of a large-scale network failure, given the information of how long an outage will be tolerated. If the network operators are prepared to tolerate the loss and impairment of calls for a time after the failure, they can at least be sure that the bandwidth manager will eventually release all the sessions, the pinholes will be closed and at some point in the future, the network will be able to carry the new calls that users will be attempting to establish. The benefit of the bandwidth manager is that this will happen even though the network failure that triggered the disturbance is ongoing. Clearly, this offers significant benefits over the case where no bandwidth manager is available, because such a network would not be able to adjust its CAC to the available bandwidth and would still be overloaded potentially, resulting in a wide-scale packet loss and loss of calls. The longer the disturbance period the network operator will tolerate, the less processing power needed in the bandwidth managers, Call Agents and SBGs. However, this comes at a cost to the end-user experience.

Another solution that can be adopted to ease the processing overload is for the Call Agents to make aggregate reservations. This might allow a single reservation to carry sufficient capacity for, say, ten or a hundred calls. Therefore, although the bandwidth manager must potentially be able to tear down each reservation in the event of a failure, it holds far fewer reservations because each reservation maps many end-to-end calls. Clearly, such an approach, while sacrificing some network efficiency, will achieve significant benefits in scalability and failure processing.

The other major challenge for a bandwidth manager is that not all calls are simple to reserve bandwidth for. As described in Chapter 2, SIP allows calls to fork, resulting in multiple related call set-ups, only one of which will mature. If the bandwidth manager reserves resources for each reservation, there is a danger that calls will fail because multiple reservations are made even though the end result can only be one call. While reservations must be made between each end point for every SIP fork, if the end-to-end route happens to share some common network resource, it is not necessary to perform multiple reservations on that resource. This is the case because only one of the sessions will mature and therefore only one reservation is required.

Another example of this problem is the case of three-way calling where customers are placed on hold at a media server in the network. Consider the example of a three-party call where the media flows are as shown in Figure 8.15.

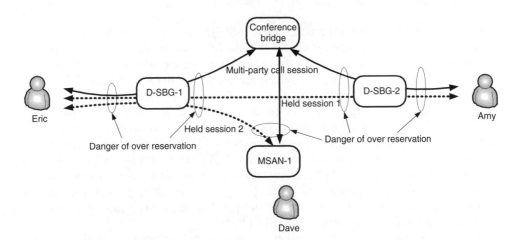

Figure 8.15 Media flows in an example three-party call service

In this case, it can be seen that calls are established between the media server and each of the end-users but because of the way three-party calls are set up, Eric has called both Amy and Dave and placed them on hold before placing them into conference. This creates a situation where Eric has three call legs that must be reserved by call control yet only one call can be active at any one time. Since Eric may have a single access link over which all calls are being set up, the bandwidth manager should be able to identify that these are related calls; otherwise, it will reserve three times the bandwidth over the access that it needs.

The solution to this is to allow the reservation requests from the Call Agent to the bandwidth manger to carry a token that identifies linked reservations. If the bandwidth manager sees two reservations with the same token, then it can determine that the reservations are linked and hence can be shared. The Diameter protocol provides a mechanism to achieve this with the SIP-Forking-Indication AVP. If an AA-Request is received by the bandwidth manager with a session Id and a SIP-Forking-Indication AVP set to the value "SEVERAL_DIALOGUES" and the session Id is known to it, then it invokes special handling. Instead of treating an AA-Request with a known session Id as a request to modify an existing session, it treats it as a new session set-up but it only reserves bandwidth on paths where the existing session is not currently holding a reservation. This solution ensures that whichever path becomes active, bandwidth will be available for the media but the minimum set of reservations are made.

8.10.9 A Call Set-up with Guaranteed QoS Using Bandwidth Managers

In order to set up a call with guaranteed QoS in the network, all the elements that make up the MSF architecture for bandwidth management must come together. This means that pinholes must be allocated at the D-SBG, SIP signalling must be sent end-to-end negotiating QoS and bandwidth managers must be queried for resources. Figure 8.16 shows the call set-up signalling for two subscribers in the same network where the calling party connects to the network with a SIP phone and the called party has an analogue phone that is connected to the network by an H.248-based Access Gateway.

The key points in this call set-up are as follows:

1. When the initial INVITE (1) arrives at the S-SBG, it creates a pinhole for the media flow at the D-SBG by sending an H.248 Add (2). This creates a context with two ephemeral terminations, one facing the calling party and the other on the network side of the D-SBG. Once the reply has been received, the S-SBG forwards the INVITE to the subscriber's Call Agent (4) modifying the SDP as appropriate.
2. The originating Call Agent determines the next hop Call Agent and forwards the INVITE. This terminating Call Agent determines that the subscriber is locally connected to one of its access gateways and sends an Add (6) to the access gateway. The Add instructs the access gateway to create an H.248 context for the call and place an ephemeral termination (i.e. RTP end point) and the called party's analogue line into the context. The Reply (7) contains the local descriptor of the calling party which includes the IP address and port on which they wish to receive media. This information is sent back through the network in the SIP 183 (8,9,10).
3. When the terminating Call Agent receives the local descriptor, it has sufficient information to reserve bandwidth for the media being sent towards the calling party. It does

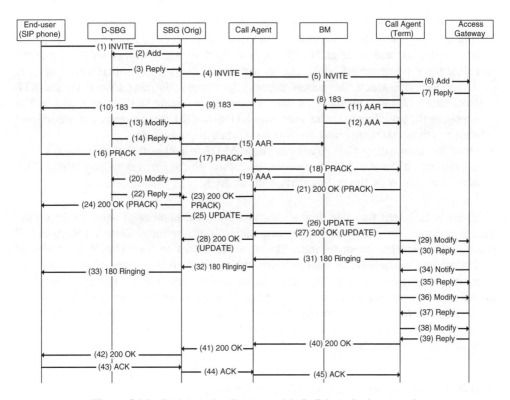

Figure 8.16 End-to-end call set-up with QoS in a single network

so by sending an AAR (11) to the bandwidth manager and is informed by the returned AAA (12) that resources have been reserved.

4. When the originating S-SBG receives the 183, it now knows the IP address and port on which the called party wishes to send and receive media. It sends an H.248 modify command to the D-SBG instructing it to set up the NAT bindings (13). It then requests bandwidth for the media flow to the called party from the bandwidth manager by sending it an AAR (15) and is informed that bandwidth is available in the returned AAA (19).

5. Once the originating S-SBG has determined that bandwidth is available, it follows the SIP precondition signalling sequence described in Section 8.10.6. It also instructs the D-SBG to open the media path for the call with an H.248 modify (22). At this stage, an end-to-end path does not exist because it has not been cut through by the access gateway; however, all the required bandwidth is reserved in the network and the media path is ready to be cut through.

6. The terminating Call Agent is informed via the SIP UPDATE (26) that bandwidth has been reserved for the call; it informs the originating Call Agent that it too has reserved bandwidth in the OK to the update (27). It then sends an H.248 modify to the access gateway (29) instructing it to deliver the Calling Line Identity (CLI) and to alert the subscriber. It also instructs it to apply a ring tone to the ephemeral termination; this results in the calling party hearing the ring tone generated by the access gateway.

7. When the called party answers the phone, the access gateway autonomously performs ring trip (to protect the called party from the acoustic shock of having a ringing phone next to the ear) and sends an H.248 Notify to the Call Agent (34) to inform it that the called party has gone off-hook. The terminating Call Agent now performs a number of actions; it instructs the access gateway to remove the ring tone from the RTP termination (36) and instructs the access gateway to enable full duplex media and to configure the jitter buffer for the incoming RTP flow (38). At this point, a bi-directional media path is established and the two parties can communicate.
8. Once the terminating Call Agent has received confirmation that the media has been cut through, it sends a 200 OK message back through the network to start billing. This is acknowledged by the calling party with an ACK.

It should be noted that this flow shows just one of a number of ways to set up the call. It assumes, for example, that the originating user's terminal does not support SIP preconditions which necessitates the S-SBG to act as a Back-to-Back User Agent and send an Update into the network to enable bandwidth reservation to be performed. If the end-users supported preconditions, then they might generate the UPDATE themselves. It is also possible to vary the times at which the pinholes are opened and while in this example the S-SBG and terminating Call Agent each reserve bandwidth for the outgoing flows, an alternative approach would be for the terminating Call Agent to reserve bandwidth for incoming and outgoing flows.

8.10.10 End-to-end QoS, Spanning Multiple Networks

Where a call spans multiple networks, the bandwidth reservation is performed in a "leap by leap" manner with each network being responsible for the reservation of resources across its core packet transport. Signalling between the two networks is likely to be handled by a pair of SBGs deployed back to back (each one representing the edge of its owning operator's managed domain). The behaviour of these SBGs and the Network Network Interface (NNI) that they form depends upon the support of the RFC3312 preconditions in the forward direction network. Figure 8.17 illustrates the case in which both the originating network and the next hop network both support RFC3312 preconditions.

In this model, the SIP INVITE is immediately passed over the NNI from the originating SBG which then begins the bandwidth reservation for the media it will transmit into its managed domain (the bandwidth for the media it will receive will be reserved by the Call Agent of the originating party).

The additional messages and preconditions in the SDP are passed in the NNI between the originating and destination side SBGs. However, each SBG assumes that the bandwidth is available on the NNI link and so it can set the preconditions to indicate that resources are available at any time.

The egress SBG (in the originating network) does not propagate the SIP UPDATE, indicating that the resources in its own network have been reserved until both the UPDATE from the Call Agent (indicating that forward direction media resources are reserved in the originating network) and the AAA response has been received from the bandwidth manager (indicating that backward direction media resources are reserved in the originating network).

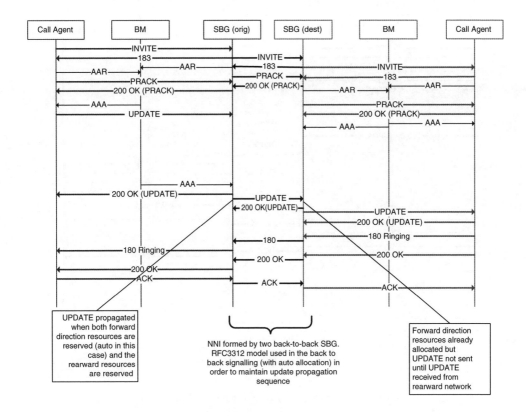

Figure 8.17 NNI for parallel bandwidth reservation

The ingress SBG (in the destination network) performs resource reservation in the normal manner, issuing an AAR to the bandwidth manager for the media resources it will transmit. The UPDATE is not sent to the terminating Call Agent, however, until both the resources have been reserved (the bandwidth manager returns an AAA response indicating this) and an UPDATE has been received from the peer SBG through the NNI (indicating that resources have been reserved in the originating network).

Only when the resources are reserved in both networks will the destination be alerted. This approach, where the SIP preconditions are supported end to end, allows the bandwidth reservation in all the networks in the path to be made in parallel.

Where the forward direction network does not support the SIP precondition model for bandwidth reservation, a different behaviour of the SBG is required; this is shown in Figure 8.18 below.

As the forward direction network is known to not support the SIP preconditions, the egress SBG in the originating network does not propagate the SIP INVITE to the peer (ingress) SBG across the NNI until the resources in the originating network have been reserved. The INVITE is only propagated after the UPDATE from the Call Agent has been received and the AAA response has been received from the bandwidth manager. The reason for withholding the INVITE until resources are reserved is the same as for using preconditions in the first place, to prevent the destination party from being alerted

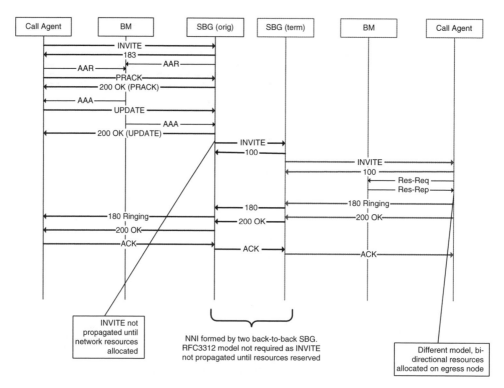

Figure 8.18   NNI for sequential bandwidth reservation

prior to resources being allocated. Since the forward direction network does not support the preconditions, there is nothing to stop it from alerting the user prior to the originating network's resources being allocated (as there is no mechanism to convey this information).

Note that in the example the destination network is shown performing a single bi-directional bandwidth reservation from the terminating Call Agent.

As can be seen in the signalling sequence of Figure 8.18, this approach leads to a sequential reservation of bandwidth which adds to the post-dial delay experienced by the caller (post-dial delay is the time taken between initiating the call and receiving an alerting indication such as a ringback tone).

8.10.11 Supporting IMS-based Networks

The provision of end-to-end QoS reservation becomes more challenging when we consider roaming end points that the IMS architecture enables. This can be seen in the example shown in Figure 8.19. This is an extreme example of a roaming case; however, these types of calls do happen in real networks today.

In this example, Subscriber A with a home network in the United States is roaming in Australia and calls Subscriber B with a home network in the United Kingdom who is also roaming in Australia. The thinner line represents the SIP-signalling path between the two end points, which needs to be routed through the home networks of both subscribers in order to pass through the Serving Call Session Control Function to which the subscribers

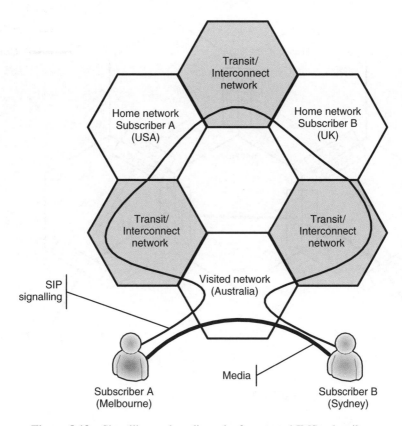

Figure 8.19   Signalling and media paths for roamed IMS subscribers

are registered (see Chapter 5 for a fuller description of IMS roaming). Clearly, it is considerably more efficient if the media does not follow the same path but can take the most direct path between Subscriber A and Subscriber B.

There are, however, implications in supporting an optimal media path between the roaming subscribers. Consider the general case shown in Figure 8.20.

The challenging aspect of this generic scenario is that the optimal routing of the media is through an IP Transit network (as the roamed-to networks have no direct interconnection) but the SIP signalling does not pass through this network; instead, it takes a path which traverses the home networks of both the subscribers. As the SIP signalling is the mechanism used to perform bandwidth reservations, the question arises of how the IP Transit network in this scenario can perform bandwidth reservation for the media flow across it.

The MSF is working on solutions to this problem which would allow the per-session bandwidth reservation model to be used in the case of roaming subscribers with optimal media routing. In developing potential solutions to the problem, the MSF has applied the following four guiding principles.

1. A single bandwidth reservation mechanism will be used for all cases.
2. Changes and additions to the SIP protocol will be avoided (or at least minimised).

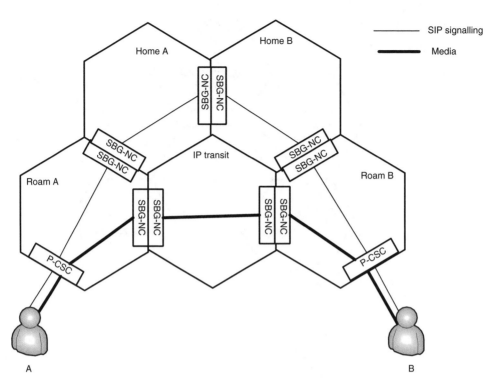

Figure 8.20   IMS roaming, generic case

3. Entities in one network are not assumed to have any knowledge of the internal structure of other networks.
4. Networks are free to route media across themselves as they see fit, according to current network load conditions.

The MSF is working through a number of potential solutions. All the current solutions entail sending SIP signalling with the media path to enable bandwidth reservation; this signalling is either a separate (to the main SIP-signalling path) correlated SIP transaction or an extension of the original SIP-signalling path. At the time of writing, this work is still at an early stage and other solutions may well emerge.

8.11 Protecting the Network from Application Layer Overload

This chapter has looked in detail at how QoS mechanisms may be used in IP and MPLS networks to engineer the network, and how the application layer can ensure that mission-critical services, such as PSTN voice, or high-value videoconferencing sessions, can almost always get through despite significant network overload. However, guarantee-ing QoS in the traffic plane is only part of the problem when it comes to supporting key national infrastructure over packet networks. A major concern is the robustness of the control plane due to focussed overload events. These may be caused by televoting or topical TV and radio phone-ins as well as emergencies. In the NGN, more worryingly, they may also be caused by malicious DoS attacks launched on the infrastructure. In any case, the NGN must be able to take action to ensure that the control plane continues to function and the overload is managed.

8.11.1 Principles of Control Plane Overload

Overload of the control plane will occur when the quantity of requests for some service reaching the application servers exceeds their capacity to handle the requests. In this case, the application server must graciously reject requests that it cannot handle while still operating as close to its capacity as possible. Unfortunately, the nature of overload is such that there are limits beyond which overload starts to impact the application server's ability to process sessions such that its overall capacity begins to reduce. This leads to the type of behaviour shown in Figure 8.21 taken from Ref. 24.

As can be seen, as the volume of offered requests rises, the processed requests rise linearly until the application server starts to invoke overload protection and reject requests. As overload protection starts to kick in, the number of processed requests continues to rise until it reaches a point at which the quantity of requests being rejected means that it starts to impact the processing capacity of the application platform; this point gives the peak capacity of the platform. Beyond this point, the effort involved in rejecting ever more requests means that the processing capacity of the platform will decline progressively until it eventually reaches zero capacity. This type of behaviour is seen in real-world network and implies that it is not just sufficient to rely on an individual computing platform's ability to cope with overload but that mechanisms must be provided in the network to allow for the throttling of requests to a given application server at some point prior to their arrival at the application server.

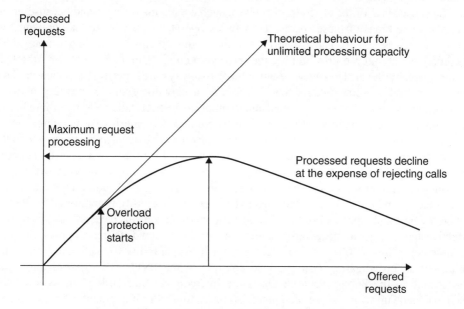

Figure 8.21   Behaviour of an application server under load

8.11.2 Control Plane Overload Control in the PSTN

The performance of the PSTN under overload largely depended on how rigorous the individual network operator and territory was in addressing the problem of network congestion. Within the United Kingdom, the major telecommunications operators have a good record in protecting against overload and they have deployed a number of mechanisms in the PSTN to assist in this, aided to an extent by the hierarchical nature of the PSTN-switched network.

Within the PSTN, individual local exchanges and concentrators may become congested owing to too many calls originating on them. In this case, call restriction is invoked which seeks to reject calls at source using mechanisms such as delaying the application of dial tone or preventing a line from returning to idle following a call. The latter approach is particularly useful for dealing with scenarios where individual subscribers are retrying rapidly because they receive a network busy tone. This overuse of the redial button is a classic problem with the PSTN and explains why overload events often have very large traffic spikes.

In addition to overload at the originating exchange, which might be caused by an incident in an area, the PSTN is also susceptible to focussed overload at destination exchanges. The most common reasons for such an event would be a phone-in, a televote or a telephone competition, which is hosted from a particular exchange. In this case, the large volume of calls arriving at the local exchange overloads it and might reduce or remove its ability to process originating calls. This is particularly problematic because it means that a single user may effectively prevent other customers from accessing the network simply by overloading their local exchange. Therefore, the approach taken in the PSTN is to restrict or gap calls to the destination number that is overloading the network and to perform this restriction as close to the source of the call as possible. Ideally, this would imply gapping at every originating exchange in the network, but in practise gapping is likely to be done at the first trunk switch that the call hits when it enters the core of the PSTN. Because of the sharp traffic spikes, such events may trigger and because of the very short rise times of these spikes, the network operator may wish to take preventative action before the event. The network operator would do this by making a note of an upcoming event and pre-provisioning the restriction controls in advance to limit the load to an acceptable level. However, because there is always a danger that a phone-in show may be missed or some other event may occur, a network operator may use network management systems to allow manual intervention to restrict calls to a given number or even run software to dynamically identify problem numbers and invoke gapping.

In some cases, however, gapping and restriction by network management may not do the job and in this case, the next line of defence for the PSTN is to use the congestion control mechanisms available in the call set-up signalling. These controls allow an overloaded exchange to send information to peer exchanges that it is congested and to request that they reduce the quantity of calls that they send. This type of capability is supported by PSTN-signalling protocols such as ISUP, although such overload controls have not always proved effective and they do not necessarily interoperate well between different suppliers. Because of these concerns, in 2001 the then UK Network Interoperability Consultative Committee published an investigation into overload controls in the PSTN. This investigation was carried out by the Public Network Operators Interconnect Standards Committee (PNO-ISC) and came up with a set of recommendations as to how to implement overload control in PSTN switches using ISUP signalling [25].

A well-behaved congestion control implementation in an application server should enable it to control the rate of requests that it receives as it enters overload. This can be done by explicitly signalling to its client nodes that it has entered congestion, thus causing them to reduce the volume of traffic that they send. The actual mechanism for signalling congestion is protocol dependent and a number of approaches have historically been taken including signalling congestion when accepting or rejecting a session request or sending explicit messages indicating congestion levels back to client nodes. In addition, there are variations on either approach based on whether a single level of congestion is signalled or whether multiple levels of congestion are signalled. In the PSTN, ISUP used the Automatic Congestion Control (ACC) mechanism, [26–28], which explicitly signals the exchange congestion level (level 1 or 2) in the release message sent by the congested exchange when a call is cleared.

One of the key findings of the PNO-ISC study into PSTN overload controls [25] was that it was critically important how the client exchange behaved on receiving an indication of congestion. Because the ISUP specifications had left this detail deliberately vague (to allow vendors a free hand to implement their own internal controls), it was found that this led to a wide variance in how well different vendor's equipment behaved during overload control. It was possible to get a "bad fit" of the two implementations when multi-vendor scenarios were considered. This meant that the performance was worse than for the single vendor case because the vendors had unwittingly chosen incompatible overload protection mechanisms.

For example, the PNO-ISC found that implementations where the client exchange just assigned a fixed restriction level on the basis of the received ISUP ACC level from the congested exchange (e.g. level 0 allow all calls, level 1 allows 50% of the calls and level 2 allows 10% of the calls) were subject to both undercontrol and overcontrol. This situation was made worse because Ref. 27 endorsed such an approach and recommended the running of a timer in the client exchange. While this timer ran, it was proposed that the client exchange did not reduce its level of restriction; this meant that even if the overloaded exchange stopped sending ACC levels in the release messages, it was a fixed period of time before the client exchange could again increase its traffic rates. The implications of this were that if the overloaded exchange could not process calls adequately when its client exchanges operate at a restriction level of (in this example) 50%, then it would continue to get overloaded until it sent a release with a level 2 ACC message. At this point, it might be progressively starved of traffic (because all its client nodes start rejecting 90% of their calls) and eventually its buffers might empty and leave it to idle waiting for timers to expire at the client nodes. Once these timers expired, the client nodes would again start sending traffic at a level which would rapidly drive the congested node back into congestion and the whole process would start again.

As a result of these findings, the PNO-ISC proposed that good implementations of ACC should not map ACC message parameters directly to restriction levels but should use them instead to derive an appropriate restriction level (e.g. on the basis of the frequency of arrival and the contained restriction levels). This general implementation is shown in Figure 8.22.

This approach means that the client node monitors the rate of arrival, and overload levels, seen in the session rejects from the overloaded node. The client node then calculates an appropriate level of restriction to apply to session set-up requests received from its

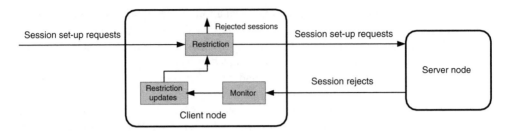

Figure 8.22   Overload control as proposed in ND:1115:2001/08 PNO-ISC/INFO/015

peers that are destined for the congested server node. This means that the client node may have a large number of internal restriction levels that it can apply to the incoming sessions even though only two levels of restriction may be indicated by the ISUP ACC parameters received in the session rejects. This allows for much finer-grained control and with appropriate operational tuning, enables efficient loading of the server node even if it is being overloaded by a number of client nodes. Implementations that follow this approach should converge to a steady level of restriction rather than oscillating between overcontrol and undercontrol. This basic principle may be reused for similar protocols in the NGN, for example SIP, although it should be noted that at the time of writing, SIP does not have a well-defined mechanism for signalling congestion to its peer nodes.

8.11.3 An Overview of Control Plane Overload Control in the NGN

Although the processing power has increased, the problem of overload has not gone away in the NGN, because network operators have chosen to consolidate their call control architecture into fewer larger and more centralised Call Agents. As for the PSTN, Call Agents may be driven into overload by originating calls from the SIP User Agents and H.248 access gateways they serve or by terminating calls originated by other Call Agents but destined for SIP User Agents and H.248 access and trunking gateways that they serve. An added complication in the NGN is that the media plane is now split from the control plane and overload mechanisms that were previously internal to a PSTN switch must now be instantiated across the interface between a media gateway and a Call Agent. However, the biggest danger to the NGN is that because it is an IP-based infrastructure, it is potentially exposed to the very worst sorts of DoS attacks, launched at it by hacked or malicious PC-based SIP terminals. Figure 8.23 shows the NGN architecture and highlights those areas that are at the greatest risk of overload from traditional telephony terminals and from new PC-based SIP clients.

Therefore, to provide a complete solution for control plane overload protection, the network operator must consider the following:

1. Overload protection at access and trunking gateways and its impact on the serving Call Agent.
2. Overload protection at the SBG which may also be a P-CSCF and its impact on the Call Agent or S-CSCF.

If the network operators cannot solve overload control in these two areas, then it is not possible for them to support critical infrastructure such as the PSTN on their NGN.

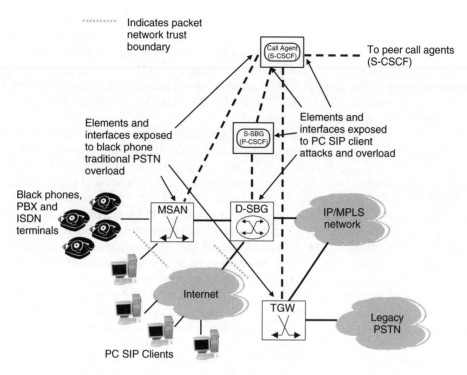

Figure 8.23 Interfaces and control plane elements at risk of overload

8.11.4 Congestion Control Mechanisms Required for Black Phones and Access Gateways

To protect an access gateway from overload, a number of capabilities must be supported.

1. The access gateway must be able to protect itself from overload caused by a local mass call event, for example, all its subscribers going off-hook at the same time.
2. The access gateway must be able to protect itself from overload caused by terminating calls and to instruct its controlling Call Agent to reduce the number of calls being sent to it.
3. The access gateway must be able to react to requests from its controlling Call Agent to reduce the number of call attempts being submitted to the Call Agent. This protects the Call Agent from overload caused by all its access gateways increasing their traffic because of some local, regional or national event.

Access gateways can employ a range of mechanisms to protect themselves from overload caused by too many of their users attempting to set up calls at once. These mechanisms are limited because the access gateway itself has very little real knowledge of what a user is attempting to do, seeing only the line events. While the Call Agent is capable of a more intelligent reaction to overload, by the time the events have been signalled to the Call Agent, it is likely that the access gateway has already undertaken significant processing and will therefore be driven to overload. A typical solution to this problem is for the access gateway to simply deny power to its lines, thus ensuring that the end-user does not obtain a dial tone and therefore cannot congest the access gateway by attempting to set up a new call. These simple solutions come into conflict with regulatory requirements, such as the need to prioritise key workers and government agencies and to give priority to emergency calls (if the volume of emergency calls are not so high that they themselves congest the network). Sophisticated implementations of overload control in access gateways will therefore ensure that these types of users and calls are given priority when they invoke internal overload controls.

To protect access gateways from overload caused by terminating calls from the Call Agent, the H.248 protocol provides two packages, H.248.10 [29] and H.248.11 [30] (see Chapter 4). It is generally believed that H.248.11 is a more effective solution as it allows for finer grain control and was indeed written specifically to address the shortcomings of H.248.10. The H.248.11 package provides an explicit event for the access gateway to signal that it is in overload. This event is sent in a NOTIFY Request message and is called "MG_Overload". Unlike ISUP ACC which was vague in documenting how network nodes should react to overload, H.248.11 requires that the Call Agent implements some variation of a leaky bucket algorithm. This allows restriction by the Call Agent to be increased on the basis of the frequency of received "MG_Overload" events and to slowly reduce the restriction as the frequency of these events reduces. In this way, a Call Agent should converge to a stable restriction level that maximises throughput while avoiding the wild oscillations in traffic load that were the hallmark of poor ISUP ACC implementations.

In order to protect the Call Agent from overload, the access gateway must detect that the Call Agent is in overload and autonomously restrict calls for its Call Agent by locally applying a call-gapping function, while ensuring that priority is given to emergency and governmental calls. This may be achieved by implementing a very simple proxy Call Agent in the access gateway that merely analyses digits and gaps calls on the basis of whether these digits map to defined high-priority strings, for example 999, 911 or 112. This approach violates the principle of H.248 that the Call Agent is always master and the access gateway should not carry out autonomous actions; this caused some debate in the H.248 standards community. However, it is now generally recognised that it is not possible to protect the Call Agent without some sort of autonomous action being taken by the access gateway and that this is a special case. Given this basic mechanism, it is still necessary for the Call Agent to signal to the access gateway to allow it to determine the level of restriction to be applied. This requires a new H.248 package which, at the time of writing, was being defined by ETSI TISPAN as the Notification Rate Package [31]. This package will be published some time in 2006.

One additional area to note when considering overload control on access gateways is that the size of the gateway is important. Overload control works best if a significant number of users are controlled by a single access gateway and performs worst for very small access gateways of a very few lines. Therefore, network operators should seek to maximise the size of access gateways when designing their network. This implies a very sparing use of the concept of virtual access gateways (see Chapter 4) as this may create a large number of small gateways and expose the network to additional risk of overload.

For ISDN terminals, the overload controls are simpler in that the Q.931 signalling is backhauled by the access gateway to the signalling gateway that is resident in the network. This signalling gateway may be co-located with the Call Agent or may be a separate entity, and it provides a natural place for overload control to be implemented.

8.11.5 Trunking Gateway Overload Protection Mechanisms

The trunking gateway is a much simpler network element than the access gateway because it does not have to handle analogue signalling interfaces. As described in Chapter 4, its role is to simply establish RTP bearers on demand from a Call Agent and cross connect them to TDM bearers from the PSTN. Therefore, although the trunking gateway can potentially enter overload because it receives too many cross connect requests from the Call Agent, it is able to use the mechanisms of H.248.11 to request the Call Agent to reduce the quantity of traffic it is being asked to handle. Because the trunking gateway is not itself a source of traffic in the same way that access gateways are, there is no requirement to protect the Call Agent against overload generated by the trunking gateway. The Call Agent/signalling gateway must, however, be able to request that its peer exchange reduce traffic directed towards it and must, in turn, be able to respond to a request by its peer nodes to reduce its traffic. However, as the Call Agent/signalling gateway effectively uses ISUP to communicate with its peer on the PSTN side, a well-implemented version of ISUP ACC and its associated exchange controls will be more than adequate.

8.11.6 A Framework for SIP Overload Control

Overload control becomes significantly more difficult when a network operator chooses to allow SIP clients that are implemented on open computing platforms, such as a PC, to connect to their network. As discussed in Chapter 3, these types of platforms offer great potential for DoS attacks by allowing compromised machines to be coordinated with other machines and allowing them to set up or tear down a large number of calls simultaneously. In order to protect the network from overload, therefore, the network operator may chose to use a SBG at the edge of the network. This allows the D-SBG to restrict the size of the signalling pinholes allowed per user and permits the S-SBG to effectively aggregate the signalling from a large number of end-users before sending it into the core and the network operator's Call Agents. Furthermore, because the S-SBG is a trusted network element (unlike the end-user's SIP clients), it can be relied upon to respond to requests to slow down its rate of signalling requests on demand from a network Call Agent. Effectively, this makes both elements of the SBG the first line of defence against SIP overload; Chapter 3 provides a description of how the SBG can be implemented in such a way as to protect the network against overload.

Although the SBG provides the first line of defence, there is still a risk that the Call Agent may become overloaded by too many S-SBGs sending too much traffic to it. Therefore, in order for a stable SIP network to be built, it requires a congestion control mechanism along the lines of the ISUP ACC solution. However, at the time of writing, no such congestion control mechanisms existed for SIP, although there was at least one early IETF draft that was looking at the subject in more detail. In the absence of such a mechanism, the basic nodal functionality used to support ISUP ACC might be adopted, and it could monitor for overload on the basis of the error messages from SIP received in response to INVITES that indicate congestion. As the number of SIP errors due to congestion increase from a given node, the monitor function could cause the restriction level to be re-calculated and restriction to be applied. Similarly, as the number of SIP errors due to congestion decreases from the node, the restriction level could be gently wound down, allowing for a controlled and gradual re-ramping of traffic. A well-implemented

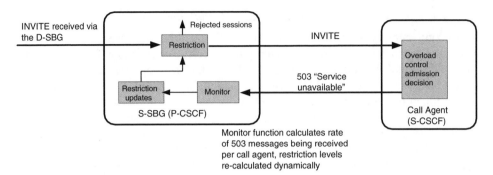

Figure 8.24   A Call Agent overload control mechanism for SIP access via an SBG

mechanism should avoid the dangers of over- and undercontrol of traffic and look to converge on a sustainable throughput rate. One candidate SIP error message that fits well with this solution is "503 Service Unavailable" as this is sent when a server encounters overload. This solution is shown in Figure 8.24.

The SIP 503 message also allows the user of a Retry-After header which sets a time period in which the Call Agent (or in this case, the S-SBG) receiving the 503 message should not send any new requests to the congested SIP Call Agent. This header is optional and is probably of limited use (and may actually be harmful) given the dangers of over- and undercontrol faced by ISUP ACC implementations that used similar crude timing mechanisms.

8.11.7 A Protocol-independent Approach (GOCAP)

It has been shown that PSTN protocols such as ISUP support signalling and nodal behaviour that can provide protection against control plane overload. It has also been shown that some NGN protocols such as H.248 support overload control signalling and others such as SIP and Diameter currently do not. While it is possible to extend all NGN protocols to support overload control mechanisms and to improvise solutions that at least minimise the impact of weak (or absent) overload controls in NGN protocols, it was recognised in ETSI TISPAN that it might be more efficient to define a generic overload control protocol. Such a protocol could be implemented by all nodes and be independent of the control plane protocol. This would mean that interfaces at a node using SIP, H.323, MGCP, H.248 and the Diameter protocols could use the same single generic overload control protocol to signal overload; and furthermore, this protocol would be optimised for the purpose of rapidly communicating overload levels.

To this end, ETSI TISPAN has proposed the Generic Overload Control Application Protocol (GOCAP) [32] which is in the very early stages of standardisation. At the time of writing, it is too early to say whether such an approach will gain widespread acceptance in the NGN or fall by the wayside. It seems likely that this will depend on how quickly (if at all) the NGN protocols such as SIP and Diameter adopt their own overload control solutions.

8.12 Summary

This chapter has introduced some of the mechanisms that enable differentiated QoS to be implemented in the converged multimedia network. Differentiated levels of QoS are required in order to support real time and conversational services, as well as non-real-time and BE services on a common network. Such service differentiation allows multiple service offerings while avoiding over-provisioning the network, thus maximising the operator's return on investment.

The NGN is based on IP, and this chapter shows that there are two elements to this: The first is to ensure appropriate per-packet treatment by the underlying network and the second is to add on top of this mechanisms such as admission control to allow the applications to limit the quantity of traffic they send into the network at each level of priority.

The most common set of mechanisms for ensuring the correct prioritization and treatment of packets in the network is DiffServ, which defines a set of per-hop behaviours for packet treatments with associated classes of service, including EF suitable for real-time services, AF for non-real-time services and a DF class, typically implying BE treatment. DiffServ can be combined with MPLS in the network to enable TE to ensure that links and network elements achieve optimal utilisation while achieving the required QoS during periods of congestion.

The chapter explained how QoS can be provided for business Layer 2 VPN services using a combination of MPLS and DiffServ treatments. Admission control is used to reserve resources through the MPLS network for the tunnel LSP, while DiffServ marking of the LSP packets enables appropriate QoS treatment in the PEs and core LSRs. It is also important to provide QoS for emerging video services, which, alongside voice and high-speed Internet, are increasingly being delivered by converged networks. Video services bring new requirements for resource management, in particular huge bandwidth and scale (in terms of numbers of subscribers). These are resolved by taking a hierarchical approach to per-subscriber and per-service queuing.

To support PSTN services over a packet network, additional mechanisms are layered on top of the network QoS mechanisms. There are a number of approaches that can be taken for this ranging from simple over-provisioning to implementing a full bandwidth manager–based QoS solution. This chapter examined the options and showed that for many network solutions simple mechanisms may work well, but for large-scale PSTN support, bandwidth management–based solutions provide additional robustness in the face of network failure. The chapter went on to examine bandwidth management–based solution in depth by considering the MSF solution for bandwidth management. It showed how SIP and Diameter signalling can be combined to allow the SIP Call Agents to request QoS from bandwidth managers; it also showed that for packet networks that support PC-based SIP clients the SBG plays a key role in ensuring that the network can guarantee QoS. The chapter also looked at some of the emerging solutions for QoS over multiple domains and, in particular, how the MSF bandwidth management architecture can support IMS roaming with media path optimisation.

Support for QoS in the network will still not guarantee service if the control plane, that is, the application servers and Call Agents that provide the service intelligence, becomes overloaded because of excessive session requests. This chapter looked at how the PSTN was protected from overload and showed that the same approach can be taken to protect the NGN. It concluded by looking at the key network elements that are vulnerable to overload and showed how they can be protected by the implementation of the appropriate overload controls.

References

1. R. Braden, D. Clark, S. Shenker. IETF RFC 1633. "Integrated Services in the Internet Architecture: An Overview". June 1994.
2. S. Blake, D. Black, M. Carlson, E. Davies, Z. Wang, W. Weiss. IETF RFC 2475. "An Architecture for Differentiated Service". December 1998.
3. S. Shenker, C. Partridge, R. Guerin. IETF RFC 2212. "Specification of Guaranteed Quality of Service". September 1997.
4. J. Wroclawski. IETF RFC 2211. "Specification of the Controlled-Load Network Element Service". September 1997.
5. B. Davie, A. Charny, J.C.R. Bennett, K. Benson, J.Y. Le Boudec, W. Courtney, S. Davari, V. Firoiu, D. Stiliadis. IETF RFC 3246. "An Expedited Forwarding PHB (Per-Hop Behaviour)". March 2002.
6. J. Heinanen, F. Baker, W. Weiss, J. Wroclawski. IETF RFC 2597. "Assured Forwarding PHB Group". June 1999.
7. J. Heinanen, R. Guerin. IETF RFC 2697. "A Single Rate Three Colour Marker". September 1999.
8. J. Heinanen, R. Guerin. IETF RFC 2698. "A Two Rate Three Colour Marker". September 1999.
9. Y. Bernet, P. Ford, R. Yavatkar, F. Baker, L. Zhang, M. Speer, R. Braden, B. Davie, J. Wroclawski, E. Felstaine. IETF RFC 2998. "A Framework for Integrated Services Operation over Diffserv Network". November 2000.
10. K. Nichols, B. Carpenter. IETF RFC 3086. "Definition of Differentiated Services Per Domain Behaviors and Rules for their Specification". April 2001.
11. D. Awduche, J. Malcolm, J. Agogbua, M. O'Dell, J. McManus. IETF RFC 2702. "Requirements for Traffic Engineering over MPLS". September 1999.
12. ATM Forum Specification. AF-AIC-0178.001. "ATM-MPLS Network Interworking, Version 2.0". August 2003.
13. ETSI ES 282 003. "Telecommunications and Internet converged Services and Protocols for Advanced Networking (TISPAN); NGN Release 1: Functional Architecture; Resource and Admission Control Subsystem (RACS)". March 2006.
14. S. Walker, P. Drew. MSF-ARCH-002.00-FINAL "MSF Release 2 Architecture". January 2005. Available at http://www.msforum.org.
15. C. Gallon, O. Schelén. MSF-TR-ARCH-005-FINAL002.00-FINAL "Bandwidth Management in Next Generation Packet Networks". August 2005. Available at http://www.msforum.org.
16. T. Taylor, S. Walker. MSF-ARCH-003.00-FINAL. "MSF Release 3 Architecture". June 2006. Available at http://www.msforum.org.
17. J. Evans. MSF White Paper "Network Engineering to Support the Bandwidth Manager". May 2006. (available at http://www.msforum.org).
18. G. Camarillo, W. Marchall, J. Rosenberg, (Editors). IETF RFC 3312. "Integration of Resource Management and Session Initiation Protocol (SIP)". October 2002.
19. O. Schelén, A. Torger, J. Johansson. MSF-IA-NRCP.001-FINAL. "Implementation Agreement for a Network Resource Control Protocol (NRCP)". September 2004.
20. E. Lundgren, U. Bodin, O. Schelén. MSF-IA-DIAMETER.001-FINAL. "Implementation Agreement for a Diameter Interface to the Bandwidth Manager". July 2006.
21. ETSI TS 183 017. "Telecommunications and Internet Converged Services and Protocols for Advanced Networking (TISPAN); Resource and Admission Control; DIAMETER protocol for session based policy set-up information exchange between the Application Function (AF) and the Service Policy Decision Function (SPDF)". Protocol Specification. March 2006.
22. P. Calhoun, J. Loughney, E. Guttman, G. Zorn, J. Arkko. IETF RFC 3588. "Diameter Base Protocol". September 2003.
23. ETSI ES 283 018. "TISPAN NGN Release 1; RACS; H.248 Profile for the Ia Interface". March 2006.
24. I. Jenkins. MSF Technical Report MSF-TR-ARCH-007-FINAL. "NGN Control Plane Overload and its Management". January 2006.
25. NICC Document ND:1115:2001/08 PNO-ISC/INFO/015. "ISUP Overload Controls". August 2001.
26. ITU-T Recommendation Q.764. "Signalling System No. 7 ISDN User Part signalling procedures". December 1999.
27. ITU-T Recommendation Q.542. "Digital Exchange Design Objectives – Operations and Maintenance". March 1993.

28. ITU-T Recommendation E.412. "Network Management Controls". March 1998.
29. ITU-T Recommendation H.248.10 (07/2001). "Gateway Control Protocol: Media Gateway Resource Congestion Handling Package". 2001.
30. ITU-T Recommendation H.248.11 (11/2002). "Gateway Control Protocol: Media Gateway Overload Control Package". 2002.
31. ETSI ES 283 039-4. "Telecommunications and Internet converged Services and Protocols for Advanced Networking (TISPAN); NGN Overload Control Architecture; Part 4: Adaptative Control for the MGC". 2006.
32. Draft ETSI ES 283 039-2. "NGN Congestion & Overload Control; Part 2: Core GOCAP and NOCA Entity Behaviours". February 2006.

Index

Authentication, Authorization and Accounting (AAA), 176, 177
AAR. *See* Authenticate and Authorize Request (AAR)
Abort-Session-Answer (ASA), 312
Abort-Session-Request (ASR), 312
AC. *See* Attachment Circuit
ACC. *See* Automatic Congestion Control (ACC)
Access circuit, 101, 106, 116
Access Gateway (AGW), 10, 14, 22, 23, 28, 32, 34, 74, 92, 93, 116, 284–289, 292–294, 297, 298, 301, 314, 316, 324–327
Access network, 1, 2, 6, 18, 19, 27, 93–96, 98, 101, 149–151, 157–159, 173, 174, 180, 181, 198, 278
Access signalling, 101, 106
Access types
  ADSL (Asymmetric Digital Subscriber Line), 10, 232
  Cable, 2, 9, 18, 65, 77, 116, 178, 276
  DSL (Digital Subscriber Line), 2, 4, 10, 11, 14, 18–20, 22, 75, 77, 93, 150, 276, 277
  Fixed, 1, 4–6, 9, 13, 92, 114, 150, 151, 156, 159, 174, 175, 228, 268, 276, 285, 291, 300, 306, 309, 323
  GSM, 4, 11, 150–155, 174, 175, 178, 206, 208, 209
  Mobile, 1, 2, 4–6, 9, 11–14, 17, 19, 21, 24, 36, 55, 56, 92, 111, 114, 150–160, 163, 168, 170, 171, 173–179, 185, 193, 202, 206, 208, 209, 228, 276, 283, 300

UMA (Unlicensed Mobile Access), 11, 12, 174, 175
Universal Mobile Telecommunications System (UMTS), 4, 150–156, 158, 159, 170, 179, 184, 185
WiFi, 4, 13, 94, 174
WiMAX, 2, 4, 150, 151, 176
wireless, 1, 2, 5, 6, 9, 11, 13, 150, 151, 159, 174–177, 182, 206
WLAN (Wireless Local Area Network), 13
Access-Resource and Admission Control Function (A-RACF), 180
Adaptive Modulation and Coding (AMC), 158
Adaptive Multi-Rate (AMR), 156
Admission control, 269
AGI. *See* Attachment Group Identifier
AGW. *See* Access Gateway (AGW)
AH. *See* Authentication Header (AH)
AINI. *See* ATM Internet Network Interface
AIS. *See* Alarm indication Signal
Alarm Indication Signal (AIS), 262, 263
Alliance for Telecommunications Industry Solutions (ATIS), 22, 162, 165–167, 181, 182, 185
AMC. *See* Adaptive Modulation and Coding (AMC)
American National Standards community (ANSI-41), 151
AMR. *See* Adaptive Multi-Rate (AMR)
Analogue, 9–11, 16, 24, 27, 74, 92, 93, 99–106, 112, 114, 116–118, 129, 138, 140, 141, 151, 179, 314, 327
Application Function (AF), 180
Application Server (AS) 2, 8, 13, 23, 24, 36, 115, 163, 164, 167, 168, 171, 194–198,

Application Server (AS) (*continued*)
    208–213, 215–225, 228, 267, 268, 289,
    290, 299, 300, 306, 307, 321, 323, 329
A-RACF. *See* Access-Resource and Admission
    Control Function
Architecture, 1, 5–7, 9, 14–24, 36, 37, 91, 92,
    99, 103, 108–111, 115, 142, 145, 146,
    149, 150, 152, 156, 160, 163, 165, 166,
    177, 180–186, 200, 205, 206, 210, 211,
    213, 223, 228, 231–233, 243–247, 251,
    253, 254, 263, 267, 271, 272, 276, 277,
    280, 283–286, 288–296, 298–300, 303,
    306, 310, 314, 318, 324, 329
ARPU. *See* Average Revenue Per User
    (ARPU)
ARQ. *See* Automatic Repeat Request (ARQ)
ASBR. *See* Autonomous System Border
    Routers (ASBRs)
Assured Forwarding (AF), 272
Asynchronous Transfer Mode (ATM), 4, 18,
    106, 118–121, 135, 136, 141, 155–157,
    184, 231, 232, 241, 244–252, 254–264,
    278, 279
ATIS. *See* Alliance for Telecommunications
    Industry Solutions (ATIS)
ATM. *See* Asynchronous Transfer Mode
    (ATM)
ATM Forum, 246, 248, 257
ATM Inter-Network Interface (AINI), 257
ATM PW, 247, 254
ATM virtual trunks, 257
Attachment Circuit (AC), 244, 249, 250,
    253–255, 260–263
Attachment group identifier, 250
Attachment identifier, 250
Attribute line, 67, 69
Audio/Video Transport Working Group (AVT),
    183
Authenticate and Authorize Request (AAR),
    307
Authentication Header (AH), 85
Automatic Congestion Control (ACC), 323
Automatic Repeat Request (ARQ), 155
Autonomous System (AS), 235
Autonomous System Border Routers (ASBRs),
    235, 280
Availability, 5, 6, 13, 92, 144, 179, 181, 185,
    203, 236, 238, 242, 244, 256, 259
Average Revenue Per User (ARPU), 4, 191

AVT. *See* Audio/Video Transport Working
    Group (AVT)

Bandwidth, 4, 5, 19
Bandwidth management, 9, 22, 23, 32, 267,
    277, 280, 281, 283, 286, 289–297, 299,
    300, 302, 303, 305–314, 316, 317, 329
Bandwidth manager, 23, 32, 288, 289
Base Station Controller (BSC), 1, 152
Base Station Subsystem (BSS), 152
Base Transceiver Station (BTS), 152
Basic Call State Model (BCSM), 15, 24–26,
    28–31, 33, 39, 77
    O-BCSM (Originating BCSM), 24–26,
        31–34
    T-BCSM (Terminating BCSM), 24, 33–35
BCSM. *See* Basic Call State Model (BCSM)
BE. *See* Best Effort (BE)
Bearer Independent Call Control protocol
    (BICC), 106, 111
Best Effort (BE), 271
BGCF. *See* Breakout Gateway Control
    Function (BGCF)
BGF. *See* Border Gateway Function (BGF)
BGP. *See* Border Gateway Protocol (BGP)
BICC. *See* Bearer Independent Call Control
    protocol (BICC), 1
Billing, 10, 19–21, 25, 33, 64, 65, 82, 92, 120,
    154, 156, 157, 161, 167, 182, 210, 214,
    227, 316
Bits, 27, 100, 101
Bluetooth, 13, 174
BNG. *See* Broadband Network Gateways
    (BNG)
Border Gateway Function (BGF), 181
Border Gateway Protocol (BGP), 235, 236,
    244, 245, 253, 256
BRAS. *See* Broadband Remote Access Server
    (BRAS)
Breakout Gateway Control Function (BGCF),
    165
British Telecom National User Part (BTNUP),
    27
Broadband, 1, 2, 4, 9, 11, 13, 14, 18–21, 23,
    77, 82, 150, 151, 174–176, 182, 231,
    232, 263, 267, 276, 277
Broadband Network Gateways (BNG), 277
Broadband Remote Access Server (BRAS), 18,
    277
Broadcast TV, 267, 277

BSS. *See* Base Station Subsystem (BSS)
BTNUP. *See* British Telecom National User
      Part (BTNUP)
BTS. *See* Base Transceiver Station (BTS)
Buffer, 71, 123

Cable Operator, 9
CAC. *See* Connection Admission Control
      (CAC)
Call Agent, 15, 19–25, 27–30, 32–39, 41–43,
      45, 46, 54–58, 60–66, 68, 73, 77,
      86–88, 90–95, 97, 108, 110, 122–124,
      143, 156, 199, 210–214, 222–226, 282,
      284–289, 291–294, 297–309, 311–318,
      324–329
Call control, 7, 10, 14, 15, 19, 24, 25, 27–30,
      33, 65, 67, 69, 71, 73, 75, 77, 81, 86, 90,
      92, 93, 98, 106, 108, 111, 149, 156, 164,
      186, 188, 192, 200, 202, 204, 215, 286,
      288, 292, 293, 298–300, 306, 314, 324
Call counting, 284, 288
Call gapping, 21
Call Processing Language (CPL), 194, 195
Call routing, 15, 21, 24, 26–28, 32, 36–39, 41,
      60–64, 76, 77, 94, 112, 119, 183, 211,
      213, 281, 285, 300
Call session control, 6
Call Session Control Function (CSCF), 159
Call-gapping, 282, 326
Calling Line Identifier (CLI), 34, 191
Capabilities Exchange Request, 307
CAPEX. *See* Capital expenditure (CAPEX)
Capital expenditure (CAPEX), 3, 5
CBS. *See* Committed Burst Size (CBS)
CDMA2000, 158
CDMA2000 1x, 158
CDMA2000 1xEV-DO, 158
CDMA2000 1xEV-DV, 158
CE. *See* Customer Edge (CE)
Cell Loss Ratio (CLR), 155
CGI. *See* Common Gateway Interface (CGI)
Charging, 10, 64, 156, 177, 182, 186, 187,
      200, 202, 210, 214
CIR. *See* Committed Information Rate (CIR)
Circuit domain, 11, 183, 184
Classification, 234, 249
CLI. *See* Calling Line Identifier (CLI)
CLR. *See* Cell Loss Ratio (CLR)
Codec, 16, 19, 36, 39, 44, 46, 48, 65–72, 77,
      100, 101, 111, 112, 116–121, 125, 126,

      132, 134–138, 156, 186, 187, 285, 305,
      310
G.711, 19, 44, 48, 65, 66, 68, 70, 100, 101,
      119, 121, 136, 156, 310
G.723.1, 44, 68, 112, 116, 136
G.729, 19, 44, 65, 66, 68, 70, 112
Cold standby, 240
Colour, 234, 272
Committed Burst Size (CBS), 272
Committed Information Rate (CIR), 272
Common Gateway Interface (CGI), 195
Common Object Policy Service (COPS), 169
Common Object Request Broker Architecture
      (CORBA), 200
Common Open Policy Service (COPS), 89
Companding, 100, 123
   A-law, 68, 100, 123, 136
   μ-law, 44, 48, 68, 70, 100, 123
Connection Admission Control (CAC), 9, 23,
      55, 274, 277, 279, 283, 285, 287–289,
      293, 295, 296, 313
Connectivity Verification (CV), 237, 238
Contributing Sources (CSRC), 72
Control Plane Interworking, 256
Controlled Load Service, 271
Convergence, 1–3, 5, 6, 14, 24, 92, 150, 151,
      159, 231, 232, 237, 239, 251, 263, 264,
      269
Core network, 1–3, 5, 6, 9, 12, 15, 17–19, 23,
      88, 92, 106, 149, 150, 152–156, 159,
      160, 168, 169, 175, 181, 185, 186, 194,
      200, 227, 228, 232, 239, 251, 254,
      256–259, 264, 267, 275, 278, 279, 284
CPL. *See* Call Processing Language (CPL)
CSCF. *See* Call Session Control Function
      (CSCF)
CSRC. *See* Contributing Sources (CSRC)
Current status, 304
Customer Edge (CE), 244, 245, 247, 254
CV. *See* Connectivity Verification (CV)

Data Link Connection Identifier (DLCI), 241
Data path Session Border Gateway (D-SBG),
      22, 23, 86, 89–92, 94, 97, 291, 292,
      294, 297–299, 301, 306, 309, 310,
      313–315, 325, 327, 328
Data services, 1, 4–6, 9, 10, 17–20, 23, 67, 71,
      72, 75, 77, 81–86, 89, 99, 101, 110, 117,
      143, 144, 149–159, 162, 168, 174, 175,
      177–179, 182, 194, 202, 204, 208, 209,
      227, 228, 269, 275, 276, 278, 300, 306

Default Forwarding (DF), 272
Denial of Service Attacks (DoS), 90
Desired Status (DS), 304
Detour
    LSP, 241, 242
DF. *See* Default Forwarding (DF)
DHCP. *See* Dynamic Host Configuration
    Protocol (DHCP)
Dial pulse signalling, 102
Dial tone, 20, 102, 103, 105, 124, 125, 138,
    139, 322, 326
Diameter Protocol, 306
Differentiated Services Framework (DiffServ),
    233, 234, 263
    behaviour aggregates, 234
    classification, 234, 249
    colour, 234
    DP (drop precedence), 234
    forwarding class, 234
    per-hop behaviours, 234
    PSC (Per-hop behaviour Scheduling Class),
        234
DiffServ. *See* Differentiated Services
    Framework
DiffServ-aware traffic engineering, 275
Digest authentication, 55, 95
Digital Subscriber Line Access Multiplexer
    (DSLAM), 18, 20, 89, 276, 277
Digital Video Broadcasting Handheld
    (DVB-H), 177
DLCI. *See* Data Link Connection Identifier
    (DLCI)
DNS. *See* Domain Name System (DNS)
Domain Name System (DNS), 18–20, 28, 29,
    39, 62, 76, 122, 124, 160, 163, 183, 187
DoS. *See* Denial of Service Attacks (DoS)
Downlink, 6, 152, 153, 157, 158, 178, 310
DP. *See* Drop Precedence (DP)
Drop Precedence (DP), 234
DS. *See* Desired Status (DS)
D-SBG. *See* Data path Session Border
    Gateway (D-SBG)
DSLAM. *See* Digital Subscriber Line Access
    Multiplexer (DSLAM)
DTMF. *See* Dual Tone Multiple Frequency
    (DTMF)
Dual Tone Multiple Frequency (DTMF), 44,
    48, 69, 102, 110, 112, 115, 116, 121,
    138, 140, 146, 192, 202, 217, 220
DVB-H. *See* Digital Video Broadcasting
    Handheld (DVB-H)

Dynamic Host Configuration Protocol (DHCP),
    58, 60, 160

EBS. *See* Excess Burst Size (EBS)
EDGE. *See* Enhanced Data Rates for GSM
    Evolution (EDGE)
EF. *See* Expedited Forwarding (EF)
Egress remarking, 279
Emergency services, 4, 21, 23, 26, 27, 30, 33,
    56, 57, 91, 92, 132, 161, 185, 186, 268,
    283, 285, 286, 288, 291, 326
Encapsulating Security Payload (ESP), 85
End-to-end delay, 268
Enhanced Data Rates for GSM Evolution
    (EDGE), 154
ENUM (telephone number mapping), 16,
    27–29, 76, 77, 124, 163, 183
ERO. *See* Explicit Route Object (ERO)
ESP. *See* Encapsulating Security Payload
    (ESP)
Ethernet, 2–4, 19, 20, 82, 83, 86, 89, 93, 184,
    231, 232, 236, 239, 244, 246, 247,
    251–257, 260, 261, 263, 276–279
ETSI. *See* European Telecommunications
    Standards Institute (ETSI)
European Telecommunications Standards
    Institute (ETSI), 14, 22, 23, 90, 103, 108,
    126, 141, 151, 158, 160, 162, 165–167,
    174, 178–181, 185, 192, 199, 201–205,
    289, 291, 302, 303, 306, 307, 326, 328
Excess Burst Size (EBS), 272
EXP bits, 234, 275
Expedited Forwarding (EF), 271
Experimental (EXP), 234
Explicit Route Object (ERO), 274
Extended PNNI, 257, 259
Exterior gateway routing protocol, 235
    eBGP (exterior BGP), 235

Facility bypass, 241, 242
Facsimile, 100, 101, 104, 110, 121, 137, 138
Fast re-route, 14, 237, 239, 240, 242, 244, 296
FDD/TDMA. *See* Frequency Division
    Duplex/Time Division Multiple Access
    (FDD/TDMA)
Feature interaction, 24, 25, 28, 211, 221
Feature servers, 25, 28, 194
Features, 15, 20, 24, 28, 31, 33, 34, 37, 38, 93,
    97, 102, 115, 120, 133, 150, 151, 154,
    178, 286

call forwarding, 24, 25, 54, 119, 120, 204, 207

call hold, 33, 70, 102, 113, 129, 131, 219, 221, 226, 313, 314

call screening, 34, 204, 216–225

call waiting, 20, 34, 129, 211, 217–225

caller display, 20, 34, 56, 57, 93, 139, 191

CCBS (Call Completion to Busy Subscriber), 24, 25

conference calling, 24, 28, 33, 116, 121, 128, 163, 202, 205, 211, 313, 314

MCID (Malicious Call Identification), 34

Fibre to the Home (FTTH), 276

Fixed Network Operator, 13

Flowspec object, 271

Forwarding class, 234, 271

FQDN. See Fully qualified domain name (FQDN)

Frame Relay, 4, 184, 232, 241, 244–247, 249, 251, 252, 254–257, 262, 263, 278

Framework SCF, 200, 201

Frequency Division Duplex/Time Division Multiple Access (FDD/TDMA), 152

FTTH. See Fibre to the Home (FTTH)

Fully Qualified Domain Name (FQDN), 16, 32, 40

Gates and Pinholes, 90

Gateway GPRS Support Node (GGSN), 153

General Packet Radio Service (GPRS), 150, 153

Generalised ID FEC, 249

Generic Overload Control Application Protocol (GOCAP), 328

Generic Routed Encapsulation (GRE), 243, 246, 252

GERAN, 150, 160

GGSN. See Gateway GPRS Support Node (GGSN)

Global System for Mobile Communications (GSM), 150, 151

Gateway MGW (GMGW), 156

Gateway MSC (GMSC-S), 156

Go interface, 169

GOCAP. See Generic Overload Control Application Protocol (GOCAP)

GPRS. See General Packet Radio Service (GPRS)

GPRS Tunnel Protocol-Control Plane (GTP-C), 157

GPRS Tunnelling Protocol-User Plane (GTP-U), 157

Gq interface, 169, 306

Gq prime interface, 306

GR.303 access signalling, 27

Graceful restart, 239, 243

GRE. See Generic Routed Encapsulation

GTP-C. See GPRS Tunnel Protocol-Control Plane (GTP-C)

GTP-U. See GPRS Tunnelling Protocol-User Plane (GTP-U)

Guaranteed QoS, 231, 290, 300, 314

Guaranteed Service, 271

GUI, 194

H.248 (Gateway Control Protocol), 7, 10, 14, 23, 24, 29, 32, 34, 72, 73, 89, 90, 93, 97, 98, 111, 116, 117, 121, 122, 126–130, 132–142, 146, 156, 157, 165, 167, 183, 284, 289, 292, 306, 307, 314, 316, 324, 326–328

H.248 packages
Stimulus Analogue Line Package, 14, 23

H.320, 101, 104

H.323, 36, 108, 109, 116, 127, 194, 328

H.324, 104

Hashes, 84

Hierarchical Virtual Private LAN Service, 253

High Speed Circuit Switched Data (HSCSD), 152

High Speed Downlink Packet Access (HSDPA), 158

High Speed Uplink Packet Access (HSUPA), 158

HLR. See Home Location Register (HLR)

Home Location Register (HLR), 152

Home Subscriber Server (HSS), 162, 208

Homogeneous layer 2 VPN, 254

Hot Spot, 175

Hot-standby, 238, 240

HSCSD. See High Speed Circuit Switched Data (HSCSD)

HSDPA. See High Speed Downlink Packet Access (HSDPA)

HSS. See Home Subscriber Server (HSS)

HSUPA. See High Speed Uplink Packet Access (HSUPA)

HTML, 194

H-VPLS, 253

IANA. *See* Internet Assigned Numbers
  Authority (IANA)
iBGP. *See* Interior BGP (iBGP)
ICMP. *See* Internet Control Messaging
  Protocol (ICMP)
I-CSCF. *See* Interrogating CSCF (I-CSCF)
IETF. *See* Internet Engineering Task Force
  (IETF)
iFC. *See* Initial Filter Criteria (iFC)
IGP. *See* Interior gateway routing protocol
  (IGP)
IGW. *See* International gateway exchange
  (IGW)
IMS. *See* IP Multimedia Subsystem (IMS)
IMSI. *See* International Mobile Subscriber
  Identity (IMSI)
IM-SSF. *See* IP Multimedia Service Switching
  Function (IM-SSF)
IN. *See* Intelligent Network (IN)
INAP. *See* Intelligent Network Application
  Part (INAP)
Indication AVP, 314
Infrastructure, 2, 3, 5–7, 11, 14–16, 21–23,
  30, 65, 93, 97, 179, 185, 191, 206, 227,
  228, 231, 232, 243, 244, 246, 252, 253,
  257, 263, 264, 276, 282, 286, 299, 300,
  320, 321, 324
Initial Filter Criteria (iFC), 206
Integrated Services Digital Network (ISDN),
  14, 23, 24, 26, 27, 29, 30, 76, 92, 101,
  104, 105, 109, 115, 136, 142, 145, 146,
  156, 182, 184, 192, 193, 325, 326
Integrated Services Framework (IntServ), 233
Intelligence, 6, 15, 16, 27, 37, 86, 92, 288,
  293, 298, 329
Intelligent Network (IN), 24–27, 30, 31, 33,
  34, 77, 182, 184, 191–193, 198, 199,
  205, 208–210, 228
  SCP (Service Control Point), 18, 20, 22, 25,
    33, 192, 193, 211
  SSP (Service Switching Point), 192, 198,
    199
Intelligent Network Application Part (INAP),
  26–28, 192, 193, 198, 199, 208, 209,
  211
Intelligent Peripheral (IP), 192
Interconnect, 10, 11, 17–19, 22, 23, 299, 300
Interconnection with peer network, 10, 11,
  17–19, 22, 23, 29, 30, 32, 76, 77, 86,
  90, 92, 93, 95, 181, 284, 286, 292–294,
  296, 299–301

Inter-exchange, 16, 32, 101, 106, 107, 110,
  112, 116, 123, 138, 140, 141, 146
Interior BGP (iBGP), 235, 270
Interior gateway routing protocol (IGP), 235,
  237, 239, 245
  iBGP (interior Border Gateway Protocol),
    235, 244
  IS-IS (Intermediate System-Intermediate
    System), 235
  OSPF (Open Shortest Path First), 233, 235,
    245
International gateway exchange (IGW), 17
International Mobile Subscriber Identity
  (IMSI), 153
International Telecommunication Union
  Telecommunication Standardisation
  Sector (ITU-T), 16, 22, 29–31, 35, 36,
  65, 75, 76, 100, 102, 104, 106, 108, 111,
  113, 114, 119, 121, 126, 127, 139, 142,
  181, 183, 185, 238, 246
Internet, 232, 233, 235
Internet Assigned Numbers Authority (IANA),
  141, 306
Internet Control Messaging Protocol (ICMP)
  Ping, 237
  traceroute, 237
Internet Engineering Task Force (IETF),
  14–16, 29, 36, 65, 67, 68, 70, 71, 74,
  76, 77, 89, 108–110, 112–114, 121,
  126, 142, 144, 146, 159, 163, 169, 170,
  181–185, 187, 194, 232, 236–238, 240,
  243, 244, 246, 248, 250, 251, 271, 273,
  293, 306
Internet Key Exchange (IKE), 85
Internet Protocol (IP), 159
Internet Protocol Security (IPSEC)
  VPNs, 245
Internet Service Provider (ISP), 11, 17, 20–22,
  179
Interrogating CSCF (I-CSCF), 160, 162, 173
Interworking, 6, 7, 35, 92, 108, 112–114, 134,
  142, 146, 154, 160, 170, 182, 183, 185,
  186, 211
Interworking LSP, 246
IntServ. *See* Integrated Services Framework
  (IntServ)
INVITE, 302, 305, 314, 316, 317
IP address, 19, 28, 39, 40, 44, 48, 58, 62, 63,
  66, 67, 70, 75, 87–90, 120, 157, 160,
  161, 168, 169, 175, 269, 271, 272, 305,
  307, 314

IP Multimedia Service Switching Function (IM-SSF), 198, 199, 208
IP Multimedia Subsystem (IMS), 1, 6, 7, 9, 14, 24, 64, 92, 106, 149, 150, 154, 158–168, 170, 173, 177, 179–182, 184–186, 206–209, 212, 214, 217–221, 223, 228, 267, 291, 306, 318–320, 329
IP Telephony (IPTEL), 183
IP Virtual Private Networks (IP VPNs), 4, 278
IP VPNs. See IP Virtual Private Networks (IP VPNs)
IPsec, 85
IPSEC. See Internet Protocol Security (IPSEC)
IPTEL. See IP Telephony (IPTEL)
IPv4, 66, 67, 87, 160, 176, 181, 295
IPv6, 66, 160, 168, 176, 181, 295
ISDN. See Integrated Services Digital Network (ISDN)
ISDN User Part (ISUP), 14, 24, 26, 27, 29, 30, 32, 34, 56, 106–115, 119, 142, 145, 146, 156, 157, 160, 165, 183, 184, 192, 193, 199, 322–324, 326–328
    IAM (Initial Address Message), 34, 56, 107, 113, 119
ISP. See Internet Service Provider (ISP)
ISUP. See ISDN User Part (ISUP)
ITU-T. See ITU-T (International Telecommunication Union Telecommunication Standardisation Sector)

JAIN. See Java for Advanced Intelligent Networks (JAIN)
Java for Advanced Intelligent Networks (JAIN), 194, 198
Jitter, 71, 73

Key Distribution, 85

L2TP. See Layer 2 Tunnelling Protocol
L2TP Access Concentrator (LAC), 18
L2TP Network Server (LNS), 18, 20
L2TPv3, 246
Label-inferred PSC LSP (L-LSP), 234, 275, 279
LAC. See L2TP Access Concentrator (LAC)
Lawful intercept, 19, 64, 93
Layer 2 mediation. See SPVC-PWE3 interworking
Layer 2 Tunnelling Protocol (L2TP), 243, 245, 251

Layer 2 Virtual Private Networks, 245
Layer 2 VPN. See Layer 2 Virtual Private Networks
Layer 3 Virtual Private Networks, 244
Link Management Interface (LMI), 262, 263
L-LSP. See Label-inferred PSC LSP (L-LSP)
LLU. See Local Loop Unbundling (LLU)
LMI. See Link management interface
LNS. See L2TP Network Server (LNS)
Load balancing, 240, 241
Local descriptor, 135, 137, 314
Local exchange, 10, 14, 16, 17, 19, 24, 27, 56, 101–104, 110, 115, 138, 276, 281, 282, 322
    Class Five Office, 16
    DLE (Digital Local Exchange), 17, 18
Local Loop Unbundling (LLU), 10, 11
Local Management Interface (LMI), 262
Local protection, 241, 242
Location, 11, 13, 15, 21, 38, 39, 54, 58, 62, 76, 86, 153, 154, 157, 162, 168, 171, 175, 185, 187, 193, 202, 204, 217, 300
Loose routing, 274

MAC. See Medium Access Control (MAC)
MAC Ping, 260
MAC trace, 260
Management Information Base (MIB), 84, 293
Maximum Transmission Unit (MTU), 247, 249, 253, 254
MBMS. See Multimedia Broadcast/Multicast Service (MBMS)
MD5, 84
Media access control, 252
Media Component Description, 307, 309
Media Gateway (MG), 11, 19, 23, 32, 65, 90, 108–110, 112, 116–121, 124–126, 128–130, 132, 134–141, 146, 156, 182, 183, 324, 326
Media Gateway Control (MEGACO), 7, 183
Media Gateway Control Function (MGCF), 159
Media Gateway Control Protocol (MGCP), 24, 32, 72, 86–88, 90, 92, 93, 108, 116, 121–126, 128–130, 132, 133, 139, 140, 146, 287, 292, 328
Media Gateway Controller (MGC), 11, 108–110, 112, 113, 116–121, 125, 126, 128–130, 132, 134–141, 145, 146, 156
Media Resource Function (MRF), 159, 166
Media Server, 13, 36, 64, 90, 97, 217, 222, 227, 313, 314

Media sub-component, 309, 310
Medium Access Control (MAC), 83, 155, 252, 253, 260
MEGACO. *See* Media Gateway Control (MEGACO)
MG. *See* Media Gateway (MG)
MGC. *See* Media Gateway Controller (MGC)
MGCP. *See* Media Gateway Control Protocol (MGCP)
MIB. *See* Management Information Base (MIB)
MMUSIC. *See* Multiparty Multimedia Session Control (MMUSIC)
MNO. *See* Mobile Network Operator (MNO)
Mobile, 1, 2, 4, 13, 177
Mobile Country Code (MCC), 153
Mobile Network Operator (MNO), 9, 11, 13
Mobile Subscriber Identity Number (MSIN), 153
Mobile Switching Centre (MSC), 152
Mobile Virtual Network Operator (MVNO), 9, 13
Mode, 134
Modem, 18, 99, 104, 110, 133, 138, 276
Modified defect loop, 263
MP-BGP, 245
MPLS. *See* Multiprotocol Label Switching (MPLS)
MPLS traffic engineering, 14
MRFC. *See* Multimedia Resource Function Controller (MRFC)
MRFP. *See* Multimedia Resource Function Processor (MRFP)
MSAN. *See* Multi-Service Access Node (MSAN)
MSF. *See* MultiService Forum (MSF)
MTU. *See* Maximum Transmission Unit (MTU)
Multimedia, 1, 2, 4, 5, 7, 11, 12, 16, 22, 24, 36, 71–75, 77, 81, 86, 101, 106, 111, 121, 133, 149, 150, 152, 154, 158, 159, 163, 167, 168, 177–179, 183, 186, 187, 202–205, 208, 231, 273, 278, 289
Multimedia Broadcast/Multicast Service (MBMS), 179, 186, 187
Multimedia Resource Function Controller (MRFC), 166
Multimedia Resource Function Processor (MRFP), 166
Multiparty Multimedia Session Control (MMUSIC), 183

Multiprotocol Label Switching (MPLS), 14, 22, 23, 32, 82, 184, 231–233, 239–64, 267, 269, 273, 275–279, 282, 284, 287–289, 292–294, 296–298, 312, 320, 325, 329
BFD (Bidirectional Forwarding Detection, 238, 261
conservative label retention, 236
CoS (class of service), 234
CR-LDP (Constraint-based Routing Label Distribution Protocol), 236
E-LSP (LSP with PSC inferred from experimental bits), 234
experimental bits, 234
FEC (Forwarding Equivalence Class), 232, 235–238, 249, 250
label, 232–238, 243, 247–250, 253, 259, 263
label binding, 235, 236, 238
LDP (Label Distribution Protocol), 235, 236, 238, 249, 250, 253, 257, 259, 260, 263
LER (Label Edge Router), 232–235, 237, 238, 240, 241, 263
liberal label retention, 236
L-LSP (LSP with PSC inferred from labels), 234
LSP (Label Switched Path), 232–247, 256, 259, 263
LSP Ping, 237, 238, 261
LSP Trace Route, 238
LSR (Label Switching Router), 232–244
LSR self test, 238
Multi-segment Pseudo wire, 250
Multi-Service Access Node (MSAN), 14, 22, 23, 313, 325
MultiService Forum (MSF), 16, 22, 24, 28, 41, 65, 142, 210–217, 221–226, 228, 267, 289, 291–296, 299, 300, 303, 306, 307, 309, 310, 314, 319, 320, 329
Multi-service interworking, 254, 255, 262
MVNO. *See* Mobile Virtual Network Operator (MVNO)

NAT. *See* Network Address Translation (NAT)
Network Address Translation (NAT), 7, 19–21, 44, 46, 75, 78, 86–89, 92, 96, 97, 154, 180, 181, 306
Network engineering, 273, 277
Network management, 3, 13, 322
Network Termination Equipment (NTE), 276

Network Time Protocol (NTP), 73
Next Generation Network (NGN), 4, 81, 156, 267
NGN. *See* Next generation network (NGN)
Non-stop routing, 239
Notification Rate Package, 326
NTE. *See* Network termination equipment (NTE)
N-to-one mode, 248
NTP. *See* Network Time Protocol (NTP)
Number analysis, 24, 26

OAM. *See* Operations, Administration, and Maintenance (OAM)
OAS SCS. *See* Open Service Access Service Capability Server(OSA SCS)
Offline Charging, 177
Offline TE, 274
One-to-one mode, 248
Online Charging, 177
Online TE, 273
Open Service Access (OSA), 199
Open Service Access Service Capability Server(OSA SCS), 208
Operating Expenditure (OPEX), 3, 5, 160
Operational Support Systems (OSS), 20, 34, 92, 227, 293, 296
Operations, Administration, and Maintenance (OAM), 6, 237, 238, 248–250, 260–264
    alarm suppression, 237
    defect indication, 237, 238, 262, 263
    loopback, 237, 238
    path trace, 237
    performance monitoring, 237
OPEX. *See* Operating expenditure (OPEX)
ORBA. *See* Common Object Request Broker Architecture (CORBA)
OSA. *See* Open Service Access (OSA)
OSS. *See* Operational Support Systems (OSS)
Overload, 11, 21–23, 30, 91, 93, 116, 122, 123, 141, 268, 281–284, 291, 312, 313, 320–329
Over-provisioning, 9, 233, 268, 269, 273, 275, 283, 329

P routers, 242–244, 247, 249, 252
PABX. *See* Private Automatic Branch Exchange (PABX)
Packet Data Convergence Protocol (PDCP), 157

Packet Data Gateway (PDG), 176
Packet Data Protocol (PDP), 154
Packet domain, 11
Packetisation time, 69, 71
Packet-Switched Network (PSN), 243, 244, 246, 247, 251, 253, 255, 258–263
Packet-TMSI (P-TMSI), 154
PAN. *See* Personal Area Network (PAN)
Parlay, 24, 194, 199–204, 208–212, 227, 228
Parlay X, 24, 194, 201, 211
Path message, 274
Path protection, 240
Payload types, 44, 68
PBX. *See* Private Branch Exchange (PBX)
P-CSCF. *See* Proxy Call Session Control Function (P-CSCF)
PDCP. *See* Packet Data Convergence Protocol (PDCP)
PDF. *See* Policy Decision Function (PDF)
PDG. *See* Packet Data Gateway (PDG)
PDP. *See* Packet Data Protocol (PDP)
PDU mode, 248
Peak Information Rate (PIR), 273
Peer-to-peer, 6, 10, 15, 16, 19, 23, 24, 27, 36, 65, 67, 73, 77, 81, 86, 105, 200, 282–284, 286–288, 291, 299
PEP. *See* Policy Enforcement Point (PEP)
Per-hop behaviour Scheduling Classes (PSCs), 271
Personal Area Network (PAN), 174
PHY (Physical) Layer, 155
PINT. *See* PSTN/Internet Interfaces (PINT)
PIR. *See* Peak Information Rate (PIR)
PKI. *See* Public Key Infrastructure (PKI)
Plain Old Telephone Service (POTS), 4, 18–20, 22, 23, 159
PNNI. *See* Private Network-Node Interface
Point to Point Protocol (PPP), 18, 19, 157
Point-to-Point Protocol (PPP), 246
Policing, 86, 249, 278
Policy Decision Function (PDF), 169, 306
Policy Enforcement Point (PEP), 169
POTS. *See* Plain Old Telephone Service (POTS)
PPP. *See* Point to Point Protocol (PPP)
Private Automatic Branch Exchange (PABX), 13
Private Branch Exchange (PBX), 30, 76, 92, 103, 105, 109, 142, 146, 159, 325
Private key, 84, 85

Private Network–Network Interface (PNNI), 257

Private Network-Node Interface, 257–260

Private User Identity, 164, 165, 168

Protection, 5, 6, 46, 81, 83, 86, 236–241, 244, 252, 259, 263, 274, 276, 281, 321, 323, 324, 328

Protocol Type Indicator (PTI), 247

Provider Edge (PE), 239, 243–254, 259–263

Proxy Call Session Control Function (P-CSCF), 92, 160, 161, 218, 221, 324

PSC. *See* Per-hop behaviour Scheduling Classes (PSCs)

Pseudo wire (PW), 236, 246–256, 257–263, 264

PSN. *See* Packet-switched network (PSN)

PSTN. *See* Public Switched Telephone Network (PSTN)

PSTN emulation, 115, 139, 146

PSTN simulation, 115, 146, 180

PSTN/Internet Interfaces (PINT), 183

PTI. *See* Protocol Type Indicator

P-TMSI. *See* Packet-TMSI

Public key, 84, 85

Public Key Infrastructure (PKI), 84

Public Switched Telephone Network (PSTN), 1, 4, 5, 7, 9–11, 13–15, 17–30, 32–35, 37, 39, 56, 57, 65, 71, 74–77, 81, 90, 92, 93, 97, 99–101, 103, 105, 107–121, 123, 125, 127, 129, 131, 133, 135, 137, 139, 141–146, 151, 152, 155–157, 159, 163, 165, 166, 180, 182–184, 191, 198, 228, 267, 277, 281–285, 287, 289, 291, 292, 294–299, 301, 311, 320, 322–325, 327–329
  emulation, 7
  simulation, 7

Public User Identity, 164, 165, 168

PW. *See* Pseudo Wire

PW ID FEC, 249

PW segment, 251

Q.931 access signalling, 24, 26, 27, 29, 104–107, 109, 136, 142, 145, 146, 183, 326

QoS. *See* Quality of Service

Quad Play, 2, 5, 7, 9, 174

Quality of Service (QoS), 4, 5, 7, 9, 11, 14, 19, 21–23, 55, 89, 149, 150, 155–158, 162, 169–172, 175, 180, 184–186, 202, 231,

233, 234, 236, 244, 245, 249, 252, 258, 263, 267–271, 273, 275–284, 286–290, 300, 302–306, 311, 314, 316, 318, 320, 328, 329

bandwidth, 4–6, 19, 23, 36, 38, 39, 65, 71, 72, 89, 91, 92, 99, 100, 108, 112, 136, 144, 150, 151, 161, 169, 174, 178, 185, 202, 231, 232, 234–236, 240, 242, 248, 251, 253, 267, 268, 270, 271, 273–283, 285–291, 293–300, 302, 303, 305–307, 309, 310, 312–314, 316–320, 329

delay, 4, 9, 11, 19, 21, 71, 73, 74, 155, 156, 215, 236, 237, 241, 268–272, 274, 280, 290, 318

jitter, 11, 19, 71–75, 112, 236, 241, 249, 270, 271, 274, 316

loss, 4, 11, 21, 72, 73, 75, 138–140, 236, 237, 268–272, 274, 278, 280, 285, 312, 313

Queuing, 279

R2, 22, 291

R3, 291

RAB. *See* Radio Access Bearer (RAB)

RACS. *See* Resource and Admission Control Subsystem (RACS)

Radio Access Bearer (RAB), 155, 171

Radio Access Network (RAN), 150, 187

Radio Access Network Application Part (RANAP), 156, 157

Radio Network Controllers (RNCs), 154, 155

RADIUS. *See* Remote Authentication Dial-In User Service (RADIUS)

RAN. *See* Radio Access Network (RAN)

RANAP. *See* Radio Access Network Application Part (RANAP)

RCC. *See* Routing Control Channel

RCEF. *See* Resource Control Enforcement Function (RCEF)

RCU. *See* Remote Concentrator Unit (RCU)

Real-time, 6, 7, 9, 23, 29, 71, 86, 156, 183, 191, 228, 270, 278, 282, 290, 296, 329

Real-Time Control Protocol (RTCP), 71, 72
  extended reports, 74
  reports, 73

Real-Time Streaming Protocol (RTSP), 7

Real-time Transport Protocol (RTP), 7, 11, 16, 25, 44, 45, 48, 51, 66–75, 77, 87–91, 94, 106, 112, 125, 136, 181, 183, 188, 249, 309, 310, 314, 316, 327

AVP (Audio-Video Profile), 45, 48, 51,
    66–72, 87, 136
Receiver Report (RR), 73
Record Object (RRO), 274
Redirection, 54
Redundancy, 238, 239, 243
Registration Procedures, 58
Regulation, 5, 14, 26, 27, 56, 93, 115, 291
Release 99, 11, 12, 151, 152
Remote Authentication Dial-In User Service
    (RADIUS), 18, 151, 174, 306
Remote Concentrator Unit (RCU), 17
Remote descriptor, 135, 137
Residential Gateway (RGW), 20, 21, 37, 75
Resiliency, 238, 252, 259
Resource and Admission Control Subsystem
    (RACS), 180
Resource Control Enforcement Function
    (RCEF), 181
Resource ReserVation Protocol (RSVP), 233,
    235, 236, 263
Restoration, 6, 236, 239, 252, 256, 263
RESV, 274
RFC 2547, 244, 245, 253
RGW. See Residential Gateway (RGW)
Ringing, 46, 63, 102, 105, 117–119, 121, 138,
    140, 172, 215, 222, 224, 292, 305, 316
Ringing tone, 33, 63, 102, 103, 105, 119–121,
    125, 130, 316
RNC. See Radio Network Controllers (RNCs)
Route distinguisher, 245
Route pinning, 239
Routing, 6, 32, 84, 86, 92, 106, 107, 122, 144,
    145, 153, 154, 157, 232–237, 239–241,
    243, 245, 252, 256–261, 269, 270,
    273–275, 280, 282, 293, 295–300, 312,
    319
    loose, 274
Routing Control Channel (RCC), 257–259
RR. See Receiver Report (RR)
RRO. See Record Object (RRO)
RSA, 84
RSVP. See Resource ReserVation Protocol
    (RSVP)
RSVP with Traffic Engineering (RSVP-TE),
    235, 236
RSVP-TE. See RSVP with Traffic Engineering
RTCP. See Real-time Control Protocol (RTCP)
RTP. See Real-time Transport Protocol (RTP)
RTSP. See Real-Time Streaming Protocol
    (RTSP)

S/MIME. See Secure MIME
SAI. See Source Attachment Identifier
Satellite Digital Multimedia Broadcast
    (S-DMB), 177, 178
SBC. See Session Border Controller (SBC)
SBG. See Session Border Gateway
Scalability, 6, 97, 236, 237, 252, 278, 283,
    288, 300, 313
SCE. See Service Creation Environment (SCE)
SCFs. See Service Capability Features (SCFs)
S-CSCF. See Serving Call Session Control
    Function
SCTP. See Stream Control Transmission
    Protocol (SCTP)
SDH. See Synchronous Digital Hierarchy
    (SDH)
S-DMB. See Satellite Digital Multimedia
    Broadcast
SDP. See Session Description Protocol (SDP)
Secure MIME (S/MIME), 94, 96
Secure URI, 95
Security, 7, 22, 30, 38, 75, 77, 81–86, 90,
    92–98, 114, 157, 161, 184, 186, 187,
    291, 299
    authentication, 10, 18, 55, 56, 58, 78,
        94–96, 98, 107, 153, 162, 200
    DoS (denial of service), 7, 81, 144, 268,
        299, 321, 324, 327
    theft of service, 7, 81, 93
Sender Report (SR), 73
Sender template specific, 242
Sequence number, 247
Service Agility, 8
Service Assurance, 236
Service Broker, 22, 24, 28, 211–217,
    221–226, 228
Service Capability Features (SCFs), 10, 200,
    201
Service Control Function, 193, 208
Service Creation Environment (SCE), 193
Service Data Unit (SDU) mode, 248
Service Delivery Platforms, 226
Service Independent Building Blocks (SIBBs),
    193, 205
Service Interworking, 254
Service Level Agreement (SLA), 5, 6, 236,
    237, 239, 283
Service Logic Gateway, 211
Service Nodes, 193
Service Orchestration, 205, 206, 217

Service Point Trigger, 207

Service Policy Decision Function (SPDF), 180, 291

Service Provider, 2–4, 7–10, 13, 18, 19, 21, 87, 116, 149, 174, 194, 198, 205, 206, 231, 232, 234, 243–246, 251–254, 256, 257, 259, 264, 268, 276, 277, 279, 280

Service Switching Function, 193, 208

Service Velocity, 8

Service-based Policy Control (SBP), 180

Services in the PSTN/IN Requesting InTernet Services (SPIRITS), 184

Serving Call Session Control Function (S-CSCF), 160, 163, 164, 167, 208, 217

Serving GPRS Support Node (SGSN), 153

Session Border Controller (SBC), 19–21, 23, 30, 38, 44, 57, 64, 75, 81, 86–89, 92, 94, 96–98, 160, 283, 285–289, 291, 299

Session Border Gateway (SBG), 11, 23, 30, 44, 81, 83, 85–87, 89–97, 284, 291, 292, 298–301, 306, 307, 312, 313, 315–318, 320, 324, 327–329

Session Description Protocol (SDP), 7, 16, 36, 39, 42, 44, 46, 48, 49, 57, 65–70, 75, 77, 87, 88, 94, 96, 106, 109, 111, 112, 120–122, 125, 127, 134, 135, 137, 171–173, 180, 186, 187, 207, 219, 226–228, 285, 302–305, 307, 314, 316

    attribute (a=) line, 45, 48, 51, 66, 68–70, 136, 302, 304, 305

    conf (precondition negotiation) attribute, 302, 304, 305

    connection data (c=) line, 44, 48, 51, 66, 67, 69, 70, 87, 135, 136

    curr (precondition negotiation) attribute, 302, 304

    des (precondition negotiation) attribute, 302, 304

    fmtp attribute, 44, 45, 48, 51, 66, 69

    inactive attribute, 69, 70

    media type and transport (m=) line, 44, 45, 48, 51, 66–70, 87, 135–137

    ptime attribute, 45, 66, 68, 69

    recvonly attribute, 69, 70

    rtpmap attribute, 44, 45, 48, 51, 66, 68–70, 136

    sendonly attribute, 69, 70

    sendrecv attribute, 44, 45, 66, 69, 70, 302, 304

Session Initiation Protocol (SIP), 1, 7, 10, 11, 14–16, 24, 28, 29, 32, 34–70, 72,

75–77, 86–98, 105, 106, 108, 109, 112–121, 124–127, 134–137, 146, 159–165, 167, 168, 170, 171, 173, 180, 183, 184, 186–188, 194–199, 206–217, 219–222, 224, 225, 228, 267, 284, 285, 287, 289, 291, 292, 299, 300, 302–307, 312–320, 324, 325, 327–329

1xx (provisional) responses, 47

2xx response, 46

3xx response, 54

4xx response, 52, 53

100 Trying response, 38, 39, 45–47, 55, 62, 113, 302

180 Ringing response, 38, 46, 47, 49, 55, 61–63, 87, 88, 113, 120, 125, 302, 303, 315, 317, 318

200 OK response, 34, 38, 46, 50–53, 55, 58–60, 96, 113, 120, 124, 125, 164, 172, 173, 226, 302, 305, 315–318

ACK request, 38, 52, 54, 105, 107, 113, 120, 125, 144, 172, 315–318

Allow header field, 43, 47, 51, 150

B2BUA (Back-to-Back User Agent), 37, 38, 57, 64, 87, 89, 94, 198, 208, 212, 299, 316

branch parameter magic cookie, 40

BYE request, 38, 52–55, 116, 120, 125, 215, 220, 221, 224, 225

Call-ID header field, 41, 45, 47, 49–53, 57, 59, 60

CANCEL request, 52, 55

confirmed dialogue, 46

Contact header field, 42, 48, 49, 51, 52, 54, 57–60, 62

Content-Length header field, 42, 45, 48–54, 59, 60

Content-Type header field, 42, 51, 96

CSeq header field, 42, 45, 47, 49–54, 57, 59, 60, 95

Date header field, 45, 47, 50, 51, 53

dialogue, 39–43, 46–53, 55, 60, 62–65, 95, 198

early dialogue, 47, 52

forking, 11, 41, 54–56, 62, 167, 168, 313, 314

From header field, 40, 41, 45, 47, 49–53, 55–57, 59, 62, 96

History header field, 54

INVITE request, 32, 34, 37–42, 44–47, 49–52, 54–58, 60–63, 66, 87, 88, 93, 95, 96, 113, 116, 120, 125, 136,

167, 171, 172, 188, 194–197, 206, 208–210, 212–224, 226, 302, 305, 314–318, 328

Max-Forwards header field, 39, 53, 57

NOTIFY request, 43, 47, 51, 116, 206, 217, 219, 220, 326

offer-answer model, 36, 46, 48, 49, 65, 66, 68–70, 83, 107, 125, 128, 134, 136–138, 151, 172–174, 178, 187, 286, 300, 303, 305, 327

P-Asserted-Identity header field, 56, 57, 94, 114

P-Charging-Vector header field, 64, 65, 210, 214

PRACK (Provisional ACK) request, 38, 46, 49, 50, 63, 187, 302, 303, 315, 317, 318

precondition signalling for QoS, 267, 302

Privacy header field, 57

Proxy Server, 7, 32, 37, 39, 40, 52, 55, 60, 62, 64, 82, 92, 160, 164, 167, 176, 177, 194–196, 198, 309, 326

RAck header field, 50

Record-Route header field, 39, 60, 62–64, 173

redirect server, 38, 54

redirection, 11, 76, 195

REFER request, 43, 47, 51, 188

REGISTER request, 58–60, 93

registrar, 7, 38, 39, 58, 59, 62, 90, 91, 95, 96

registration, 10, 56, 58, 59, 76, 90, 95, 108, 141, 161, 163, 164, 173, 188, 208, 214, 219

re-INVITE, 46, 63, 64, 219–221, 226

Request-URI field, 39–41, 60, 62, 114, 187, 196, 212, 213, 216, 222, 223

Require header field, 42, 43, 47, 59

Route header field, 60, 61, 63, 64, 208, 210, 212–214, 219, 274

RSeq header field, 47, 50

Session-Expires header field, 42, 46

SIP: URI, 32, 39–42, 48, 57, 60, 62, 63, 76, 88, 95, 124, 163, 165, 167

SIP-I (SIP with encapsulated ISUP), 14, 29, 94, 113, 114

SIPS: URI, 95

Supported header field, 43, 59, 307

target URI, 61–63

tel: URI, 76, 79, 165, 183

To header field, 41, 45, 47, 49–53, 55–57, 59, 62, 96, 114

UA (User Agent), 7, 32, 37, 39, 40, 42, 43, 45–47, 49–60, 62–64, 66–70, 87, 88, 92, 93, 95, 96, 164, 194

UAC (User Agent Client), 37

UAS (User Agent Server), 32, 37

UPDATE request, 43, 46, 47, 51, 302, 303, 305, 315–318

Via header field, 39, 40, 45, 47, 49–53, 57–63

Session Initiation Protocol Project INvestiGation (SIPPING), 183

Session Termination Answer, 312

Session Termination Request (ST-Request), 311

SG. *See* Signalling Gateway (SG)

SHA1, 84

Shaping, 249, 269

Shared Secrets, 84

SIBBs. *See* Service Independent Building Blocks (SIBBs)

Signalling, 234–236, 247, 250–252, 256–262, 264

Signalling Gateway (SG), 11, 108–110, 118, 145, 146, 155, 160

Signalling Interworking, 257

Signalling network, 101, 144, 146

Signalling path Session Border Gateway (S-SBG), 22, 23, 86, 89, 91, 92, 94, 291–294, 297–301, 305–307, 312, 314, 316, 325, 327, 328

Signalling System No. 7 (SS7), 11, 21, 24, 26, 29, 30, 93, 106, 107, 142, 144–146, 183

Signalling Transport (SIGTRAN), 14, 29, 30, 109, 110, 142–144, 146, 183

DUA (DPNSS User Application Layer), 30

IUA (ISDN User Adaptation Layer), 14

M2PA (MTP2 User Peer to peer Adaptation Layer), 30, 145, 146

M2UA (MTP2 User Adaptation Layer), 29, 30, 145, 146

M3UA (MTP3 User Adaptation Layer), 29, 30, 145, 146

SIGTRAN. *See* Signalling Transport (SIGTRAN)

SIMPLE, 184

Simple Network Management Protocol, 84, 296

Simple Object Access Protocol (SOAP), 201

Single segment PWs (SS-PWs), 251, 252

SIP. *See* Session Initiation Protocol (SIP)

SIPPING. *See* Session Initiation Protocol
    Project INvestiGation (SIPPING)
SLA. *See* Service Level Agreement (SLA)
SLF. *See* Subscription Locator Function (SLF)
SOAP. *See* Simple Object Access Protocol
    (SOAP)
Soft Permanent Virtual Connections (SPVCs),
    256, 257
Softswitch, 156, 199, 227
SONET. *See* Synchronous Optical Network
    (SONET)
Source Attachment Identifier (SAI), 250
Spanning Tree Protocol (STP), 252, 253
S-PE. *See* Switching PE (S-PE)
Special Resource Function, 192
Special service circuits, 99, 101
SPIRITS. *See* Services in the PSTN/IN
    Requesting InTernet Services (SPIRITS)
SPVC-PWE3 interworking, 259, 260
SPVCs. *See* Soft Permanent Virtual
    Connections
SR. *See* Sender Report (SR)
SS7. *See* Signalling System No. 7 (SS7)
S-SBG. *See* Signalling path Session Border
    Gateway
SS-PWs. *See* Single segment PWs
SSRC. *See* Synchronisation source (SSRC)
Standards, 6, 16, 22, 24, 25, 30, 33, 36, 103,
    108, 109, 113, 121, 150, 176, 179–182,
    184, 185, 193, 199, 206, 228, 326
Status signalling, 262
STP. *See* Spanning Tree Protocol
Stream Control Transmission Protocol (SCTP),
    7, 29, 30, 141–145, 307
Strict routing, 274
Subscriber Identity Module, 151
Subscription Locator Function (SLF), 162, 165
Supervision, 25, 101–103, 106, 140, 146
    hook-flash, 102
    off-hook, 102, 103, 117, 119–121, 124,
        125, 138–140, 316, 325
    on-hook, 102, 103, 120, 121, 124–126, 130,
        138–140
SVC. *See* Switched Virtual Connections
Switched Virtual Connections (SVC), 256
Switching PE (S-PE), 251, 252
Symmetric RTP, 75, 77, 88
Synchronisation source (SSRC), 72
Synchronous Digital Hierarchy (SDH), 231,
    232, 237, 239

Synchronous Optical Network (SONET), 231,
    239

Tag
    in preconditions, 304
    in user part of SIP Route header field, 219,
        223
    parameter of SIP From or To header field,
        40, 41, 45, 47, 49–53, 55, 57, 59
    used in calls to SIP CGI server, 197
TAI. *See* Target Attachment Identifier
Target Attachment Identifier (TAI), 250
TCP. *See* Transmission Control Protocol (TCP)
TD-CDMA. *See* Time Division CDMA
    (TD-CDMA)
TDM. *See* Time Division Multiplex (TDM)
TE. *See* Traffic Engineering (TE)
Telecoms & Internet converged Services and
    Protocols for Advanced Networks
    (TISPAN), 22, 90, 180, 289
Telephone line, 2, 5, 9, 10, 24, 25, 32, 34, 56,
    67, 74, 93, 101–104, 117–121,
    123–126, 129, 138–140, 156, 228, 276,
    283, 314, 322, 325
    local loop, 101
Telephone numbers, 16, 32, 75–77, 183
Temporary Mobile Subscriber Identity (TMSI),
    154
Terminating PE (T-PE), 251, 252
TGW. *See* Trunking Gateway (TGW)
Third parties, 5, 25, 94, 194, 299
3GPP. *See* 3rd Generation Partnership Project
    (3GPP)
3rd Generation Partnership Project (3GPP), 6,
    11, 12, 24, 36, 106, 111, 114, 126,
    150–152, 154–156, 158, 160–162,
    167–170, 173, 174, 176, 177, 179, 181,
    184–188, 198, 199, 206–209, 291, 302,
    306, 307
Threat model, 81, 83, 96, 97
Three-colour marker, 272, 273
Time Division CDMA (TD-CDMA), 158
Time Division Multiplex (TDM), 3, 4, 10,
    14–17, 19, 21–24, 29, 30, 32, 34, 39,
    77, 152, 153, 191, 199, 231, 246, 249,
    256, 257, 281, 302, 327
Time Division Synchronous Code Division
    Multiple Access (TD-SCDMA), 158
TISPAN. *See* Telecoms & Internet converged
    Services and Protocols for Advanced
    Networks (TISPAN)

TLS. *See* Transport Layer Security (TLS)
TMSI. *See* Temporary Mobile Subscriber Identity (TMSI)
ToS. *See* Type of Service bits (ToS)
T-PE. *See* Terminating PE (T-PE)
Traffic contract, 234, 269, 272, 278
Traffic engineered, 14, 267, 296
Traffic Engineering (TE), 233–237, 239, 244, 252, 256, 263
Traffic Management, 234
Transit exchange, 16, 17, 19, 24, 27, 32, 297
    Class Four office, 17
    XIT, 17
Transmission Control Protocol (TCP), 7, 29, 57, 122, 141–144, 232, 233, 235, 270, 273, 274, 307
Transport Layer Security (TLS), 92, 94, 95
Trigger point, 25, 30, 33, 192, 206
Triple Play, 2, 5, 7, 9, 174, 276
Trunking Gateway (TGW), 19–23, 284, 287, 289, 292, 294, 297, 298, 301, 325
21st Century Network, 24
Type of Service bits (ToS), 234, 244

UA. *See* User agent (UA)
UDP. *See* User Datagram Protocol (UDP)
UE. *See* User equipment (UE)
ULTRA TDD, 158
UMA. *See* Unlicensed Mobile Access (UMA)
UMA Network Controller (UNC), 174
UMAN. *See* Unlicensed Mobile Access Network (UMAN)
UNC. *See* UMA Network Controller (UNC)
UNI. *See* User-Network Interface (UNI)
Universal Mobile Telecommunications System (UMTS), 4, 150–156, 158, 159, 170, 179, 184, 185
Universal Resource Identifier (URI), 32, 54, 57, 62, 64, 76, 95, 165, 183, 216
Universal Terrestrial Radio Access Network (UTRAN), 150
Unlicensed Mobile Access (UMA), 11
Unlicensed Mobile Access Network (UMAN), 175
Uplink, 6, 152, 153, 157, 158, 310
URI. *See* Universal Resource Identifier (URI)
User agent (UA), 7
User Datagram Protocol (UDP), 7, 29, 40, 42, 44, 45, 47, 49–53, 59, 67, 71, 121, 122, 141, 144, 157, 183, 233, 270, 274, 310

User equipment (UE), 104, 155, 159
User-Network Interface (UNI), 257
UTRAN. *See* Universal Terrestrial Radio Access Network (UTRAN)

V5 access signalling, 23, 24, 26, 27, 29, 103, 104, 139, 142, 146
Value-added services, 8, 24, 27, 30, 36, 37, 164, 191, 193, 195, 197, 199, 201, 203, 205–207, 209, 211, 213, 215, 217, 219, 221, 223, 225–228
VC. *See* Virtual Channel
VCC. *See* Virtual Channel Connection (VCC)
VCCV. *See* Virtual Circuit Connectivity Verification (VCCV)
VCI. *See* Virtual Channel Identifier (VCI)
Video on Demand (VoD), 10, 11, 18, 20, 277
Video services, 2, 5, 6, 10, 11, 16, 19, 36, 67, 71–74, 77, 101, 104, 121, 134–137, 156, 159, 174, 177–179, 182, 183, 205, 232, 267, 268, 276–278, 282, 283, 286, 289, 329
Virtual Channel (VC), 241, 247, 248
Virtual Channel Connection (VCC), 241, 248
Virtual Channel Identifier (VCI), 248, 250, 259
Virtual Circuit Connectivity Verification (VCCV), 260, 261
Virtual Leased Lines (VLLs), 277
Virtual Local Area Network (VLAN), 252, 256
Virtual Path Identifier (VPI), 248, 250, 257–259
Virtual Private LAN Services (VPLS), 4, 246, 252–254, 260, 264, 276, 278
Virtual Private Network (VPN), 175, 231, 243–247, 250, 253–256, 261–264
    IP VPN, 236, 245
    layer 2 VPN, 236
Virtual Private Wire Service (VPWS), 246, 253
Virtual routing and forwarding (VRF), 245
Virtual tunnel, 275, 285
Visitor Location Register (VLR), 154
VLAN. *See* Virtual Local Area network
VLL. *See* Virtual Leased Lines (VLLs)
VLR. *See* Visitor Location Register (VLR)
VoD. *See* Video on Demand
Voice mail, 10, 54, 62, 139, 221, 222, 224–226
Voice on net service, 20–23
Voice over IP (VoIP), 15, 20, 22, 23, 33, 36, 38, 43, 65, 74, 75, 77, 86, 87, 183, 270
Voice services, 2–7, 9–11, 14–17, 19–24, 27, 28, 36, 38, 65, 71–75, 77, 81, 82, 86,

Voice services (*continued*)
 90, 99–101, 106, 108–112, 115, 116,
 121, 137, 139, 149–152, 155, 156, 158,
 159, 174, 175, 182, 191, 203, 205, 227,
 228, 232, 257, 267, 268, 270, 276, 278,
 281–287, 291, 310, 320, 329
VoIP. *See* Voice over IP
VPI. *See* Virtual Path Identifier (VPI)
VPLS. *See* Virtual Private LAN Services
 (VPLS)
VPN. *See* Virtual Private Network (VPN)
VPWS. *See* Virtual Private Wire Service
VRF. *See* Virtual routing and forwarding

WAG. *See* Wireless LAN Access Gateway
 (WAG)

Wallclock, 73
W-CDMA. *See* Wideband Code Division
 Multiple Access (W-CDMA)
Wideband Code Division Multiple Access
 (W-CDMA), 158
WiFi, 4, 13, 94, 174
WiMAX. *See* Worldwide Interoperability for
 Microwave Access (WiMAX)
Wireless LAN (WLAN), 149, 150
Wireless LAN Access Gateway (WAG), 176
WLAN. *See* Wireless LAN (WLAN)
Worldwide Interoperability for Microwave
 Access (WiMAX), 151

X.509, 84, 85
XML, 194, 195